# The Victorians: A Botanical Perspective

Luís Manuel Mendonça de Carvalho
Editor

# The Victorians: A Botanical Perspective

Volume 1

🐎 Springer

*Editor*
Luís Manuel Mendonça de Carvalho
Biosciences Department (ESAB)
Beja Polytechnic University
Beja, Portugal

ISBN 978-3-031-68758-7        ISBN 978-3-031-68759-4   (eBook)
https://doi.org/10.1007/978-3-031-68759-4

This Springer imprint is published by the registered company Springer Nature Switzerland AG
The registered company address is: Gewerbestrasse 11, 6330 Cham, Switzerland

If disposing of this product, please recycle the paper.

*Hic liber dedicatus est
Maria Águeda Conceição D'Andrade
(1924–2011), a very Victorian lady who
loved tea, ferns, violets and books.*

# Introduction

The Victorian era, spanning from 1837 to 1901, witnessed unprecedented advancements in science and technological innovation that reshaped society, culture, and the world at large. It was marked by a profound advance in botany, as botanical explorations were at the forefront of scientific inquiry. Simultaneously, expeditions to far corners of the globe, sponsored by wealthy patrons and scientific institutions, yielded a wealth of botanical specimens previously unknown to Western science. Pioneering naturalists and botanists, such as Joseph Dalton Hooker, Charles Darwin, Alfred Russel Wallace, and Richard Spruce, embarked on voyages that expanded the known boundaries of botany and revolutionized our understanding of biodiversity and, above all, of the evolution of life on Earth.

The systematic cataloging and classification of plant species became a hallmark of Victorian botany. Institutions such as the Royal Botanic Gardens, Kew, played a central role in these efforts, amassing vast collections of plant specimens and botanical knowledge. The development of taxonomic systems laid the groundwork for modern botanical classification, providing a framework that continues to shape our understanding of plant diversity. Botanical gardens have also emerged as centers of botanical education and public engagement in science, providing spaces for both scientific research and leisure and promoting botanical knowledge among the public.

The legacy of Victorian botanical gardens lives on as they are vital resources for studying plant biodiversity and ecosystem dynamics, as well as opportunities for public engagement and environmental advocacy. The enduring fascination with plants and horticultural displays cultivated during the Victorian era persists in contemporary gardening trends and even in botanical tourism.

The Victorian era witnessed a strong desire to cultivate exotic plant species from around the world, and wealthy collectors often competed to acquire the rarest and most prized specimens. Hence, botanical gardens and private estates became showcases for these botanical treasures, with glasshouses and conservatories constructed to house delicate plants from distant and exotic lands.

The development of new horticultural techniques, such as plant hybridization and propagation, spurred innovation in Victorian gardening practices. Pioneers such as the gardener Joseph Paxton—designer of the Crystal Palace—, experimented with novel methods for cultivating and displaying plants on a grand scale, resulting in elaborate bedding schemes reflecting the Victorian desire for order, symmetry, and botanical spectacle.

This period coincided with the height of the British Empire, and botanical exploration played a crucial role in imperial expansion, with subsequent changes in landscapes and disruptions of natural ecosystems due to the introduction of new crops, such as tea, rubber, and quinine, which transformed economies and lifestyles both in the UK and abroad.

The Victorian legacy includes the civil rights struggles of women; workers' rights, including children's rights; labor reforms; challenging religious doctrines; the prevention of cruelty to animals; improvements in public health and treatment of diseases; postal reforms; and many other legacies in biology, physics, chemistry, and engineering that reinforce scientific enlightenment and social progress. The scientific advancements and discoveries made during this period laid the foundation for our world, shaping the trajectory of social justice movements and advocacy efforts in human civilization and influencing virtually every aspect of our contemporary life. Both formal and informal scientific societies, dedicated to the advancement of knowledge across various scientific disciplines, served as forums for scientists and intellectuals to exchange ideas, present research findings, and engage in innovation and interdisciplinary collaborative projects. The professionalization of science within Victorian society also led to the establishment of codes of conduct and ethical guidelines for science, emphasizing the importance of honesty and accountability.

This volume presents challenging perspectives for areas in which plants were relevant players for the Victorians: Bruce Hunt discusses the use of gutta-percha and the development of electrical measurements in Victorian Britain; Catherine Higgs presents a detailed history of cocoa, Cadbury, and forced labor in the São Tomé and Príncipe Islands; Helen Elletson reveals the beauty, imagination, and order of William and May Morris' flowers; Jonathan Smith introduces us to the world of Charles Darwin and Victorian Botany Culture; Luke Keogh and Angela Kreutz explain the crucial role of the Wardian Case in the global transport of plants; Meegan Kennedy reveals the connection between Mid-Victorian Botany and Microscopy; Sara Albuquerque and Ângela Salgueiro give us glimpses of the colonial collections at the 1862 London Exhibition; Vera Gonçalves and Filipe Barroso explain how botany was connected with the development of photography; Vicky Albritton and Fredrik Albritton Jonsson evoke the desire for a return to Nature and a simple life; and, finally, Francisca Fernandes and Luís Mendonça de Carvalho take us into a journey through the history of violets.

# Acknowledgments

The authors express their deepest gratitude to Springer editorial staff, especially to Eric Stannard, Natalie Muñoz, Karthiga Barath, and Arun Siva Shanmugam.

# Contents

# Chapter 1
# Insulation for an Empire: Gutta-Percha and the Development of Electrical Measurement in Victorian Britain

Bruce J. Hunt

**Abstract** In the second half of the nineteenth century, British engineers and businessmen built the world's first truly global telecommunications system, a vast network of submarine telegraph cables that reached almost every significant port in the world. Almost all of these cables were insulated with gutta-percha, a natural plastic derived from the latex of *Palaquium gutta* trees found in Malaya, Sumatra, and Borneo. Cable telegraphy, one of the most technologically advanced industries of the day, was wholly dependent on axe-wielding tribesmen who gathered gutta-percha in the forest. Once it was shipped to London, gutta-percha became one of the most intensively studied substances in the world, its mechanical and electrical properties subjected to minute investigation by leading British physicists and engineers, with far-reaching consequences for the development of electrical science. The story of gutta-percha illustrates important patterns in the appropriation and transformation of tropical materials and their integration into imperial economic and technological systems.

**Keywords** Gutta-percha · Telegraph · Cable · Electricity · Measurement · Singapore · Malaya · Borneo

In the second half of the nineteenth century, British engineers and businessmen built the world's first truly global telecommunications network, a vast array of submarine cables that snaked through Europe's coastal waters, reached across the Atlantic to

This is an updated version of a chapter previously published in *Semaphores to Short Waves* (ed. Frank A. J. L. James). Proceedings of a Conference on the Technology and Impact of Early Telecommunications Held at the Royal Society for the Encouragement of Arts, Manufactures and Commerce, on Monday 29 July 1996. Organized by the British Society for the History of Science, the Newcomen Society, and the RSA.

B. J. Hunt (✉)
The University of Texas at Austin, Austin, TX, USA
e-mail: bjhunt@austin.utexas.edu

North America, ringed the coasts of South America and Africa, and extended onward to India, Australia, China, and Japan. Now, for the first time, London could communicate almost instantaneously with New York, Bombay, Sydney, and Hong Kong. Such a revolutionary development could not help but affect the way commerce and diplomacy were conducted, the way news was disseminated, and not least, the way the British Empire was run. Science, too, felt the effects, as the enormous task of building and maintaining a global cable network exerted a powerful influence on the pace and direction of British electrical research and helped shape some of its most characteristic features in the mid-Victorian era.

The great majority of these cables were insulated with gutta-percha. By examining the history of this now largely forgotten tropical substance, and by exploring how it came to be the focus of some of the most advanced electrical research of the mid-nineteenth century, we will be able to shed useful and perhaps surprising light on the role cable telegraphy played in the operation of the Victorian British Empire and the development of electrical physics, as well as in the exploitation of the forests of Southeast Asia and the Malay Archipelago.

Gutta-percha is a milky white substance derived from the latex of certain trees (*Palaquium gutta*, formerly known as *Isonandra gutta*) that are native to Malaya, Borneo, and Sumatra (Godfrey 2018; Obach 1898). Though often loosely referred to as "sap," this latex, like that of rubber trees and many other plants, has a quite distinct nature and function: rather than distributing nourishment, as sap does, the latex provides protection against insects and other pests (Fig. 1.1). Gutta-percha is a natural plastic, easily shaped at temperatures above about 40 °C and becoming

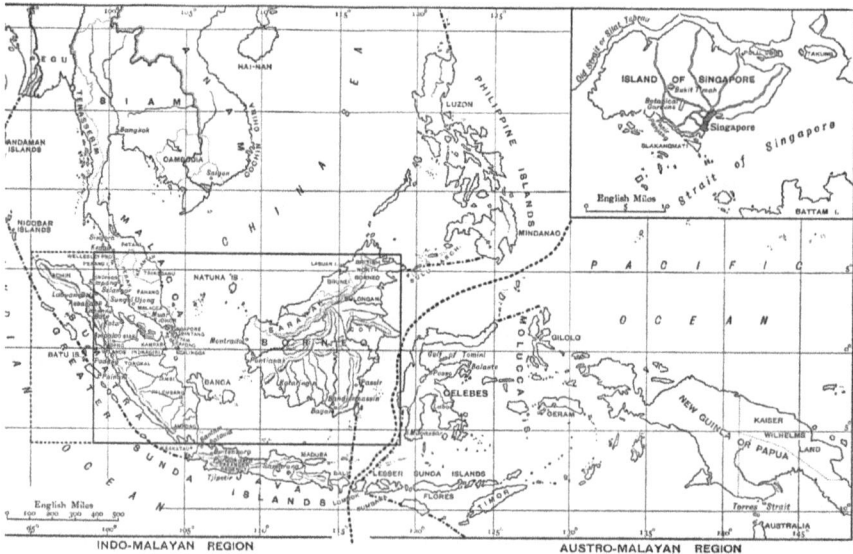

**Fig. 1.1** Map showing (in box) the range of the trees from which gutta-percha is derived. (From Obach 1898)

hard and yet remaining flexible as it cools. It is chemically similar to rubber, but not springy. Malayans had long gathered gutta-percha from the forest and fashioned it into whips, tubs, knife handles, and other useful objects, but it remained almost unknown in the West until the 1840s. Thereafter it soon became an important item of European commerce, with dire effects on the gutta trees, which were virtually wiped out over much of their range. In particular, Europeans found that gutta-percha was a good electrical insulator, and it soon became the insulating material of choice for use on the new submarine telegraph cables they began to lay after 1850. Throughout the rest of the nineteenth century and well into the twentieth, the growth and maintenance of the global cable network depended on access to supplies of gutta-percha. A long cable—across the Atlantic, say, or from Suez to Bombay— required hundreds of tons of high-grade gutta-percha, and British firms' effective control of the Singapore-based gutta-percha trade reinforced Britain's nearly total domination of the cable industry.

It may seem jarring to realize that one of the most advanced "high tech" industries of the second half of the nineteenth century, submarine telegraphy, relied so heavily on an exotic tropical material gathered mainly by axe-wielding Malayans. But the story of gutta-percha in fact illustrates an important way in which science and technology functioned in the Victorian Empire: they provided the motive and means for appropriating and transforming materials taken from distant parts of the world, and so helped draw those materials and their producers into the imperial economy. Moreover, in the case of gutta-percha, this process formed a self-reinforcing loop: Britain's imperial and commercial power gave it favored access to Malayan gutta-percha supplies, and so facilitated the construction of a cable network that, in turn, greatly strengthened the Empire and British commerce—including British control of the gutta-percha trade.

The global cable network was often referred to in the nineteenth century as the "nervous system" of the British Empire and of world commerce; as such, it constituted a key means by which modern technology extended its reach around the world (Headrick 1991; Hunt 2021). Insulated by a sheath of gutta-percha, the British metropolis was, in effect, able to extrude itself throughout the world, creating for itself a way to draw in information, send out commands, and exercise power almost instantaneously wherever a cable could be run.

Gutta-percha played a major role in the construction of the world cable system, and also in the production of new scientific knowledge. The importance of gutta-percha to the cable industry made it the object of enormous attention; great resources, both intellectual and material, were brought to bear on it, so that in the second half of the nineteenth century, this latex from Malayan trees became one of the most intensively studied and tested substances in the world.

For several decades after the 1850s, knowledge of the properties of solid dielectrics (i.e., electrical insulators) largely consisted of knowledge of those of gutta-percha; many techniques for the measurement of very high resistances and of electrostatic induction were first worked out on, and largely applied to, the gutta-percha used in submarine cables. Important scientific work by Michael Faraday (1791–1867), William Thomson (1824–1907, later Lord Kelvin), Fleeming Jenkin

(1833–1885), James Clerk Maxwell (1831–1879), and many others was closely tied to the study of gutta-percha, and such characteristic features of Victorian electrical theory as its emphasis on activity in the dielectric or "field" surrounding charges and currents grew in part from the demands and opportunities presented by cable telegraphy, and so indirectly from the use of gutta-percha. The path from the Malayan forest to Maxwellian field theory is a long and twisting one, but it is well worth exploring.

## First Use

We have no record of when Malayans first discovered gutta-percha, but they had been using it to make whips, hats, tubs, and knife handles long before Europeans came on the scene. A few gutta-percha items apparently made their way to Europe in the seventeenth century, but the real "discovery" of gutta-percha in the West dates from the 1840s. This was the heyday of "economic botany," Europeans' deliberate effort to gather and study plant specimens and materials from throughout the world for possible commercial exploitation (Brockway 1979; Drayton 2000; Headrick 1996). A Portuguese engineer, Jose d'Almeida, first described gutta-percha and its uses to the Royal Asiatic Society of London early in 1843, but the new material first came to wide notice a few months later when an East India Company surgeon, William Montgomerie, published a paper about it and sent samples to London, where they were displayed at the Society of Arts (Bright 1898, p. 248; Obach 1898, p. 85). Small quantities of gutta-percha were soon being shipped from Singapore to London, and British manufacturers of rubber goods began to expand their lines by offering molded gutta-percha items such as boot soles, pump buckets, bottle stoppers, and golf balls. (It is now used mainly by dentists to stop up root canals.)

About this time, William Siemens (1823–1883) came across some samples of gutta-percha on the London market and sent them to his brother Werner (1816–1892) in Berlin "as curiosity" (Siemens 1966, p. 51). Werner Siemens was intrigued by the new material, particularly when he found it to be a good electrical insulator, and by 1846, he was wrapping copper wires in thin sheets of gutta-percha for an experimental underground telegraph line. He soon found, however, that the seams did not hold well, so he developed a press, resembling a macaroni-making machine, that could extrude a continuous sheath of gutta-percha around a wire. This worked very well, and Siemens used such seamless gutta-percha covered wire on the Berlin–Frankfurt underground telegraph line he built for the Prussian government in 1848. (The Prussian authorities feared that an overhead line would be destroyed by the "turbulent populace.") Unfortunately, Siemens tried to make the gutta-percha more durable by treating it with sulfur, on analogy with the vulcanization of rubber. The sulfur in the treated gutta-percha reacted with the copper wire, corroding it and breaking down the insulation. The consequent failure of the line (exacerbated by the voracity of the German rats, which dug down and nibbled the gutta-percha) discredited underground lines and gutta-percha insulation in Germany for years to come

(Löser 1969). Siemens's firm later returned to the production of gutta-percha-insulated telegraph lines, but mostly at William Siemens's London operation and mostly for submarine cables.

The electrical uses of gutta-percha were first brought to public notice in Britain by Faraday in a short note in the *Philosophical Magazine* early in 1848. Independently of Siemens, Faraday had also found that gutta-percha was a good electrical insulator and sought to bring its utility for making the handles and supports of electrostatic devices to the attention of "working philosophers, both juvenile and adult" (Faraday 1848). He did not mention using it to insulate wires, but British telegraphers, who for several years had been making do with tarred hemp and the like, quickly took up the new material and found it to be far superior to anything they had tried before. By 1849, they were stringing gutta-percha-insulated wires along the walls of damp railway tunnels and laying them beneath the streets of London and other cities. Willoughby Smith (1828–1891) of the dominant British manufacturer, the Gutta Percha Company, soon developed improved techniques for processing and applying gutta-percha for electrical purposes (Galton et al. 1861, p. 149; Smith 1891, pp. 1–3).

Gutta-percha proved very useful for underground telegraph lines, but it was in submarine telegraphy that it really came into its own. The new material had come onto the British market at a propitious time. Several plans had already been proposed for underwater cables across rivers and seas, particularly the English Channel, and a few experimental attempts had been made, but none of the available insulating materials worked very well; tarred hemp did not seal out the water well enough, and unvulcanized rubber broke down and was difficult to work with (Bright 1898, pp. 2–5). Under these circumstances, gutta-percha appeared to be a godsend. Early in 1850, the Gutta Percha Company received an order from two enterprising brothers, Jacob and John Watkins Brett, for 25 nautical miles of copper wire covered with gutta-percha to a diameter of half an inch (Smith 1891, pp. 1–2). The Bretts had long proposed laying a telegraph cable across the English Channel, and now, having lined up financial backers, secured a concession from the French government, and finally found a suitable insulating material, they were set to carry their plan into practice. After many difficulties, their engineer, Charlton Wollaston, laid the cable in August 1850, only to see it fail almost immediately. The Bretts and their backers persevered, however, and in September 1851 they laid a new cable (its "core" of copper wires covered with gutta-percha now protected by an outer armoring of iron rope) from Dover to Calais. Despite rather primitive testing methods—Wollaston reportedly sometimes used his *tongue* to test for the flow of current—this first Channel cable worked well and proved very profitable (Bright 1898, pp. 6, 11–13).

The success of the 1851 cable (due, in retrospect, more to luck than real expertise) set off a flurry of other cable-laying projects around the British Isles and in the Mediterranean, all using gutta-percha as insulation. Several of these early cables failed badly, but by 1856 enough had succeeded to encourage the American entrepreneur Cyrus Field (1819–1892) and a group of British investors to contemplate laying a cable across the Atlantic—a huge leap beyond anything previously attempted. The cable snapped while being paid out on the first attempts in 1857 and June 1858, but in August 1858 the first Atlantic cable was, to great rejoicing, finally successfully laid from Ireland to Newfoundland. The cable had been badly made and roughly handled, however, and it failed after just a month of fitful service.

The spectacular failure of the Atlantic cable, followed in 1860 by the costly collapse of the British government-backed Red Sea cable, threatened for a time to tag long-distance cable telegraphy as a great *failed* technology, with some blaming the supposed deficiencies of the gutta-percha insulation. But a thorough investigation in 1859–1860 by a joint committee of the British government and the Atlantic Telegraph Company absolved gutta-percha and pointed to haste and improper handling as the real culprits in the failures (Galton et al. 1861; Hempstead 1989; Hunt 2021, pp. 97–143). This verdict, together with renewed successes in laying shorter cables, helped restore public confidence in submarine telegraphy, and the successful completion in 1866 of two cables across the Atlantic ushered in a new era of maturity in the cable industry.

The success of the 1866 Atlantic cables soon set off a boom in submarine telegraphy, and by 1875 cables had been laid to India, China, Japan, and Australia and along the coasts of Africa and South America. Virtually all of these were manufactured and laid by British firms; indeed, most were controlled by one man, John Pender (1815–1896), known as the "cable king" (Ash 2018). Originally a cotton merchant in Glasgow and Manchester, Pender had been an investor in the first Atlantic cable and played an important part in renewing the effort after the 1858 failure. In 1864, he engineered the merger of the Gutta Percha Company with the iron rope makers Glass, Elliot to form the mammoth Telegraph Construction and Maintenance Company (Telcon), which dominated cable manufacture for many years and virtually monopolized the gutta-percha trade. Pender later formed the Eastern Telegraph Company and its many affiliates (chief ancestors of Cable & Wireless), and from the 1870s until his death in 1896 he effectively controlled both the operation and manufacture of most of the long cables in the world. He was also a Member of Parliament and (with the disingenuous motto "Telegraphs know no politics") worked closely with successive British governments to promote cable interests. Pender served the Empire by providing secure and reliable communications throughout the world, and the Empire served Pender by providing a ready market, and sometimes substantial subsidies, for the services he sold (Barty-King 1979, pp. 69–89; Coates and Finn 1979, pp. 170–71).

As the world cable network grew in the late nineteenth century (from about 15,000 nautical miles in 1866 to over 200,000 by 1900), so did the demand for gutta-percha. Rubber was its only real rival as cable insulation; when specially vulcanized and carefully applied in a process devised in the 1860s, rubber could be made to work satisfactorily, and was often used in warm tropical areas where gutta-percha tended to soften (Bright 1898, pp. 158–60). But the superior reliability and ease of manufacture of gutta-percha (and the backing of the huge Telcon combine) gave it the edge, and rubber-insulated cores never constituted more than a small fraction of the total mileage of submarine cables.

The long dominance of gutta-percha as an insulator resulted in part from the technological conservatism that overtook the cable industry once it achieved maturity around 1870. With huge capital investments literally sunk on the seabed, cable firms were understandably reluctant to experiment with untried materials and designs. Once they had found a type of cable that worked reasonably well, as copper insulated with gutta-percha manifestly did, they saw little reason to run the large risks involved in trying something new, particularly when the prospective benefits (slightly faster signaling speeds or lower manufacturing costs) were relatively small. Cable engineers and manufacturers instead emphasized rigorous quality control and the incremental refinement of established techniques. Innovation in cable telegraphy was confined mainly to the relatively inexpensive and easily replaceable sending and receiving instruments, with the design and manufacture of the cables themselves remaining largely unchanged from the 1860s until the 1920s or (in some respects) even the 1950s, when gutta-percha finally gave way to synthetic polyethylene insulation (Finn 2009; Noakes 2013).

Britain imported only about 200 pounds of gutta-percha in 1844, before its electrical use began, but this rose rapidly to 700 tons in 1848, 1500 tons in 1865, and over 2700 tons in 1873. British imports (the great bulk of the world market, brought in mainly through Singapore) peaked at about 5300 tons in 1890 and again at 6600 tons in 1900 before falling substantially, and averaged around 1800 tons per year from 1850 to 1900 (Obach 1898, p. 97; Godfrey 2018, p. 114). When one considers that a mature tree yielded only about a kilogram of gutta-percha, and was usually felled to extract it, one begins to grasp the devastating effects the hunt for gutta-percha had on the forests of Malaya (Headrick 1987; Tully 2009). In a series of lectures on gutta-percha at the Royal Society of Arts in 1897, Eugen Obach (of Siemens's London cable factory) said that the opening of the European market in the 1840s set off "almost a craze" for gutta-collecting among the Malayans, and that a huge number of old-growth trees, "probably hundreds of thousands, were ruthlessly destroyed during the first four or five years." There were virtually no large gutta trees left on Singapore by 1850, and they soon became scarce throughout much of southern Malaya (Obach 1898, p. 12).

Gutta trees were soon found on Borneo and Sumatra, however, and production of gutta-percha continued to rise through the rest of the century, though with substantial ups and downs and amid growing concern about the reliability of long-term supplies (Obach 1898, p. 97; Bright 1898, pp. 258–59; Godfrey 2018, pp. 76–82, 114). Moreover, as the old-growth trees were felled and harvesters turned to younger and less suitable trees, the quality of the gutta-percha offered for sale declined dramatically (Fig. 1.2). Indeed, cable engineers later in the century envied the high quality and low price of the gutta-percha available in the early days, remarking that tests on part of the original 1850 Channel cable that had been pulled up after 25 years' submersion "served to shew what excellent stuff it was composed of" (Bright 1898, pp. 10n, 13).

From the 1880s on, efforts were made to secure more reliable supplies of gutta-percha or find substitutes for it. Balata, a gutta-like substance derived from a South American tree (*Manilkara bidentata*), came into some use after 1890. Efforts were also made to find ways to gather gutta-percha without felling the trees, either by tapping them or by extracting latex from their leaves. Tapping proved ineffective, however, since the vesicles that held the fluid latex were not interconnected, and the

**Fig. 1.2** Kayans gathering gutta-percha in the Borneo forest. (From Hose and McDougall 1912)

leaf-based method had a low yield and was difficult to carry out (Bright 1898, pp. 259–60; Obach 1898, p. 19). In the 1880s, the French, chafing at British control of the world cable system, tried to secure their own supplies of gutta-percha by setting up plantations in Indochina, but through mismanagement, most of the trees soon died (Obach 1898, pp. 15–17; Godfrey 2018, pp. 69–70). The Dutch (aiming more at increasing their sales to the British than at breaking into the cable industry themselves) began to grow gutta trees on plantations in Java in the 1880s, but since it took 10 or 20 years before a tree yielded usable latex, such plantation methods only slowly replaced the harvesting of wild trees. By the 1920s, however, a steady supply of gutta-percha was being produced by plantations in Malaya and the Dutch East Indies, many of them controlled by Telcon (Bright 1898, p. 259; Brown 1927, p. 29; Headrick 1987, p. 14).

## Gutta-Percha and Electrical Measurement

Gutta-percha presents a striking example of a characteristic pattern of colonial trade: a natural material is extracted from the periphery of an empire, brought to its center, transformed in a way not available to those in the periphery, and then shipped back to the periphery (either the source region or another) and sold or used in some other way to reinforce the power of those at the center. One need only think of "Lancashire" cotton being sold in Egypt or India (by John Pender, among many others) to grasp the dynamics of the process. What makes the case of gutta-percha especially striking is that it served not only as a commodity in this cycle of imperial commerce and power, but also, once it had been transformed into cable insulation and laid on the seabed, as a crucial part of a technology that enabled the imperial system to sustain itself and expand its reach.

One of the keys to turning gutta-percha from mere coagulated latex into a pillar of British imperial power was the gathering of detailed knowledge of its electrical properties. In the 1850s and 1860s, British physics laboratories and cable company testing rooms became centers for the study, measurement, and standardization of gutta-percha (as well as copper and other materials) in preparation for using them in cables. By gathering and codifying this knowledge, scientists and engineers were able to make gutta-percha into a reliable component of a sophisticated technological system. In the process, they also transformed the practice of electrical measurement in ways whose consequences were to reach far beyond cable telegraphy.

Only the simplest electrical tests were used on the earliest cables, as the story of Wollaston's tongue suggests. As long as enough current got through to produce readable signals, no one worried very much about electrical niceties (Jenkin 1866, p. 462). But the failure of several early cables led engineers to adopt more elaborate testing procedures, and to begin using galvanometers, standardized resistance coils, and other instruments to locate breaks and incipient faults in the wire or insulation. Since even a tiny flaw could grow until it put an entire cable out of action, manufacturers had a strong incentive to emphasize strict quality control and to develop ever more searching techniques of electrical measurement and fault detection.

By the 1860s, British cable factories had become important arenas for electrical measurement, particularly of resistance. As Jenkin noted in his 1873 textbook, *Electricity and Magnetism,* "In telegraphy the measurement of resistance plays a very important part, regulating the choice of materials and enabling the electrician to test the quality of the good supplied" (Jenkin 1873, p. 229). Concern with quality control and fault detection drove cable engineers to push the limits of available measuring techniques. Indeed, Thomson, who was intimately familiar with both cable factories and university physics laboratories, later declared that:

> Resistance coils and ohms, and standard condensers and microfarads, had been for ten years familiar to the electricians of the submarine-cable factories and testing-stations, before anything that could be called electric measurement had come to be regularly practised in almost any of the scientific laboratories of the world (Thomson 1884, p. 150).

It was no accident that the cable engineers Latimer Clark (1822–1898) and Charles Bright (1832–1888) were among the first to call for a connected system of electrical units and standards, as they did at the 1861 meeting of the British Association for the Advancement of Science. When, at the same meeting, Thomson and Jenkin took the lead in forming the British Association Committee on Electrical Standards, one of their chief aims was to serve the needs of the cable industry while also forging closer links between that industry and the community of electrical physicists. The set of units and standards the British Association committee established—substantially the system of ohms, amps, and volts that we still use today—strongly reflected its roots in cable telegraphy (Schaffer 1992; Smith and Wise 1989, pp. 687–95; Hunt 1994; Hunt 2021, pp. 144–80).

At first, cable manufacturers and contractors paid little attention to the electrical qualities of the materials they used; as Willoughby Smith later said of the copper used in the first Channel cable, "To its electrical conditions and quality no attention was given, for the simple reason that all copper wire was credited with the possession of equal value in these respects" (Smith 1891, p. 2). But it soon became clear that different samples of copper and gutta-percha could in fact differ widely in their electrical properties. In 1857, Thomson and some of his Glasgow students used a Wheatstone bridge to compare samples of wire made for use in the first Atlantic cable. They found to their surprise that some of these supposedly identical "pure" copper wires conducted current only about half as well as others. Thomson quickly published a paper drawing attention to this variation in the conductivity of commercial copper, later traced to the effect of small impurities, especially arsenic, and stressed the importance of using only high conductivity copper in telegraph cables (Thomson 1857; Hunt 2021, pp. 151–53). He also persuaded the Atlantic Telegraph Company (on whose board of directors he sat) to specify that only high conductivity copper be accepted for use in a new length of cable then being ordered, and the Gutta Percha Company soon began testing all wire sent to it and rejecting any that did not meet the new standard of conductivity (Blake-Coleman 1992, 150–73).

Testing of gutta-percha followed a somewhat similar pattern, as engineers and manufacturers identified different grades of raw material and realized that different processing methods could significantly affect the conductivity of the final product. But such differences in the conductivity of gutta-percha turned out to be of far less practical importance than those in copper. Researchers found that the conductivity of gutta-percha varied widely with temperature (by about a factor of 10 between 5° and 25 °C), making the measurement of small variations between samples both difficult and pointless (Clark 1868, p. 89; Jenkin 1860, p. 415). Moreover, engineers found that very high insulation did not improve the performance of a cable; indeed, signaling was actually somewhat better on cables with a little leakage across the gutta-percha sheath. As long as the insulation was free of serious defects, particularly ones that might grow and become fatal, there was no need to use gutta-percha of especially low conductivity. As one expert noted in 1898, conductivity tests were a useful way to check the general quality of gutta-percha samples, but from an electrical point of view, "any gutta-percha would probably offer a sufficient specific resistance, and usually much more is obtained than is really necessary" (Bright 1898, p. 326). Using more gutta-percha than was really needed for electrical purposes was another example of the conservatism of late Victorian cable engineering; extra gutta-percha was applied more as a hedge against possible breakdown than as a way to improve the speed or clarity of signaling.

Beyond considerations of quality control, however, knowledge of the electrical properties of gutta-percha was important for the quantitative understanding of signaling rates and the efficient design of cables. From the late 1850s on, great effort and ingenuity were devoted to measuring both the conductivity and specific inductive capacity (permittivity) of gutta-percha, with significant effects on the development of electrical measuring techniques and on electrical science as a whole.

The most important early study of the electrical properties of gutta-percha was that undertaken in 1859 by Jenkin, then a young engineer at the Birkenhead cable works of Robert Stirling Newall (1812–1889) and later a leading figure in submarine telegraphy. Jenkin had been given the task of testing coils of cables being readied for use in the ill-fated Red Sea project, and he brought to bear the best available techniques of electrical measurement (Jenkin 1861, p. 264; Thomson to Stokes, 8 Feb. 1860, in Wilson 1990, 1: 254–55). He also met and began to correspond with Thomson, inquiring about improved measuring techniques and instruments, appropriate electrical units, and the like. The collaboration between Jenkin and Thomson proved very fruitful, and later ripened into a lucrative patent partnership (Smith and Wise 1989, pp. 699–705; Cookson and Hempstead 2000, pp. 67, 73–77).

Before 1859, measurements of cable insulation had sought only to compare the leakage across the gutta-percha sheath of one cable with that across the sheath of another. In the most direct method, a large battery and galvanometer were connected to one end of the core of a cable while the other was kept insulated with its outer covering grounded (usually by immersing the cable in water). The galvanometer then indicated how much current was leaking through the gutta-percha. Another method was to connect the insulated core to a battery and charge it to a high potential, as measured on an electrometer; the core was then disconnected, with both of its ends kept insulated, and its charge allowed to decay by leakage through the insulation. By measuring the time it took for the potential to fall to half its initial value (ranging from a few seconds to several hours), one could calculate the rate of leakage and the relative conductivity of the insulation (Clark 1861, pp. 298–300).

Using mainly the first of these methods, Jenkin set out in 1859 to measure the conductivity of gutta-percha not just against that of another insulator, but against that of copper itself. As he noted in his published report, he treated gutta-percha as a conductor "offering a resistance to the electric fluid similar to that offered by any so-called good conductor, such as copper or iron" (Jenkin 1861, p. 464). Jenkin was seeking to erase any strict dividing line between insulators and conductors and instead treat all materials as part of a continuous range of conducting dielectrics—an approach deriving ultimately from Faraday, and that was to become central to Maxwellian electrical theory (Buchwald 1985, p. 28; Hunt 1991b, pp. 62–63). But while the electrical conductivity of different materials might all lie on a single continuum, it is one with an extremely wide range; indeed, it is one of the widest of any known physical property. Jenkin's ability to put the conductivities of copper and gutta-percha on the same scale was a triumph of electrical measurement and strong testimony to the delicacy and range it had begun to achieve by the late 1850s—in large part in response to the demands and opportunities presented by cable telegraphy.

Jenkin began by using a Wheatstone bridge and some standard (though not entirely satisfactory) resistance coils to measure the resistance of the copper core of the cable. He found it to be about $25 \times 10^7$ "Thomson units" (in later terms, just under 1 ohm) per nautical mile. Jenkin then turned to the gutta-percha sheath. He would have liked to have measured its resistance directly with the same sort of bridge arrangement, using the copper core as one arm of the bridge and the gutta-percha as the other, but he had, he said, "no measuring coils of sufficient resistance for this purpose" (Jenkin 1861, p. 472). Much of the impetus for the construction of high-resistance coils a few years later derived from attempts to measure the conductivity of gutta-percha, and it was no accident that when Maxwell performed his fundamental measurements of the ratio of electromagnetic and electrostatic units in 1868, he thanked Willoughby Smith of the Gutta Percha Company for the loan of a million-ohm coil (Maxwell 1868, p. 645). Since no such calibrated coils existed in 1859, Jenkin was forced to resort to an indirect method, using a very delicate galvanometer to measure the leakage through the gutta-percha sheath and then comparing its sensitivity with that of the ordinary galvanometer he had used to measure the flow of current through the copper core. A simple calculation then yielded the ratio of the resistance of the gutta-percha to that of the copper, and so the specific resistivity of the gutta-percha, which Jenkin found ranged from $600 \times 10^{17}$ Thomson unit-feet at 27 °C to over $5600 \times 10^{17}$ at 10 °C (i.e., from about $2 \times 10^{12}$ to about $17 \times 10^{12}$ ohm-feet). Gutta-percha thus had a resistivity about $10^{20}$ times that of copper (Jenkin 1861, pp. 468–72).

Beyond a rough estimate Thomson had made in 1857 while working on the first Atlantic cable, Jenkin's measurements of gutta-percha were, Thomson said, "the only ... absolute measurements of the electric resistance of an insulating material" made up to that time (Thomson 1887, p. clvi). Even before Jenkin's work was published, Thomson praised it lavishly in the venerable *Encyclopedia Britannica*, declaring that by "experiments of a very precise character," Jenkin had "put in practice systematically for the first time" methods for measuring the specific resistance of insulating substances such as gutta-percha not simply relative to an arbitrary material standard, but in "absolute" terms based on units of length, mass, and time (Thomson 1860, p. 97). When in 1864 Thomson nominated Jenkin for election to the Royal Society of London, he singled out this paper of Jenkin's as "in one respect the most valuable; as it contains an accurate determination of a property of matter (the specific resistance of the insulator of the cables tested) in approximate absolute measure"—and there were few things Thomson valued more than an accurate determination of a property of matter, particularly one of practical value (Thomson to Stokes, 20 April 1864, in Wilson 1990, 1: 320). Jenkin's figures provided basic design information that enabled cable engineers to calculate, for cables of any length or thickness, what fraction of the total current would reach the far end through the copper and what fraction would leak out laterally through the gutta-percha. Jenkin's measurements at Birkenhead also laid the groundwork for his collaboration with Maxwell a few years later on the British Association Committee on Electrical Standards, and for the eventual adoption by cable engineers of the "absolute" units, principally the ohm, that the committee devised and disseminated (Hunt 2021, pp. 159–80).

Besides resistance, gutta-percha had another important electrical property: specific inductive capacity, or permittivity. In 1837, Faraday had discovered that introducing different dielectrics between the plates of a capacitor increased the amount of charge stored for a given difference of potential, and, taking air as his reference material, had measured the ratio of this increase for sulfur, shellac, and glass (Faraday 1838, pp. 25–37). That insulators really played such an active role in electrical phenomena was not widely accepted at the time; indeed, Thomson later said that "the existence of specific inductive capacity was either unknown, or ignored, or denied, by almost all the scientific authorities of the day" (Thomson 1887, p. clvii; see also Smith and Wise 1989, pp. 219–28). But when in 1853 Clark and other cable engineers noticed a puzzling "retardation" of signals sent along underground and submarine lines, and Faraday traced this to the effect of electrostatic induction across the insulation, the role of the dielectric began to attract more attention, at least in Britain, and the specific inductive capacity of gutta-percha became a matter of pressing concern to cable engineers (Faraday 1854; see also Hunt 1991a, 2021, pp. 14–24).

According to the theory Thomson had worked out in 1854, the maximum rate at which signals could be sent along a cable was determined by the product of its resistance and capacitance (Thomson 1855; see also Smith and Wise 1989, pp. 446–63, and Hunt 2021, pp. 26–36). Thus, other things being equal, the lower the specific inductive capacity of the insulation, the better. Determining the relative inductive capacities of different types of gutta-percha, rubber, and other insulators thus became a task of great practical importance. Moreover, if Thomson's theory was correct, one could, in principle, gauge the specific inductive capacity of a material by analyzing how a cable insulated with it responded to pulses of current. One could then also determine the ratio of electromagnetic to electrostatic units, and so shed light on some of the deepest questions of electrical theory.

In his testimony to the Joint Committee on Submarine Telegraphs in December 1859, Jenkin noted the importance of comparing the induction in cables insulated with gutta-percha and with rubber (preliminary tests suggested rubber had a substantial advantage), and declared that "experiments upon.... specific inductive capacity are very much wanted." Virtually nothing had been done along these lines since Faraday's original work on sulfur, shellac, and glass; as for measurements of gutta-percha, rubber, or other materials used to insulate cables, Jenkin said, "I know of none" (Jenkin testimony in Galton et al. 1861, pp. 137, 141).

Jenkin had, however, devised a way to extract a value for the inductive capacity by analyzing pulses sent through a cable, and at the end of a long paper on "Experimental Researches on the Transmission of Electric Signals through Submarine Cables—Part I," submitted to the Royal Society in May 1862, he remarked that "From this value [of a signalling-rate parameter] the electrostatical capacity per unit length, and the specific inductive capacity of the dielectric could be determined. These points will, however, be more fully treated of in the second part of this paper" (Jenkin 1862, p. 116). As it happened, however, no "Part II" ever appeared, though as Thomson later noted, "Jenkin had in fact made a determination at Birkenhead of the specific inductive capacity of gutta-percha"; indeed, Thomson said, "This was the very first true measurement of the specific inductive capacity of a dielectric which had been made after the discovery by Faraday of the existence of the property" (Thomson 1887, p. clvii). Moreover, as Thomson pointed out in nominating Jenkin for election to the Royal Society of London, when this was combined with measurements of the resistivity of copper, one needed only a simple electrostatic measurement "to complete the data for a comparison between electrostatic and electromagnetic measure" (Thomson to Stokes, 20 April 1864, in Wilson 1990, 1: 320).

Given that the coincidence between the speed of light and the ratio of electrostatic and electromagnetic units was then the main empirical evidence in favor of Maxwell's new electromagnetic theory of light, Jenkin's measurement was clearly of great interest and importance. It is all the more to be regretted, then, that he never published a full account of his work. Word of his results got around, however; indeed, by December 1861, Maxwell was asking Thomson: "Do you know any good measures of dielectric capacity of transparent substances? ... I think Fleeming Jenkin has found that of gutta percha[,] caoutchouc [rubber] &c. Where can one find his method, and what method do you recommend" (Maxwell to Thomson, 10 Dec. 1861, in Harman 1990–2002, 1: 696). Maxwell did not say how he first heard of Jenkin's results; although the two had been schoolmates years before at the Edinburgh Academy, their personal acquaintance apparently dates from 1862, when Maxwell joined Jenkin and Thomson on the British Association Committee on Electrical Standards. In any case, Maxwell knew enough to ask Thomson, who pointed him in the right direction and probably put him in direct touch with Jenkin. Maxwell later made good use of Jenkin's measurements, and others soon learned of them as well. In his 1868 *Elementary Treatise on Electrical Measurement,* Latimer Clark credited Jenkin for providing values for the specific inductive capacities of gutta-percha (4.2), pure rubber (2.8), and other insulating materials, and Jenkin's numbers soon spread through the textbook literature (Clark 1868, p. 148).

In 1861, Clark said of the discovery of specific inductive capacity that "At the date of Faraday's interesting researches it could little be foreseen that such an obscure phenomenon should be destined to become one day, as it has now, a consideration of high national importance, and one which has a direct and most important bearing on the commercial value of all submarine telegraphs" (Clark 1861, p. 313). It was, of course, the advent of cable telegraphy that gave this "obscure phenomenon" its "high national importance," and that made the study of dielectrics—particularly gutta-percha—and the measurement of their properties the focus of so much effort in Victorian Britain, with ramifications that reached deep into the practice and content of electrical science. Gutta-percha was one of the key materials of nineteenth-century cable telegraphy. Cables could have been insulated with rubber (but then we would just have a slightly different story to tell about a slightly different tropical material), or just possibly with some kind of tarred cloth or paper, though it seems unlikely that the latter could have been made to work very well. Without gutta-percha or something very similar to it, neither the British Empire nor British electrical physics would have developed in quite the way they actually did in the second half of the nineteenth century. And without the demands of the cable industry, the forests of Malaya, Borneo, and Sumatra would have been spared the depredations of the gutta-percha hunters.

The global system of submarine cables that grew up in the latter part of the nineteenth century served in many ways as a vast electrical laboratory. As Maxwell noted in the preface to his great *Treatise on Electricity and Magnetism* in 1873:

> The important applications of electromagnetism to telegraphy have … reacted on pure science by giving a commercial value to accurate electrical measurements, and by affording to electricians the use of apparatus on a scale which greatly transcends that of any ordinary laboratory. The consequences of this demand for electrical knowledge, and of these experimental opportunities for acquiring it, have been already very great, both in stimulating the energies of advanced electricians, and in diffusing among practical men a degree of accurate knowledge which is likely to conduce to the general scientific progress of the whole engineering profession …. (Maxwell 1873, 1: vii–viii).

Gutta-percha was an important component of this enormous telegraphic laboratory and itself the focus of considerable study. Without gutta-percha (or something very much like it), a global cable network would not have been built in the nineteenth century; the "energies of advanced electricians" in Britain would not have been stimulated in quite the directions they actually were; and theories of electromagnetic propagation and the action of the field would not have evolved quite as they actually did.

Not bad for what many took to be mere "Malayan tree sap."

# References

Ash S (2018) The cable king: the life of John Pender. Ash, London
Barty-King H (1979) Girdle round the earth: the story of cable and wireless and its predecessors. Heineman, London
Blake-Coleman BC (1992) Copper wire and electrical conductors: the shaping of a technology. Harwood, Chur
Bright C (1898) Submarine telegraphs: their history, construction, and working. Lockwood, London
Brockway LH (1979) Science and colonial expansion: the role of the British Royal Botanic Gardens. Academic Press, New York
Brown FJ (1927) The cable and wireless communications of the World. Pitman, London
Buchwald JZ (1985) From Maxwell to microphysics: aspects of electromagnetic theory in the last quarter of the nineteenth century. University of Chicago Press, Chicago
Clark L (1861) Report. In: Galton D et al (eds) Joint committee report, pp 293–335
Clark L (1868) An elementary treatise on electrical measurement. Spon, London
Coates V, Finn B (eds) (1979) A retrospective technology assessment: submarine telegraphy—the transatlantic cable of 1866. San Francisco Press, San Francisco
Cookson G, Hempstead CA (2000) A Victorian scientist and engineer: Fleeming Jenkin and the birth of electrical engineering. Ashgate, Aldershot
Drayton R (2000) Nature's government: science, imperial Britain, and the 'improvement' of the world. Yale University Press, New Haven
Faraday M (1838) Experimental researches in electricity—eleventh series. Phil Trans 128:1–41
Faraday M (1848) On the use of gutta percha in electrical insulation. Phil Mag 32:165–167
Faraday M (1854) On electric induction—associated cases of current and static effects. Phil Mag 7:197–208
Finn B (2009) Submarine telegraphy: a study in technological stagnation. In: Finn B, Yang D (eds) Communications under the seas: the evolving cable network and its implications. MIT Press, Cambridge, pp 9–24

Galton D et al. (1861) Report of the joint committee to inquire into the construction of submarine telegraph cables, British parliamentary papers, LXII.591, London

Godfrey H (2018) Submarine telegraphy and the hunt for gutta-percha. Brill, Leiden

Harman PM (ed) (1990–2002) Scientific papers and letters of James Clerk Maxwell. 3 vols. Cambridge University Press, Cambridge

Headrick DR (1987) Gutta-percha: a case of resource depletion and international rivalry. IEEE Technol Soc Mag 6:12–18

Headrick DR (1991) The invisible weapon: telecommunications and international politics, 1851–1945. Oxford University Press, Oxford

Headrick DR (1996) Botany, chemistry, and tropical development. J World Hist 7:1–20

Hempstead C (1989) The early years of oceanic telegraphy: technology, science and politics. Proc Inst Elec Engineers 186A:297–305

Hose C, McDougall W (1912) The pagan tribes of Borneo. 2 vols. Macmillan, London

Hunt BJ (1991a) Michael Faraday, cable telegraphy and the rise of field theory. Hist Technol 13:1–19

Hunt BJ (1991b) The Maxwellians. Cornell University Press, Ithaca

Hunt BJ (1994) The ohm is where the art is: British telegraph engineers and the development of electrical standards. Osiris 9:48–63

Hunt BJ (2021) Imperial science: cable telegraphy and electrical physics in the Victorian British empire. Cambridge University Press, Cambridge

Jenkin F (1860) On the insulating properties of gutta percha. Proc Roy Soc 10:409–415

Jenkin F (1861) On the insulating properties of gutta percha. In: Galton D et al (eds) Joint committee report, pp 464–481

Jenkin F (1862) Experimental researches on the transmission of electric signals through submarine cables—part I. Laws of transmission through various lengths of one cable. Phil Trans 152:987–1017

Jenkin F (1866) Submarine telegraphy. North British Review 45:459–505

Jenkin F (1873) Electricity and magnetism. Longmans, London

Löser W (1969) Der bau unterirdischer telegraphenlinien in Preussen von 1844–1867. NTM 6:52–67

Maxwell JC (1868) On a method of making a direct comparison of electrostatic and electromagnetic force; with a note on the electromagnetic theory of light. Phil Trans 158:643–658

Maxwell JC (1873) Treatise on electricity and magnetism, 2 vols. Clarendon Press, Oxford

Noakes R (2013) Industrial research at the Eastern Telegraph Company, 1872–1929. Br J Hist Sci 47:119–146

Obach E (1898) Cantor lectures on gutta percha. Society of Arts, London

Schaffer S (1992) Late Victorian metrology and its instrumentation: 'a manufactory of ohms'. In: Bud R, Cozzens S (eds) Invisible connections: instruments, institutions, and science. SPIE, Bellingham, pp 23–56

Siemens W (1966) Inventor and entrepreneur: recollections of Werner von Siemens. Lund Humphries, London

Smith C, Wise N (1989) Energy and empire: a biographical study of Lord Kelvin. Cambridge University Press, Cambridge

Smith W (1891) The rise and extension of submarine telegraphy. Virtue, London

Thomson W (1855) On the theory of the electric telegraph. Proc Roy Soc 7:382–399

Thomson W (1857) On the electric conductivity of commercial copper of various kinds. Proc Roy Soc 8:550–555

Thomson W (1860) Telegraph, electric. In: Encyclopedia Britannica, vol 21, 8th edn, pp 94–116

Thomson W (1884) Electrical units of measurement. In: Practical applications of electricity. Institution of Civil Engineers, London, pp 149–174

Thomson W (1887) Note on the contributions of Fleeming Jenkin to electrical and engineering science. In: Colvin S, Ewing JA (eds) Jenkin papers, vol 1. Longman, London, pp clv–clix

Tully J (2009) A Victorian ecological disaster: imperialism, the telegraph, and gutta-percha. J World Hist 20:559–579

Wilson DB (ed) (1990) The correspondence between Sir George Gabriel Stokes and Sir William Thomson, Baron Kelvin of Largs. 2 vols. Cambridge University Press, Cambridge

# Chapter 2
# Cocoa, Cadbury and Forced Labour in São Tomé and Príncipe, West Africa

Catherine Higgs

**Abstract** "Cocoa, Cadbury and Forced Labour" explores an early twentieth century controversy. One of the world's premier chocolate makers, the Quaker firm Cadbury Brothers Limited of Birmingham, England, spent a decade investigating whether slaves harvested the cocoa the company purchased from the Portuguese island colony of São Tomé and Príncipe. This chapter considers competing British and Portuguese colonial ambitions, the tension between capitalism, philanthropy and ethics, and the different interpretations held by British and Portuguese employers about the nature of work. Higgs examines these issues from multiple perspectives, including those of Joseph Burtt, the fellow Quaker Cadbury Brothers sent to Africa to study the conditions of labour. In a story still familiar a century after Burtt's journey, the cocoa controversy revealed the idealism, naivety and racism that shaped attitudes toward Africa, even among those who sought to improve the conditions of its workers and to support its farmers.

**Keywords** Cocoa · Cadbury · Chocolate · Slavery · Forced labour · São Tomé and Príncipe · Angola · Ghana · Côte d'Ivoire

## Introduction

In the early twentieth century the controversy over whether slaves were harvesting cocoa (*Theobroma cacao*), on the West coast of Africa became an international scandal, not only because slavery had, in theory, long been outlawed but also because one of the major purchasers of the cocoa was the chocolate maker Cadbury

C. Higgs (✉)
Department of History and Sociology, University of British Columbia Okanagan, Kelowna, BC, Canada
e-mail: catherine.higgs@ubc.ca

© The Author(s), under exclusive license to Springer Nature Switzerland AG 2024
L. M. Mendonça de Carvalho (ed.), *The Victorians: A Botanical Perspective*,
https://doi.org/10.1007/978-3-031-68759-4_2

Brothers Limited.[1] The source of the cocoa was the Portuguese island colony of São Tomé and Príncipe, which contracted most of its labourers from Angola, also a Portuguese colony. The Cadburys were Quakers, members of the Religious Society of Friends. Among the philanthropic causes they supported was the anti-slavery movement, and the protection of indigenous workers. William Cadbury helped finance the *African Mail*, the newspaper founded by the activist Edmund Morel in the early 1900s to expose the brutality of forced labour on the rubber concessions in the Congo Free State, the private territory claimed by the Belgian King Leopold II. (Higgs 2012: 2, 21).

Further complicating the cocoa controversy were British and Portuguese colonial considerations and the interests of capitalists. These included the Portuguese planters producing cocoa and the estate owners in Portuguese Mozambique on Africa's east coast earning fees by supplying labour to British-owned gold mines in the Transvaal in British-occupied South Africa. The British Foreign Office worried that pressuring Portugal to address labour concerns in São Tomé and Príncipe might lead to a decline in Mozambican miners in British Transvaal. (Higgs 12–13).

Ultimately, concern over the condition of workers prompted William Cadbury to hire Joseph Burtt to investigate the conditions of labour in Portuguese Africa. His journey coincided with the investigation of Henry Nevinson, an activist journalist writing an exposé of slavery in Portuguese Africa for the American *Harper's Magazine*. Their separate journeys along the slave route from Benguela on Angola's west coast to Kavungu on its eastern border led to a similar conclusion. Slaves laboured on the cocoa estates of São Tomé and Príncipe. (Higgs, 8, 23, 114) The long delay in Cadbury Brothers' decision to stop buying São Toméan cocoa, however, prompted charges of hypocrisy in British newspapers. The firm sued for libel. Cadbury Brothers also aligned itself with the British Foreign Office in its efforts to improve the conditions of labour in Portuguese Africa, even as it sought in the Gold Coast (Ghana), an alternate source of cocoa produced by free African farmers. (Higgs 145, 147–148, 156, 159–160) The complexity of this controversy over cocoa a century ago echoes into the present day.

## Cadbury Brothers Limited

Before Cadbury Brothers Limited became a manufacturer of fine drinking cocoa, the firm sold coffee (*Coffea*) and tea (*Camellia sinensis*) in Birmingham, England, beginning in 1824. The Cadburys were Quakers, members of the Religious Society of Friends. Excluded from politics and lacking the connections that came with membership in the Church of England, many Quakers turned to business during the

---

[1] This article is adapted from Catherine Higgs, *Chocolate Islands: Cocoa, Slavery, and Colonial Africa* (Athens: Ohio University Press, 2012) with the permission of the press. Unless otherwise noted, citations are to *Chocolate Islands*, where readers will find extensive endnote references to archival, primary and secondary sources.

Industrial Revolution and acquired wealth in mining, banking and insurance. Capitalism was acceptable, but Quaker products had to be beneficial to society and priced fairly. In many ways, this made the Cadburys atypical Victorians.

When Richard Cadbury and his brother George took over Cadbury Brothers in the early 1860s, it was nearly insolvent. They saved the firm by investing in cocoa in 1864. Two years later, they introduced "Pure Cocoa Essence", marketing it to Victorian consumers as a nutritious hot drink. (Higgs 1-2n3, n5) Cocoa brought the Cadbury family wealth, some of which they invested in good causes favoured by the Society of Friends. These included the adult school movement, prison reform and the abolition of slavery and forced labour. Drinking cocoa as an alternative to hard liquor neatly supported another Quaker cause, the temperance movement. It was no coincidence that two other large British cocoa manufacturers, J. S. Fry and Rowntree, were also Quaker family firms.

For the Cadburys, improving workers' lives was also a key initiative. In 1878, the brothers purchased the site for the new Cadbury factory at Bourneville, which included housing for foremen. (Higgs 2-3n5) In 1895, George Cadbury bought 120 acres adjacent to the factory and designed Bournville Village as a model working-class community. Intended as an antidote to the dire conditions in which many poor residents of Birmingham lived, it came complete with garden cottages and parks, but no pubs. George Cadbury donated it to the Bournville Village Trust in 1900. (Higgs 2, 3n8)

## Cocoa and Controversy

Controversy touched the cocoa that allowed the Quaker Cadburys to do good works and treat their British workers with dignity. A major supplier of the company's cocoa beans was the Portuguese island colony of São Tomé and Príncipe, off West Africa's coast. Most of the workers on the islands came from Angola, a large territory further south along the West African coast which Portugal had begun colonizing in the sixteenth century. Britain and Portugal had been allies since the fourteenth century, but Portugal was the far weaker partner. In the early 1800s, Britain pushed Portugal to end its participation in the Atlantic Slave Trade, which Britain had done in 1807. When the major European powers, including Britain, France and Germany met in Berlin in 1884–1885 to formally map the colonization of Africa in what became the 1885 Berlin Act, Portugal managed to assert its historic claims to Angola, Mozambique, Guinea and the islands of São Tomé and Príncipe and Cape Verde. Britain and Germany blocked Portugal's ambition to link Angola on the west coast to Mozambique on the east coast by claiming territory in the interior of Africa. (Higgs 10n4, 11n5)

In the late Victorian era, Britain claimed an empire that included parts of North America, Africa and Asia. Britain also had extensive business interests in Angola and maintained a consulate in Luanda, the colonial capital. In his 1894 report to the British Foreign Office, Consul Clayton Pickersgill described what he considered a slave-labour system practiced on the coffee estates of Cazengo in northern Angola. Portugal had outlawed slavery in its colonies in 1858, but like France in Madagascar and Britain in Zanzibar, it struggled to redefine newly freed slaves legally and to compel them to work in a wage economy. In the Portuguese colonies, former slaves, now called *libertos* (freedmen), were obliged to work for 20 years for their former masters. The system ended 3 years early, in 1875, and *serviçais* (singular *serviçal*), literally meaning "servants" in Portuguese, replaced the libertos. In southern Angola, *serviçais* had their 5-year contracts renewed without their knowledge by the local colonial magistrate. Investigating further, Pickersgill traveled north along the West African coast to the equator and the islands of São Tomé and Príncipe, where he found the "*serviçal* system" well entrenched. The good working conditions impressed him, but in an 1897 addendum to his earlier report, he quoted a line from a song sung by the contract labourers: "In São Tomé there's a door for going in, but none for going out". As in Angola, contracts were automatically renewed, and workers had no mobility. They were effectively slaves. The response from the British Foreign Office was silence. (Higgs 11-12n7)

There were prominent Portuguese critics of labour practices on São Tomé and Príncipe, including Júdice Biker. In the late 1890s, while working for the reform-minded governor-general of Angola, António Ramada Curto, Biker wrote a report on improving the recruitment, payment, and working conditions of African labourers. In December 1897, Biker lambasted a system that brought Angolans from the interior, signed them to 5-year contracts to work in São Tomé and Príncipe that they did not understand, and subjected them to 11½-hour workdays on coffee and cocoa plantations in a tropical climate where they suffered high death rates. They never had the opportunity to return home. Biker asked, did "the *roceiro* [plantation owner] makes their life so agreeable, dressing them and feeding them so well, instructing them, civilizing them, creating necessities for them? … Do they choose to continue to work there, renewing their contracts? … Would that this were so". In part, Biker was protesting a system that robbed Angola of labour. On the issue of working conditions on the islands, however, most Portuguese officials—like their British counterparts—chose silence. (Higgs 12n8)

## William Cadbury and Labour in São Tomé and Príncipe

Addressing the issue of whether Cadbury Brothers was purchasing cocoa harvested by forced labour would fall to Richard Cadbury's son William, who joined the family firm in 1887 at age 20. (Higgs 8n22) On an April morning in 1901, 2 months after the death of Queen Victoria, William Cadbury sat at his desk in Bournville reading a catalogue listing a cocoa estate for sale in São Tomé. Along with the estate's 6175 acres, its buildings, machinery, tools, and vehicles, two items caught Cadbury's eye—cattle valued at £420 and "200 black labourers" valued at £3555. Offering workers for sale made him suspect that the conditions of labour on São Tomé were less than ideal. In the early 1900s, Cadbury Brothers imported about

55% of its cocoa from São Tomé and Príncipe, the world's third largest exporter of cocoa after Ecuador and Brazil. The firm would not knowingly use slave-harvested cocoa in the manufacture of its chocolate. Labour, whether at the Bournville Works or on the *roças* of São Tomé, should be dignified and a worker should be free to leave his or her job. (Higgs 9n1)

Troubled by the possibility that slaves were harvesting the cocoa that went into Cadbury Brothers' drinking cocoa and chocolate candy, William Cadbury wrote to William Albright and Joseph Sturge, fellow Quakers active in the British Anti-Slavery Society. Cadbury explained: "One looks at these matters in a different light when it affects one's own interests, but I do feel that there is a vast difference between the cultivation of cocoa and gold or diamond mining, and I should be sorry needlessly to injure a cultivation that as far as I can judge provides labour of the very best kind to be found in the tropics: at the same time we should all like to clear our hands of any responsibility for slave traffic in any form". Before he criticized his Portuguese cocoa suppliers, William Cadbury wanted more evidence. (Higgs 12n9)

The gold and diamond mines to which Cadbury was referring were in what was then the British-occupied South African Republic (the Transvaal), and the British Cape Colony. The Anglo-Boer War or Second South African War had begun in October 1899. Britain claimed that the conflict was about restoring the civil rights denied to British miners in the gold-rich republic. Residents of the Transvaal, descendants of seventeenth-century Dutch settlers who harboured a deep antipathy to the British, concluded it was all about stealing their gold. George Cadbury, a committed pacifist like all members of the Society of Friends, publicly opposed the war in the pages of the *Daily News*, the newspaper he had bought for that purpose. His outspoken position hardly endeared him to the British Foreign Office. Meanwhile, British-owned mining companies in occupied Transvaal were desperate for labour, and their recruiting practices were sometimes controversial. One of their major sources of workers was the neighbouring Portuguese colony of Mozambique. Writing to Albright and Sturge, William Cadbury argued that working conditions in São Tomé were superior to those in the Transvaal mines, but he did note that in both places, "it sounds very much as if the labour was contracted on the same lines", implying that recruiters had used force to secure workers. Though reluctant to go public without confirmation, Cadbury Brothers would sanction neither slave trading nor slave labour. (Higgs 12-13n10)

The 1902 Foreign Office report on São Tomé and Príncipe noted that in 1901, 4752 contract workers (2616 men and 2136 women) had arrived on the islands. Most came from Angola. In his public report, Arthur Nightingale did not describe conditions in Angola or on the islands as slavery, nor did he comment on whether workers returned home at the end of their contracts. He did acknowledge the high death rate among serviçais, particularly on Príncipe, where sleeping sickness was endemic. This necessitated "constant fresh supplies of labourers". (Higgs 13) In his private correspondence with the Foreign Office, Nightingale described, "A repugnant traffic in human beings … that will not be abolished until some strong influence from outside is brought to bear on the matter". (Higgs 13n11)

On January 29, 1903, the Portuguese government issued a royal decree in an attempt to curb the worst abuses of the contract-labour system and, more significantly, to ensure the labour supply for the islands of São Tomé and Príncipe. Five-year contracts were strictly enforced, and 40% of a worker's monthly wage was set aside to pay for his or her repatriation. Workers' hours were reduced to 9½ hours per day, with a half day on Sunday, for a 62-hour workweek. A minimum wage was introduced, with a guaranteed increase of 10% at the end of each 5-year contract. Some critics saw the 1903 labour decree as a classic example of the Portuguese strategy of "para o Inglês ver", meaning "for the English to see". In other words, it was "window dressing", and on the ground, little would change. (Higgs 13–14 n12)

The Portuguese legation in London meanwhile went on the offensive in support of the January 1903 decree. Replying to queries from the British Anti-Slavery Society about the condition of labour in São Tomé and Príncipe in late February 1903, the Marquis of Soveral conveyed the government's position that "Portugal may justly boast of having completely suppressed the wicked traffic, which nowadays only has an existence in the imagination of certain philanthropists". (Higgs 14n13)

Portuguese indignation also revealed a fundamentally different interpretation of the meaning of work. If for the British and especially for the Quaker Cadburys work was intimately linked to notions of the dignity of labour and mobility and the freedom to leave one's job, the focus for the Portuguese was on working conditions. Contracted African labourers in São Tomé and Príncipe were housed, fed, and clothed and were paid for their work; their medical and emotional health were also attended to. Indeed, as one letter to the Lisbon newspaper *O Século* (The Century) would put it, "the servants of São Tomé are much happier than our own peasants". (Higgs 14-15n14)

Still unconvinced, William Cadbury visited Lisbon in March 1903 and spoke with several prominent Portuguese businesspeople involved in the cocoa trade on São Tomé and Príncipe, and in the recruitment of labour in Angola. They included José Constantino Dias, the Count of Vale Flôr, Francisco Mantero and António Ferreira. (Higgs 16–18) A common refrain, including from the British consul in Lisbon, Sir Martin Gosselin, was to lay the blame for the rumoured Angolan slave trade at the feet of African chiefs, whom Gosselin dismissed as "petty kings" who "treated the people as cattle and for liquor or money will readily trade their own people". This was a misrepresentation. As a general rule, the chiefs traded their enemies and only sold their own people when they had committed crimes or other offenses that challenged the society's stability. Gosselin also conveniently ignored the question of whether the chiefs would trade anyone in the absence of a demand for their labour. In William Cadbury's view, paternalistic though it may have been, African responsibility was less the issue than the Portuguese obligation as a "civilized" people to impose order and end the trade. (Higgs 19n26) A final meeting with Manuel Gorjão, the Portuguese minister of marine and colonies in Lisbon left Cadbury somewhat reassured that Gorjão would do his best to punish anyone convicted of cruelty to African workers, in keeping with the 1903 labour reforms.

Cadbury Brothers Limited also began searching for a private representative to send to Portuguese Africa. (Higgs 19-20n28)

In Lisbon, Gosselin fretted about the diplomatic implications of Cadbury's meetings with prominent cocoa planters and his connections to the British Anti-Slavery Society. He would have fretted more had he known that Cadbury sought advice about suitable representatives from H. R. Fox Bourne, the mercurial secretary of the Aborigines' Protection Society (APS). Founded in 1837, the year Victoria was crowned Queen, the APS shared the Anti-Slavery Society's mission to end slavery, but it also had a broader agenda to protect the rights of indigenous peoples in both British and foreign colonies. Fox Bourne was a thorn in the side of the British Foreign Office. He condemned the contracting of workers for the Transvaal as slave trading and was harsher still in his assessment of the Congo Free State. In June 1902, using Pickersgill and Nightingale's reports on West Africa and Consul Roger Casement's report on labour in the Congo Free State, Fox Bourne wrote to the British Foreign Office to protest the forced recruitment of workers from the Congo basin destined for Angola's coffee estates and roças in São Tomé and Príncipe. (Fig. 2.1) Gosselin's fear, conveyed to the Foreign Office, was that the cocoa controversy would "be taken up by the Press and Parliament and 'an attempt made to create a feeling against Portugal similar to that aroused against the Congo Free State.'" British South Africa remained dependent on the flow of workers from Portuguese Mozambique. (Higgs 20n29)

In 1903, the treatment of Africans in King Leopold II's Congo Free State was a cause célèbre, with the atrocities reported by the crusading journalist Edmund

Dependencia agua - Jzé S. Thomé.

**Fig. 2.1** Workers at a dependência of Água Izé roça, São Tomé, in the early 1900s. Author's collection

Morel. Leopold II claimed the territory, which was thus "free" of oversight by the Belgian government. At the 1884–1885 Berlin Conference, Belgian representatives had persuaded British, French, German and Portuguese delegates that Leopold's motives were philanthropic, and they pointed to his support of mission schools. By the 1890s, however, large parts of Leopold's African fiefdom had been sold as rubber (*Landolphia* sp.) concessions to European and American speculators, who turned them into armed camps. The severing of a worker's hand was just one of the horrific punishments imposed for failing to meet quotas in collecting raw rubber. In the late 1890s, Morel, then a clerk for a Liverpool shipping company, regularly visited Brussels and watched the delivery of cargoes of ivory and rubber from the Congo Free State. The ships returned to Africa filled with armed soldiers and few trade goods. Morel concluded that slaves laboured on the concessions in the Congo Free State. In late 1903, he founded the Congo Reform Association to expose the atrocities. Among its supporters—and also a backer of Morel's newspaper, the African *Mail*—was William Cadbury. (Higgs 21n30)

## Joseph Burtt and the Cocoa Islands

In July 1904, after consulting with Travers Buxton of the Anti-Slavery Society and Fox Bourne of the APS, William Cadbury hired Joseph Burtt, a fellow Quaker, to travel to Portuguese Africa on behalf of Cadbury Brothers Limited. (Higgs 22) Burtt had worked for nearly 20 years at the Gloucestershire Bank before leaving in 1898 to join the left-leaning Whiteway Colony. "It is hard to say", he would later recall, "if it was the horror of industrial cities or the degradation of the workers, or shame in my participation in an evil system that gave me a passionate desire to escape to some spot where I and my friends could settle and cultivate the land". (Higgs 4n11) In our age, we might assume that Burtt had a midlife crisis, but his rejection of capitalism and his search for a "practical utopia" had a long history in England. Among the late nineteenth-century experiments was George Cadburys' Bournville Village, his response to the growing recognition of poverty in Britain. (Higgs 4)

William Cadbury sent Burtt to study Portuguese in Porto, Portugal's second city, in September 1904. In October 1904, the activist journalist Henry Nevinson visited Buxton at the Anti-Slavery Society. Nevinson, a noted war correspondent, had negotiated a contract with the American *Harper's Magazine* to write a series of articles about the slave trade in Portuguese West Africa. On October 10, 1904, he wrote to William Cadbury to ask for an introduction to planters in Angola and São Tomé. Cadbury answered Nevinson politely and gave him the name of A. G. Ceffala, who ran the Eastern Telegraph Company station in São Tomé. Cadbury doubted Nevinson's seriousness, noting that he spoke no Portuguese, had no contacts in Portugal, and planned a brief tour of just 2 or 3 months. To Fox Bourne at the APS, Cadbury expressed his concern that Nevinson's "impressions will all be from the surface, and I hope what he publishes will not be of such a sensational character as to injure Joseph Burtt's chance". (Higgs 22, 23, n35, n36)

Burtt spent 8 months studying Portuguese in Porto and acquired a thorough working knowledge though not complete fluency, as he was the first to admit. When one of Cadbury's Lisbon contacts wrote to express his concern about whether Burtt was the right choice, Cadbury replied that he believed Burtt's "journey will be no less useful because he knows nothing at all about cocoa". Furthermore, Burtt had "no monetary interest in the *result* of his enquiry"—an implied criticism of Nevinson. Finally, Cadbury admitted that Burtt had been the only choice available. (Higgs 24n39)

Burtt arrived in São Tomé in June 1905 and quickly got to work. Writing to Cadbury, Burtt described Paulo Magalhães, manager of the roça Rio D'Ouro. Its owner was the Count of Vale Flôr, but like most of São Tomé's large planters, the count was an absentee landlord, with homes in Paris and Lisbon. Magalhães sent Burtt a large mule and a small boy who carried his luggage and walked behind him as he made his way along a "fine well-kept road with telephone wires running by the side", cocoa trees everywhere, groves of bamboo (*Bambusoideae*), and the occasional giant oca (*Oxalis tuberosa*), its enormous trunk "buttressed with great flanging roots". (Higgs 26n4)

On his tour of the roça Burtt got his first whiff of fermenting cocoa, an odour only slightly more tolerable than that coming from the serviçais' chicken coops. "These people are like children", Magaltrães told Burtt, adding that "if he made a tidying up clearance of the offending hovels", they would "be very sad". Whatever Burtt might have thought of the explanation, just days into his sojourn in São Tomé, he did not press his host. (Higgs 26n5)

Rio d'Ouro employed 1500 serviçais, or contract workers, and together with another 1500 workers employed on the count's other two roças, they produced 50,000 sacks of cocoa weighing 128 pounds each per year, for a total of 6.4 million pounds. Both men and women harvested cocoa, cutting the pods from the trees with machetes and then splitting them open. Women extracted the beans from the husks and placed them in baskets emptied into the small iron trucks of the estate's tramline and delivered to the fermenting sheds and then to the drying sheds.

The children on the roça occupied themselves collecting the berries dropped by rats—a chronic problem everywhere on the island—after they had eaten the sweet outer husk of the cocoa pod. Singing happily, the children came into the plantation's paved central square carrying 5-pound baskets of cocoa beans on their heads. Burtt liked Magaltrães's rapport with the children, mildly abusive though it appears to have been. After he gave each child a biscuit in reward for their work, the manager "would turn up the little flat faces and clap them with his open hand. Evidently, they looked upon this as a sort of religious rite, which though a little trying was highly satisfactory, and they would smile and blink their big eyes". After roll call (*forma*), the children and their parents went off to dinner, as did Burtt, who collapsed into bed early after a long day of watching other people work. (Higgs 26n6, 27n7)

Água-Izé, the most profitable roça on the São Tomé, took the edge of Burtt's enthusiasm. Owned by Francisco Mantero, whom William Cadbury had met in Lisbon, it was also run by a manager. Something about Água-Izé bothered Burtt, though he could not quite put his finger on it. "The moral atmosphere here seemed

different" and Burtt was not as happy. Unlike the well-maintained road that had
taken him to Rio d'Ouro, the road to Mantero's estate was full of potholes. Burtt and
the manager took a tour of Água-Izé via the roça's tramline, a narrow-gauge
Decauville railway snaking its way along the hillsides, their small car pulled by a
team of mules. Though he neglected the road to his estate, Mantero had been an
early advocate of railways as an efficient means to traverse the island's rugged ter-
rain and get cocoa to the coast for transshipment by sea. (Higgs 28, 30, 31n11)

William Cadbury wanted numbers and Burtt duly reported them. In 1905, 1 million of Água-Izé's cocoa trees were producing; 1.5 million more trees would reach
maturity within 2 years. In 1904, Mantero's combined estates on São Tomé and
Príncipe yielded 4 million pounds of cocoa, and the company paid its investors
"12% on the share capital of half a million" pounds sterling. Água-Izé employed
1668 African workers. A daily diet of dried fish, rice, beans, bananas, and breadfruit
plus meat once a week sustained the labourers. The mortality rate for 163 children
on the estate was very high (46 had died in 1904) and for adults very low (only 53
had died). Dysentery and anemia were the most commonly reported causes of death,
and occasionally, smallpox. Like many large plantations, Água-Izé maintained its
own hospital, and the staff explained the high death rate among children much the
same way Francisco Mantero had answered William Cadbury's questions in Lisbon
in 1903: it was due to "the unproductiveness of the black races, and their lack of
faithfulness to one partner". Angolan contract workers on the islands came from
polygynous societies, a reality that Portuguese planters tended to condemn as evidence of sexual promiscuity and the cause of labourers' poor health and low productivity. Burtt did encounter hospital patients suffering from syphilis, but that disease
affected both black and white workers on the island and was far down a list of illnesses—topped by fever—that reduced workers' productivity. (Higgs 31-32n12)

Burtt's next stop, the roça Caridade, sat 250 feet above sea level with a dramatic
view of the coast. Its owner was a black man, Jerónimo José da Costa. After his
unease at Água-Izé, Burtt was again charmed. Caridade sat in a verdant valley
shaded by trees and low mountains and cooled by a fresh breeze. To his surprise,
three employees just out from Lisbon had already succumbed to fever—probably
malaria—but overall, the death rate among adult workers appeared fairly low.
African workers received half a pound of meat and slightly more in rice and beans
per day to fuel 8½ hours of labour (1 hour less than the maximum dictated by the
1903 decree). (Higgs 32n13)

Costa was a *forro*, a free person born in São Tomé and Príncipe, descended from
the slaves first brought to the islands in the 1500s to cultivate sugarcane. Costa's
ownership of land made him a member of the islands' African elite. (Fig. 2.2) In
1905, Costa estimated the value of his roça at £140,000. He invested in new farming
techniques and supported a local arts and trade school in the town of São Tomé.
Though much smaller than Rio d'Ouro's 10,000 acres and less productive than
Água-Izé, with its yearly yield of 4 million pounds of cocoa, Costa's operation was
substantial by local standards. In the early 1880s, small African farms had produced
about half of the islands' coffee and 60% of its cocoa. By the early 1900s, African
farmers' share of exports had dropped to 6% as large entrepreneurs displaced

**Fig. 2.2** Two São Toméan women. Author's collection

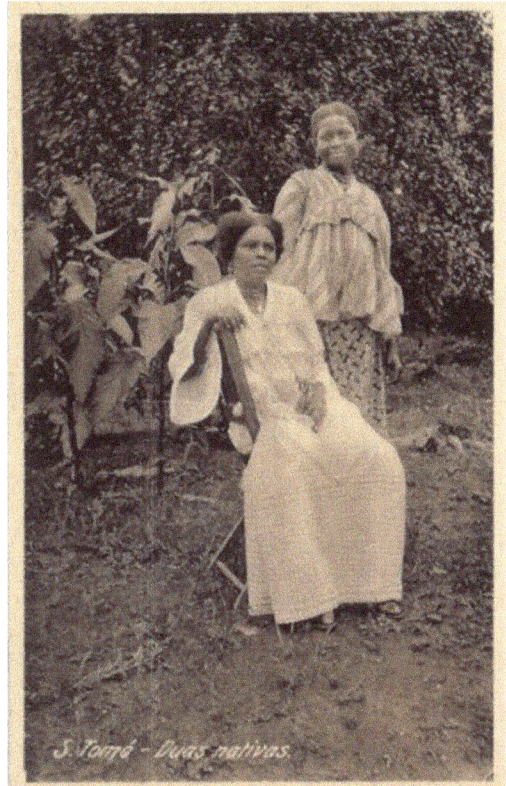

S. Tomé - Duas nativas.

smallholders—often by legal purchases but sometimes violently. They also claimed large stretches of forest to which no one had previously held title, effectively recolonizing the islands. Costa employed 300 serviçais, whose death rate was about 4.6% a year. For the 100 children on his estate the death rate was 7%, considerably lower than the 28% at Água-Izé. Costa had planted 1730 of Caridade's 2500 acres from which he expected a yield of 16,000 *arrobas*. One arroba equaled 32 pounds and in 1905 netted a market price of 14 British shillings. 528,000 pounds of cocoa, less costs of £5000, yielded a profit for the year of £6000. Demonstrating that naiveté that had worried Burtt's critics in Lisbon, Burtt found Caridade and its owner so enchanting that he thought it the ideal purchase for Cadbury Brothers. It was a prospect the firm had declined to consider in 1901 and would not consider in 1905. (Higgs 28, 37, 33n15)

Burtt also visited the model plantation Boa Entrada on São Tomé, owned by Henrique José Monteiro de Mendonça, who had in Burtt's opinion, properly embraced the European obligation to civilize and protect the African worker. Mendonça had wide-ranging business interests that included textiles and diamonds in addition to cocoa. He served on the boards of the Burnay Bank, the National Overseas Bank, and on Francisco Mantero's Principe Island Company. Boa Entrada dated from the 1880s, and Mendonça acquired its 4200 acres through marriage to the founder's daughter. He began upgrading it in the early 1900s, eventually building a hospital, a children's nursery school, a primary school and a music school. (Higgs 37, 39n22)

Burtt saw many of these improvements when he visited Boa Entrada in July 1905. A carriage ride along a well-kept road delivered him into "the large open square of the roça", where everything was "advanced and up to date; the houses of the serviçais [were] built of brick", with each family allotted an apartment. The new hospital had 36 beds and there was a separate ward for infectious cases. On the far side of the square were stables for the mules and horses and a cocoa-drying shed. In the centre was a "large shallow cemented pool of running water with an electric lamp near". A water turbine generated electricity, and the walls of the pool were serrated for washing clothes. Boa Entrada boasted 10 miles of Decauville narrow-gauge railway lines to move cocoa around the estate and down to the pier to load aboard ship. (Higgs 39n23)

## Henry Nevinson's Critique of Forced Labour

Henry Nevinson, who had completed his 6-month tour of Angola for *Harper's Magazine*, accompanied Burtt on his visit to Boa Entrada. Nevinson arrived in São Tomé on June 17, 1905, shortly after Burtt. In his diary, Nevinson described Burtt as "a big, innocent-looking man" with too "much luggage". Nevinson also recorded an American businessman's observation that "the planters had ordered everything to be carefully arranged for his visits". Nevinson left for Príncipe on June 21, returning to São Tomé 5 days later after a rough boat ride delayed by bad weather. He and Burtt did find time for a long talk "about the state of things", and Burtt reported to Cadbury that Nevinson admitted "that the people on the roças seem comfortable" though he disliked how much power the planters had. Nevinson recalled their talk, as they walked along the coast to a fishing village, rather differently. Burtt, he concluded, was "about the youngest man of 43 that could live—the mind of a youth, confused and interesting and full of dreams and theories". Burtt's time at the communist Whiteway Colony had apparently left him "despising the working man" and convinced that the "slave system" on São Tomé was good for Africans. Burtt, for all his admiration of Jerónimo Costa, the black owner of Caridade, was nevertheless a racist. "All very crude and youthful stuff", the 49-year-old Nevinson thought, "full of contradictions and very astonishing". On a whim and despite his misgivings, he joined Burtt on his visit to Boa Entrada on June 29. (Fig. 2.3) He would leave São Tomé just 2 days later, on July 1. Worn out from his long Angolan sojourn and always sensitive to injustice, Nevinson could not have enjoyed his carriage ride to Boa Entrada. The bad-tempered Portuguese driver used his whip indiscriminately on the mules, and on a black man trying to direct him. Burtt repeated stories he had heard about Brazil from Ceffala at the telegraph station—of a planter finding his wife with her black lover, killing the man, cooking him, and then feeding him to his

**Fig. 2.3** Model housing for workers, Boa Entrada roça. Author's collection

wife, who was "chained up … in a barn … till she died". Nevinson found it all very wearing. (Higgs 37, 39, 40n25)

Boa Entrada, for all of Mendonca's innovations, was an unhealthy place. It sat in the low-lying and swampier region of the island, 5 miles inland and without the benefit of the prevailing winds from the south. Death rates for black and white workers were high. Nevinson recorded a death rate of 25% for Boa Entrada's children. Twenty-seven of the 143 adult deaths between 1903 and 1905 were undiagnosed. The roça's doctor identified two major causes: endemic alcoholism among all workers, and for black workers, geophagy, or dirt eating. The doctor rejected the idea that it was a way to replace minerals missing in workers' diets, and instead saw it as a form of resistance distinctive to black workers. A more obvious form of resistance was running away, but this was risky and ineffective. Those who did find sanctuary in the forest discovered that there was little to eat besides pigeons and rats. (Higgs 41, 42, 43, 44n32)

Workers did have recourse to the legal system and to an ombudsman, the general curator of servants, or *curador*. Sometimes the workers refused to renew their contracts, a process the curador oversaw. The problem, as Burtt reported to Cadbury, was that even though workers had the technical right to complain to colonial officials, they needed permission to leave the roça to do it. If serviçais did make it into town, they discovered that the planters were more powerful than the colonial administrators and that labourers who complained were punished for "disobedience" back on the roça. Even those officials willing to listen had trouble doing so in the absence of interpreters who could speak the languages of the Angolan serviçais, many of whom were Ngangela or Wiko. (Higgs 45n33)

Burtt for his part, struggled to recognize the humanity and dignity of African workers. Forgetting for a moment that he was reporting to one of England's most successful chocolate makers, he wrote to William Cadbury:

> What I feel so much among these people is the wicked waste of opportunity of making and producing human happiness. With these gentle, patient creatures anything could be done with firm kindness. They are mere clay under the strong white hand. Our comparative omnipotence and omniscience ought to arouse a feeling of responsibility and a desire to be of use to them, but greed for money and power don't leave room for much else.

Cadbury could agree on the issue of happiness, since it was one of the goals of the reforms undertaken at the Bournville Works. There was, however, an essential distinction: management took seriously the views of workers at Cadbury Brothers. (Higgs 51-52n46) (Fig. 2.4).

Nevinson spoke no African languages and unlike Burtt, no Portuguese, but he argued against the European propensity to "think of 'black people' in lumps and blocks", stressing "that each African has a personality as important to himself as each of us is in his own eyes". The journalist still managed to objectify Africans, though in a manner less condescending and more romantic: "We profess to believe that external nature is symbolic and that the universe is full of spiritual force; but we cannot enter for a moment into the African mind, which really believes in the spiritual side of nature". (Higgs 52n47)

The second of Nevinson's articles on "The New Slave Trade" appeared in *Harper's Magazine* in September 1905. He wrote the article in Luanda in December 1904, but it began with an extract from a letter Nevinson had written in May 1905, after 2 months spent tracing the slave route in Angola's interior. After 9 weeks of walking, he was thin, ill from fever, possibly poisoned, and still 200 miles from the

**Fig. 2.4** Cocoa drying tables (Taboleiros para secar cacau). Author's collection

coast. Unsure whether he would survive, Nevinson had his report delivered to Benguela and mailed to the magazine's editor. The dramatic opening and the confused timeline prompted William Cadbury to write to Nevinson to complain about the "somewhat stagey nature" of the article and to note that it gave ammunition to Portuguese interests inclined to question the reliability of English critics. Nevinson assured Cadbury he was disgusted by the inclusion of the letter and the confusion it caused. He had informed the editors at *Harper's* that if they tried anything similar again, he would withdraw his series of articles. Nevinson's assertion that "of course it will be put right in the book" could not have reassured William Cadbury. (Higgs 55n54)

Burtt meanwhile spent 5 weeks on the smaller island of Príncipe, where he visited 13 roças and concluded that "the condition of the serviçal in Príncipe is quite as good as in S. Thomé and but for sleeping sickness carried by the tsetse fly, some of the roças would be healthy". On some of those roças, sleeping sickness increased the mortality rate to 30%. (Higgs 64)

The third of Nevinson's *Harper's* articles appeared in October 1905. The journalist flatly accused the Portuguese of running a slave trade from the Angolan hinterland to Catumbela and then to Benguela on the coast. Cadbury's reservation remained that Nevinson spoke "no Portuguese and that the Portuguese cannot be expected to accept his articles as a fair statement of the case". Cadbury was also concerned that Burtt's observations could be open to criticism, especially from Fox Bourne of the Aborigines' Protection Society. "As far as I have gathered you have made almost all your visits by appointment", Cadbury wrote to Burtt. He then reassured them both by adding: "It would be very difficult to remove all signs of illtreatment even with a week's notice, and I should suppose that you are fairly satisfied in you own mind that most … of the estate owners have dealt honestly and straightforwardly with you". (Higgs 66n17) Burtt replied that 5 months of watching serviçais working had persuaded him that "they looked orderly, well fed and contented". He remained hesitant, however, to offer "an opinion of the labour question of the islands till I had seen the method of procuring labour". (Higgs 68n20)

The 2-week trip down the west coast to Luanda aboard the *Sobo* gave Burtt the opportunity for further reflection. The interim report he sent to Cadbury acknowledged some of the concerns raised by Nevinson. São Tomé and Príncipe's contract workers were not freely recruited. Further, "practically no attempt is made to supply the serviçal with anything beyond his physical needs, and he has no life beyond the plantation". No child attended school; few workers had the opportunity to attend church. "The life of the serviçal spent under constant rule and authority does not develop character", he concluded, "and the prospect before the children born on the estate, although one of security as regards food and shelter is at its best but a mild form of slavery". (Higgs 68n21)

William Cadbury had his answer: no matter how attentive Portuguese planters were to the housing, food, and working conditions of their serviçais, the majority never left the islands. (Fig. 2.4) They lacked mobility. They were slaves. The other side of the equation—the recruiting of labour—had been described forcefully by Nevinson's articles in *Harper's Magazine*. A slave trade in Angola produced workers for São Tomé and Príncipe. Burtt might well have returned to England had Cadbury's letter of November 7, 1905, reached him. Cadbury suggested, "The firms who sent you out may feel that Nevinson's report is so thorough" that it was unnecessary for Burtt to go to Angola. When next Burtt wrote to Cadbury, it was from Luanda, "bankrupt and beautiful", a phrase lifted, tellingly, from Nevinson's September 1905 *Harper's* article. (Higgs 68–69n22)

## Burtt in Angola

Cadbury's November 1905 letter finally reached Burtt in February 1906. It included his criticisms of the fourth installment of Nevinson's series. Cadbury reiterated that Nevinson did not speak Portuguese, but Burtt thought that "his knowledge of French and keen powers of observation would in measure compensate for this". The problem for Burtt was that Nevinson had spent only 2 weeks in São Tomé and Príncipe. Burtt's contacts in Luanda's British community—diplomats, railway men, and harbor officials—assured him that Nevinson's "fears of assassination were entirely imaginary", though "common … to persons who have been overdone in Africa". Burtt felt that he owed the firm an opinion rooted in his own discreet observations of the recruiting process so that his employer would "have something more substantial than the brilliant journalism of a man unhinged with fever and fatigue". Beyond this obligation, Burtt admitted to another motive—a desire to "see the mountains and hear the lions" and, in a long tradition of British travelers, to visit Angola's interior. (Higgs 75–76, n16)

Having ignored his employer's suggestion that he return to England, Burtt continued collecting food and supplies for his journey. Most important, he needed African porters, who were in high demand to carry ivory, rubber and beeswax from the interior to the coast. (Higgs 78n17) In search of carriers, he headed south to Benguela, Angola's second city. There, he fell desperately ill, an ironic turn given his suggestion that fever and fatigue had distorted Nevinson's articles. Trapped in a hot, expensive and dirty hotel room, Burtt struggled first with a skin infection. Next, he feared that he might be coming down with sleeping sickness. He finally succumbed to malaria. After five sleepless nights, he checked into Benguela's hospital for 10 days' rest, only to fall victim to a second attack of malaria. He recovered after an injection of quinine. (Higgs 79n21)

A short trip to Catumbela, just north of Benguela, introduced Burtt to the trade in serviçais. Once a stop on Angola's slave route, it maintained a small trade in rubber and a larger one in contract workers. Burtt identified three large dealers, two smaller dealers and two contracting agents working to supply workers to planters on São Tomé and Príncipe. For *Harper's* readers, these men were slave traders. Nevinson described a Portuguese trader who openly drove his captives to the coast, "shackled, tied together, and beaten along with whips". The man lost 600 of 900 captives along the way. At Catumbela, "the slaves were rested, sorted out, dressed", and then marched the 15 miles to Benguela, "usually disguised as ordinary carriers". Burtt met with a local Portuguese judge critical of the labour-recruiting process and left convinced that "the whole city and district is eaten out with slavery and rotten to the core". First prize in a recent raffle, ostensibly a bicycle, had been a young woman. "A phrase used in the description of the bicycle when taken in a 'slang' sense", Burtt explained to Cadbury, "was very descriptive of her, and of course everyone saw through the thin disguise". (Higgs 84n30)

Burtt's Dutch host in Catumbela regretted the drain of local labour to the islands but thought the repatriation of workers from São Tomé and Príncipe a waste of time. In the absence of effective law enforcement in Angola, he argued, returned serviçais would be quickly re-contracted. Burtt countered that if the judges "were strong enough, the whole system could promptly be crushed at the ports by liberating all negroes who said they did not wish to go to the islands". The judge at Catumbela, however, had 13 cases of slavery before his court, and he thought it "hopeless to attempt to do anything through the Portuguese government". Instead, he recommended that Cadbury Brothers "make an appeal through [your] king to Don Carlos himself". (Higgs 84n31)

The judge's suggestion was an odd one because Victoria's heir, Edward VII, was a constitutional monarch who exerted no direct influence over government policy. The Portuguese constitution allowed King Carlos I more leeway. He functioned as a "moderating power", an executive meant to serve as a neutral negotiator between the branches of the government. The king's dominant personality undermined his ability to be an effective power broker. Portugal's debt complicated efforts to reassert its place on the world stage and made the king's position shakier. In the Côrtes, the Portuguese parliament, antimonarchist Republicans—still in the minority in 1906—loudly criticized government incompetence and corruption. The Catumbela judge may have been a monarchist; there were devoted royalists in the Angolan civil service, even if there were few in Lisbon. In 1906, however, Catumbela, along with the small coastal town of Novo Redondo, elected a Republican municipal council. If repatriating workers to Angola from São Tomé and Príncipe was a waste of time, so too was making appeals to a distracted and increasingly isolated Portuguese king. (Higgs 85n32)

## The Slave Route: Cocoa Is Blood

In March 1906, Burtt read "Islands of Doom", the final installment in Nevinson's series for *Harper's*. Burtt found the article "so unfair" that he could barely contain his irritation. "On page 328," he complained, "the guest who says 'The Portuguese are certainly doing a marvelous work for Angola' must have been myself". The balance of the quotation, which Burtt declined to repeat, read, "Call it slavery if you like. Names and systems don't matter. The sum of human happiness is being infinitely increased". Burtt was outraged at the implication that he supported slavery. "I need hardly tell you", he assured Cadbury "that what I did say conveyed an entirely different meaning". When he met Nevinson in June 1905, however, Burtt had just arrived on São Tomé. He was naive, but it seems unlikely—given his mission for Cadbury Brothers—that he would have openly supported slavery. Nevinson had just spent 5 months trekking through Angola. Physically and emotionally spent and convinced that slave trading continued in Angola, he was probably incapable of coming to any other conclusion during his 2-week visit to São Tomé and Príncipe. The islands of doom were the place Angolan slaves went to die. (Higgs 86-87n37)

That Burtt made the journey in reverse, first spending 6 months on the islands before going to Angola, shaped his own view of conditions on São Tomé and Príncipe, as well as his reaction to Nevinson's article. The overall death rate, he reminded Cadbury, was 11%, not the 20% claimed by Nevinson. Burtt also rejected the journalist's assertion that "the prettiest girls are chosen by the agents and the gangers [as] their concubines". White men did forge relationships with black women, but they were "chosen almost exclusively from the natives of S. Thomé". Nevinson's assertion that children were encouraged to engage in bestiality for the amusement of the planters deeply perturbed Burtt. To the contrary, he recalled that one boy had been "severely flogged" for his outrageous behaviour. Flogging, Burtt reassured Cadbury, was nevertheless rare. Workers were not routinely whipped to death as Nevinson implied, nor had Burtt found any evidence that runaways were hunted for sport. For him, the final article in the *Harper's* series was an abomination that pointed "clearly to this, that Nevinson writes authoritatively on matters of which he knows practically nothing". Despite the 6 months he himself spent on the islands, Burtt still struggled "to get at the truth amongst strangers, and if I found it difficult in 6 months it is quite impossible for Nevinson to have learnt much in as many hours, which was about the time he spent at the roças". (Higgs 87n38)

By June 1906, Burtt had finally managed to secure enough porters for the 1500-mile return journey along the slave route from Benguela on the west coast to Kavungu on Angola's eastern border. The long delay was commonplace. It had taken Dr. Ansorge, a zoologist working for the British Museum whom Burtt had met briefly in São Tomé, 4 months to plan his excursion, at a cost of £700. (Higgs 89) Burtt also asked Cadbury if he could find a medical doctor to accompany him into the interior. Cadbury found Dr. Claude Horton, who spoke no Portuguese but made up for it, in Burtt's estimation with charm, intelligence and robust good health. (Higgs 93n53)

A further delay at Benguela gave Burtt and Horton the chance to have breakfast with William E. Fay, a missionary of the American Board of Commissioners for Foreign Missions (ABCFM) stationed at Bailundo, a hundred miles to the east of Benguela. The ABCFM station opened in 1881. Fay and Wesley M. Stover arrived the following year. By 1885, Fay had helped prepare a preliminary dictionary of Umbundu. Though he had no formal medical training, he also managed the clinic at Bailundo, and he quizzed Horton enthusiastically about the latest medical advances. (Higgs 94) Stover and Fay preferred not to comment publicly on the trade in contract workers that passed by their mission. Privately, Fay told Burtt exactly what he thought. The missionary praised the kindness with which many Portuguese treated Africans, and he agreed with Burtt "that under a similar system the English would probably be worse, we having a stronger racial antipathy to the black". Nevertheless,

Fay had misgivings about Burtt's agenda in Portuguese Africa. He refused on principle to drink cocoa and directed Burtt to "tell Mr. Cadbury from me that cocoa is blood". (Higgs, 95n58)

Burtt did not and could not claim the missionary's position as his own. He confessed, however, "at times I get so wroth" over the abuses in the labour-recruiting system that "I forget my Quaker breeding". His inclination was to offer a more balanced analysis. The problem was the tremendous profit made on a worker contracted for £6 in Ambrizette and then re-contracted in São Tomé for £36. The inability of officials in Luanda to govern the hinterland effectively, the low salaries of bureaucrats in rural districts, and a broad tolerance for corruption combined to tempt many to traffic in labour. By tracing the slave route himself, Burtt hoped to acquire some understanding of how the contracting process might be improved. (Higgs 95n59)

At the end of July 1906, Burtt and Dr. Claude Horton finally left Benguela on Angola's west coast, on foot, heading to Kavungu, 750 miles to the east. At Catembula, 5 days into their journey, Burtt found a wooden ankle shackle, designed for two captives. In a scene reminiscent of Nevinson's articles, Burtt also found "a decomposing corpse, parts of the scalp & body were gone but the feet & limbs were intact". The man "lay on his back with his limbs spread out, … a small basket, a large wooden spoon, a mat & a few filthy clothes" by his side. Whether the man was a slave freed from the shackle or a free carrier Burtt could not say, though he had been told, "when a free carrier dies on the road he is buried by his relatives, if in company with them". He added, "We saw many such graves". (Higgs 96, 97n3)

Burtt and his porters reached Bailundo in mid-August 1906, where they rested for 2 days. When Nevinson visited Bailundo in early 1905, the trade in labourers destined for São Tomé and Príncipe was operating openly. For the most part, however, caravans heading west tended to be small, and overseers mixed slaves in among the free carriers. In August 1906, Burtt encountered one such group carrying rubber in the Bihé district. His porters identified 25 individuals as slaves "put here & there in the long file" and "conspicuously different in appearance from the well-nourished free carriers. The lean limbs & boney bodies told a tale of long hunger". They reminded Burtt of "a number of new arrivals on a roça at Príncipe" whom he had seen "in much the same condition". (Higgs 101–102n11)

By the time Burtt's party reached the mission station at Ochilonda in the Bihé District on August 22, 5 days after leaving Bailundo, they were all in bad shape. Sore feet hobbled the Africans, who took turns carrying an incapacitated Burtt in a hammock. Forty-three of his 50 porters quit. Burtt spent 9 days in Ochilonda recruiting 22 replacements. He also got some sense of the missionaries' efforts to block slave trading and convert Africans to Christianity. The focus of Burtt's mission, however, was narrow: to determine if slave trading continued in Angola and if those slaves ended up as serviçais on São Tomé and Príncipe. For a more dramatic sense of what might have happened to Burtt and Horton after they left Bihé on September 1, crossed the Cuanza River, and passed through the Hungry Country, William Cadbury could turn to, among other sources, Nevinson's articles in *Harper's*. (Higgs 103–105)

In September 1906, Burtt and Horton reached Cinjamba mission station, where Nevinson had ended his journey in 1905. Burtt's health had improved though he was still weak. The missionaries, as always, were gracious in sharing their stories of the ongoing slave trade. Horton wrote to Cadbury that it was still "difficult to get any first-hand evidence of the actual conditions here though there can be no doubt that it is simply slavery under another name". (Higgs 108n22) That slaves had been renamed serviçais was the conclusion of the August 1906 report Arthur Nightingale submitted to the British Foreign Office on labour conditions in Portuguese West Africa. From the Portuguese perspective, however, there was no firsthand evidence to support allegations of slavery. In September 1906, Augusto Ribeiro of the Geographical Society in Lisbon defended Portuguese labour practices in the colonies, condemning the "malevolent ignorance" of investigators motivated more by treachery than by a desire for the truth. (Higgs 107–108)

The Brethren missionary Fritz Schindler welcomed Burtt and Horton to Kavungu on October 4, 1906. The multilingual Swiss missionary was the source of much of Burtt's information about the slave route. (Higgs 109) At Kavungu, Burtt wrote to Cadbury, "We seem to be in the middle of the traffic". Just to the north, six Portuguese trading firms sat along the Angolan border with the Congo Free State, just out of sight of a Portuguese fort intended to police the border, a job it lacked the resources to do effectively. Many of the authorities of the Congo Free State were openly complicit in slave trading and in procuring the forced labour used to collect rubber. These were among the horrors that Edmund Morel's Congo Reform Association— avidly supported by William Cadbury—sought to end. (Higgs, 110, 111, 112)

On October 14, Burtt posted a long letter to Cadbury, sending it across the North Western Rhodesian border. It would take "a couple of months or so" for the letter to reach Birmingham, via mail steamer up the east coast of Africa. In fact, the last time Cadbury and his commissioner appear to have directly exchanged letters was in July 1906, when Burtt's caravan left Benguela. On October 19, Cadbury sent a letter to Burtt's Benguela address, stressing, "how important it is that your outline report and summing up should not be delayed". Cadbury asked that Horton bring a draft report with him when he returned to England in December. Burtt could "supply further details" when he reached Birmingham. (Higgs 112n34)

Burtt finished his 22-page preliminary report on December 24, 1906. (Higgs 113) It drew heavily on the more than 300 pages of letters he had written to Cadbury over the year and half since he first landed on São Tomé. Burtt still aimed for balance. Workers—most from Angola—were, he wrote, "generally, well housed and fed". They had access to medical care when ill. The 5-year contracts under which they laboured were "nominally free, and controlled by good laws, including the admirable law of 1903 relating to serviçaes". Planters, however, found it all too easy to circumvent the laws, especially if they had powerful friends in Lisbon. The average serviçal was "taken from his home against his will, ... forced into a contract that he does not understand, and never returns to Angola. The legal formalities", Burtt concluded, "are but a cloak to hide slavery". (Higgs 114n37)

For all of William Cadbury's misgivings about Henry Nevinson, Joseph Burtt's December 1906 report placed Burtt firmly in the journalist's camp. When Nevinson arrived in São Tomé in mid-1905 for a 2-week stay, exhausted after his fourth-month walk along Angola's slave route, he had been incapable of seeing the working conditions on the island as anything but slavery. Burtt spent 5 months on São Tomé, 5 weeks on Príncipe, and a year in Angola. He took Nevinson's journey in reverse and over a longer time span, but he ultimately reached the same conclusion. Africans were enslaved in Angola in a still-active trade. Shipped to the islands, they found themselves "doomed to perpetual slavery". Among Burtt's suggestions for improving the "existing conditions" in Portuguese West Africa were informing the British public—as Nevinson had done—and insisting that the Portuguese government enforce its own labour laws. (Higgs 114n39)

## Mozambican Miners in the Transvaal

The question still unanswered in December 1906 was how Mozambique fit into the humanitarian criticism of Portuguese labour policy in Africa. Burtt's third proposal in his preliminary report for improving the contracting process was to recruit "labour from the Portuguese colony on the east coast of Africa, where labour is voluntary". Subject to the same 1903 labour decree that governed Angola and São Tomé and Príncipe, Mozambique supplied tens of thousands of miners each year to Witwatersrand gold mines in the British colony of Transvaal. In October 1906, Portuguese parliamentarians asked why no one was protesting labour contracting in Mozambique when the legal process was identical to that practiced elsewhere in Portuguese Africa. In the same session, the Portuguese colonial minister, José Francisco Ayres de Ornellas, praised the transparency of recruiting across the border between Mozambique and the Transvaal. Not everyone, however, was convinced. In April 1901, William Cadbury had wondered if force was used to recruit mineworkers from Mozambique. Nevinson had no doubts: "In Mozambique, the agents of capitalists bribe the chiefs to force laborers to the Transvaal mines, whether they wish to go or not. ... We may disguise the truth as we like under talk about 'the dignity of labor' and 'the value of discipline,' but, as a matter of fact, we are on the downward slope to a new slavery". Joseph Burtt's final task as Cadbury's commissioner to Africa was to determine the truth of that charge. (Higgs 114-115n40)

In 1902–1903, private companies in Mozambique and the colonial government opened their territories to licensed recruiters form the Witwatersrand Native Labour Association (WNLA). For each worker, the Zambezia Company received a fee of 5040 reis (approximately £1), four times the annual tax it could collect from a peasant farmer working on its leaseholds. (Higgs 118) Colonial officials charged recruiters and workers for issuing contracts and registering them, and they added extra fees for passports. For each labourer who signed a 1-year renewable contract, Mozambique's government received 13 shillings, a substantial sum when multiplied by 60,000 migrant workers per year. The Portuguese government also insisted on having a resident curador in Johannesburg, who would guarantee that Mozambican miners were well treated. As Burtt had discovered in São Tomé, the curador's responsibilities could be contradictory: he was also responsible for preventing "clandestine" workers from sneaking across the border without paying their taxes and fees. If a miner did not finish his contract and pay his taxes, the curador also made sure that he completed 90 days of forced labour for the colonial government when he returned home from Mozambique. (Higgs 120n10)

In a sense, Nevinson was right: the harsh conditions of the leaseholds, the strictures imposed by local chiefs, and the unpaid labour demanded by colonial officials all "forced" Mozambican workers across the border into the Transvaal. Yet by the first decade of the twentieth century, some Africans did cross voluntarily, in order to earn cash to pay their taxes and to contract marriages. Many also indulged an increasing attraction to European clothing, foodstuffs such as tea and coffee, and other consumer goods. For those who favoured this second, rosier explanation of Mozambican labour migration, there was also "the incalculable benefit that Africans would be educated, taught the value of money, and, most important, taught how to work". (Higgs 120-121n11)

The brevity of Joseph Burtt's visit to Lourenço Marques—he stayed only 2 weeks—led him to adopt the latter view. Burtt missed the irony. He had dismissed Nevinson's negative assessment of labour conditions in São Tomé and Príncipe because the journalist had spent only 2 weeks there. In Mozambique, Burtt found the contrast with labour recruiting for São Tomé and Príncipe striking: "Not only were the natives eager to volunteer for work in the Transvaal mines, but I saw them returning to their homes when the period of work for which they had been indentured had expired". (Higgs 121n11) More research revealed unhappiness. Separated from their wives and children, Mozambican miners in the Transvaal lived in ramshackle housing provided by the mines and spent their days underground exposed to the fine dust that left many suffering from the debilitating lung disease known as miner's phthisis. (Higgs 125)

Nevinson had a point: wealthy Portuguese planters exploited Angolans on São Tomé and Príncipe. In the Transvaal, Mozambicans made mostly British mine owners rich. In Birmingham, in late January 1907, William Cadbury edited the first, limited-circulation proof of Burtt's exposé of slavery on the chocolate islands. Similar accusations made by Nevinson in 1906 had infuriated Portuguese commentators. Yet Portugal did not loudly protest the suffering experienced by African miners on the Rand at the hands of British mine-owners. Francisco Mantero's efforts to redirect Mozambican labour to Portuguese planters on São Tomé and Príncipe found little support in Lisbon, despite the wealth generated by the cocoa islands. The revenue that flowed from the Transvaal's mines in the form of fees and taxes, to Portugal and Mozambique trumped all other concerns. (Higgs 125n19)

To the question of whether the short contracts and ability to cross the border made Mozambican miners in the Transvaal free, Burtt offered a qualified yes. He remained concerned about the living conditions and high mortality rates for African miners. He understood that "one thing works into another, and that it is quite within the range of possibility that stopping the slave trade in Angola, might affect the mining interests" in the Transvaal, thereby destabilizing the British colony. Demanding that Portugal stop forcibly exporting Angolans to São Tomé and Príncipe's roças might prompt the Portuguese to do what still seemed unlikely in early 1907: close Mozambique's southern border and attempt to starve the Transvaal's British-owned mines of labour. (Higgs 130–131)

## Burtt's Report and the Case against Cadbury Brothers Limited

The report that finally emerged in mid-July 1907 was several pages shorter than Burtt's December 1906 original. The tone was more official, and the report included a list of the British and Portuguese colonial officials Burtt had met. The 1903 Portuguese decree on contract labour was explained in detail. (Higgs 136n9) The most striking difference between the two reports was the careful language in the 1907 version. As Burtt acknowledged, great care was taken to avoid "referring to the serviçais as slaves or to the serviçal system as slavery, because, approaching the matter as I did with an open mind, I have wished to avoid question-begging epi-thets". His December 1906 draft had reflected no such worries. On the islands, he had been struck by "the mental distress and hopelessness of a man separated from his family and doomed to perpetual slavery—a condition often accelerating death". Six months later, this became, "The mental distress and hopelessness of a man sepa-rated from his family, and placed in a strange environment have a highly prejudicial effect upon him". (Higgs 136n10)

In August 1907, Burtt headed back to Porto to arrange the translation of his report into Portuguese, confident that if he diligently avoided the "I'm an English Saint and you're a Portuguese sinner tone", the planters would prove responsive. In September, Nevinson addressed the African section of the Liverpool Chamber of Commerce, arguing in favour of a cocoa boycott and urging the British government to do everything in its power to end slavery in Portuguese Africa. On October 21, Cadbury responded with his own address to the chamber, presenting a synopsis of Burtt's report. He rejected the idea of a boycott, since it would rob the chocolate makers of the leverage they enjoyed as major buyers of São Toméan cocoa. What British firms declined to buy, he said, would be "very readily absorbed by other nations, who do not concern themselves with the method of production". It was a weak defense, and Cadbury knew it. Nonetheless, he found allies in his audience. John Holt, whose shipping company traded in West Africa, criticized the chamber's president, Albert Jones, for attacking the chocolate makers while serving as a repre-sentative of King Leopold's Congo Free State. The Congo Reform Association, which Cadbury supported financially had exposed labour abuse there. In Holt's view, if the chocolate manufacturers wanted to protect themselves from charges of hypocrisy as businesspeople and as philanthropists, they needed to seek out cocoa indisputably produced by free labour. That Edward Thackray, the Cadbury Brothers cocoa buyer, was in fact following Holt's suggestion remained, in October 1907, confidential. (Higgs, 137, 138n14)

In November 1907, Cadbury visited Lisbon to meet with the Portuguese planters on behalf of the English cocoa makers. Neither Fry nor Rowntree sent delegates. Cadbury thanked the planters for their invitation and the hospitality they had shown Burtt. He praised the quality of São Tomé's cocoa and said that his firm wanted to keep buying it. He conceded that workers enjoyed excellent treatment on many estates but insisted that their death rates were too high, that the recruiting methods in Angola bordered on slave trading, and that contract labourers never returned home. For Cadbury Brothers to continue buying cocoa from the islands, the firm had to be sure "that in the future it is to be produced by free labour". The planters assured Cadbury that they shared his "liberal and humane sentiment". They wanted São Toméan workers return home and "carry back to their country the accounts of the treatment they have received in the islands". Planters had already contributed £100,000 to a repatriation fund. The first workers contracted under the January 1903 law would begin returning home in 1908, taking with them severance payments of £18 each. Workers who decided to renew their contracts would receive an automatic raise of 10%. The colonial minister, Ornellas, also planned to send one of his senior officials to investigate conditions in Angola in January 1908, with the explicit "intention of replacing the present irresponsible recruiting agents by a proper government system", modelled "on the lines employed with success in Mozambique". (Higgs, 138, 139, 141n21, 142n23) Cadbury's relief proved short-lived. On February 1, 1908, King Carlos and his heir Luís Felipe were assassinated in Lisbon. The succession of the king's 18-year-old second son to the throne as Manuel II did not restore stability. In March 1908, a new Portuguese government sent Francisco de Paula Cid to investigate labour recruiting in Angola. In Luanda, *The Voice of Angola (A Voz d'Angola)* noted that Cid had made no effort to end what the newspaper considered a slave trade in serviçais while serving as governor of Angola's Benguela District in the 1890s. There was no reason to think that his current mission would do anything except "enrich half a dozen ambitious men at the expense of shaming Portugal". Horatio Mackie, the British consul in Luanda, shared the newspaper's cynicism and suggested appointing a commission that would interview contract labourers and thereby expose the abuses in the recruiting system. (Higgs 142, 143n27)

In April and May 1908, Cadbury Brothers sought and secured apologies for articles in the *Manchester Guardian,* the *Standard,* and the *Evening Standard* that had questioned why the firm was still buying São Toméan cocoa. In June, William Cadbury decided to visit São Tomé and Príncipe and Angola, to gain "a first hand insight into the exact conditions". (Higgs 143) In an editorial on September 26, 1908, just days before Cadbury planned to depart for São Tomé, the *Standard* called attention to the inherent contradictions of his business and philanthropic interests. The attack was political. The pro-imperialist Conservative Party-aligned *Standard* opposed George Cadbury's Liberal-leaning anti-imperialist *Daily News.* Surprisingly, Nevinson wrote to the *Standard* to defend the efforts by William Cadbury to expose abusive labour practices in Africa, and he noted that an editorial in the *Daily News* had condemned the practices in São Tomé in May 1908. When no apology was forthcoming, Cadbury Brothers sued the *Standard* for libel. (Higgs 145n32)

William Cadbury's visit to the islands and to Angola from October through December 1908 was disappointing. He visited several large plantations on São Tomé and Príncipe where he inquired about workers' hours, wages, their opportunities for recreation, and viewed their living quarters. The cost of producing cocoa on São Tomé, he concluded, was quite high: "Few estates made above average profits, most were mortgaged to the hilt, and many were losing money hand over fist". Colonial officials proved unreceptive. They ignored Cadbury's repeated requests to witness the contracting and re-contracting processes. (Higgs 145n34, 146)

Cadbury and Burtt left Luanda on January 11, heading north for the Gold Coast (modern-day Ghana), where free African farmers had been growing cocoa for two decades, though not on a scale large enough to meet commercial demand. In early February, Cadbury visited two farms, along with William Leslie, a cocoa buyer for Cadbury Brothers. At Mangoase, a new railroad was under construction. When finished, Cadbury thought, it might hold "the key to a vast cocoa district" in which the firm could invest, and he authorized the purchase of 14 acres for a factory site. (Higgs 148n41)

The activist and journalist Edmund Morel had told Cadbury in mid-1908 that if he went to Africa, he would end up boycotting São Toméan cocoa. Within a week of Cadbury's return to England on March 9, 1909, the firm decided to stop buying cocoa from the chocolate islands. Cadbury Brothers wrote to officials in the Foreign Office to inform them. A notice in the *Daily Mail* on March 17, 1909 announced the decision by the three Quaker cocoa firms Cadbury Brothers, Fry and Rowntree. (Higgs 148) Cadbury meanwhile decided not to publish his draft report of his trip to São Tomé and Príncipe and Angola until the firm's libel suit against the *Standard* had concluded, though he did send a copy of "Labour in Portuguese West Africa" to the British Foreign Office. (Higgs 148, 149)

British officials may have shared Cadbury's report with their counterparts in Lisbon. In April 1909, Francisco de Paula Cid presented his findings on labour recruiting in Angola to the colonial minister in Lisbon. Cid recommended that recruiting be limited to specific districts (those that Portugal had effectively occupied) and be directly supervised by "specially appointed officials … responsible to the Governor General of the Province". The repatriation of workers from São Tomé and Príncipe should be mandatory, and once they arrived in Angola, help provided to allow them to return home. Finally, the process of recruiting labour from Mozambique for the islands should continue. The official decree issued on July 17, 1909, limited contracts to 3 years for workers from Angola and set out further guidelines on recruiting, wages, working conditions, and health care, but it did not make repatriation compulsory. (Higgs 149n44)

Cadbury Brothers' week-long libel trial against the *Standard* began in Birmingham on November 29, 1909. The political divide reflected in the September 1908 *Standard* editorial continued, with the Conservative lawyer Edward Carson defending the *Standard* and the Liberal Rufus Isaacs representing Cadbury Brothers. For the firm, at issue was the newspaper's allegation that Cadbury Brothers had not tried to do anything to improve workers' lives on the islands. William Cadbury could document the company's concern with labour practices in São Tomé and

Príncipe from 1901. Beginning with his first visit to Lisbon in 1903, he had spent 6 years and thousands of pounds trying to improve labour conditions on the cocoa islands. (Higgs 151n49)

Carson, however, succeeded in refocusing the trial on slavery and away from the *Standard*'s libel of the chocolate makers. In instructing the jury, Judge William Pickford asked them to consider whether there had been "a dishonest plot to delay the matter being brought before the British public in order to enable the plaintiffs to go on buying slave-grown cocoa when they knew they ought to give it up". The jury returned its verdict within an hour, deciding in favour of Cadbury Brothers but awarding only 1 farthing (one-quarter of a penny) in damages. Pickford directed the newspaper to pay £3000 of the firm's court costs. The paltry award implied strongly that even though the jury agreed that Cadbury Brothers had been libeled it was not terribly sympathetic to the firm's position. (Higgs 152)

## Improving the Conditions of Labour

Nor had the broader issues that prompted Cadbury Brothers to begin investigating the conditions of labour in Portuguese Africa been resolved. Beginning in February 1910, labour agents began operating again in Angola. Skeptical British humanitarians remained vigilant and, from the Portuguese perspective, intrusive. Cadbury had satisfied his own conscience by boycotting São Tomé's cocoa, but along with Burtt, he remained the target of Portuguese commentators determined to assert their nation's innocence of the lingering charges of slavery. The controversy over cocoa and slavery would prove never ending. (Higgs 153n56)

Britain's Foreign Office struggled with continued Portuguese intransigence. In May 1911, Portugal passed a law limiting contracts on São Tomé and Príncipe to 2 years but reaffirming the legal obligation of Africans to work and the right of colonial officials to force them to work. The repatriation of workers remained voluntary. Disappointed British officials responded by issuing, in 1912, the first in a series of "white books" on "Contract Labour in Portuguese West Africa". The report confirmed that although working conditions on the islands were good, labourers who had completed their contracts were not returning home to Angola. (Higgs 157)

In 1913, backed by Portugal's Colonial Office, Angola's governor-general, José Norton de Matos, sent colonial officers to every estate. Serviçais received new contracts limited to 2 years at the end of which they could return home. Vagrancy remained illegal, and vagrants could be compelled to work, as was the case in most European colonies. Formally, slavery in Angola's hinterland had ended. With the Witwatersrand Native Labour Association, which funneled labourers from Mozambique to the Transvaal's mines as its model, a consortium of planters founded the Emigration Society of São Tomé and Príncipe. The society was subject to government regulation, and published a yearly statistical report. (Higgs 159n11)

In its 1914 white book, the British Foreign Office acknowledged the improvements. In 1916, the British consul H. Hall Hall confirmed from Luanda that workers were returning regularly to Angola carrying their severance payments from the repatriation fund. Those who stayed on the islands appeared to be doing so by choice. The cocoa controversy faded into obscurity, replaced by news of the Great War that had begun in August 1914. (Higgs 159-160n12)

Despite these improvements, Cadbury Brothers Limited continued to buy its cocoa from the Gold Coast (Ghana). Planters in São Tomé and Príncipe faced a range of additional challenges: soil degradation, swollen shoot disease, an infestation of thrips insects that attacked cocoa and the ever-constant high price of labour. In 1918, São Toméan cocoa exports declined 56% over the previous year. In the end, the high quality of cocoa produced by small-scale African farmers elsewhere in West Africa would out-compete planters on São Tomé and Príncipe in producing cocoa for world markets.[2]

## Conclusion

São Tomé and Príncipe's African cocoa farmers failed to regain a foothold in the competitive international cocoa market. African farmers in Ghana and the Côte d'Ivoire became the major suppliers of cocoa to the world's chocolate manufacturers. Yet allegations of slavery persisted. In 2000 and 2001, newspaper editorials and a television documentary alleged that Cadbury and the French firm Nestlé were knowingly buying cocoa harvested by child labourers enslaved on Côte d'Ivoire's cocoa plantations. Cadbury representatives replied that the company purchased the bulk of its cocoa from Ghana. The firm referred critics to the World Cocoa Foundation, established in 2000 to encourage "sustainable, responsible cocoa growing" in Africa, Asia and the Americas. Foundation initiatives included literacy and vocational-training programs for labourers. In Côte d'Ivoire, the world's largest exporter of cocoa, government officials agreed to cooperate with cocoa exporters and importers and the International Labour Organization to "eliminate child slave labour in the cocoa chain". In 2002, the London-based Anti-Slavery International (the successor to the British Anti-Slavery Society) estimated that approximately 284,000 children worked in the cocoa fields of West Africa. Of these, perhaps 15,000 (5.3%) were enslaved in Côte d'Ivoire. Less clear was whether the employers considered those young workers slaves, in societies where children commonly worked in extended family settings. Identifying the specific factors (including hunger and poverty at home) that forced children to labour on Côte d'Ivoire's plantations would prove as difficult as it had been for Burtt in São Tomé and Príncipe in 1905. Today, despite anecdotal evidence collected by a new generation of crusading journalists, African workers remain largely anonymous—and the meaning of slavery and freedom in an African context sometimes unfathomable—to the

---

[2] W. G. Clarence-Smith, "The Hidden Costs of Labour on the Cocoa Plantations of São Tomé and Príncipe, 1875–1914," *Portuguese Studies* 6 (1990): 170. See also, Higgs, 160.

predominantly Western consumers of chocolate. (Higgs 164n19) In Côte d'Ivoire, much envied in the rest of Africa for its prosperity and stability, the peace shattered in September 2002. Cocoa prices had fallen in the late 1990s; the accusations of child slavery soon followed. As the economy weakened, political, ethnic and religious tensions increased. Treaties were negotiated and broken. Yet truckers hauling cocoa through conflict zones made it safely to the coast, and cocoa beans found their way to chocolate manufacturers. A fragile peace was reached in 2007. Côte d'Ivoire's president, Laurent Gbagbo, had claimed power in 2000. He used cocoa profits to buy guns, as did his opponents. In November 2010, Gbagbo was defeated in an election deemed fair by international observers. He refused to leave office. His opponent, the victorious president-elect Alassane Ouattara, called for a ban on cocoa exports in order to starve Gbagbo of the funds required to pay the army and run the government. World cocoa prices soared. Gbagbo conceded defeat in April 2011. (Higgs 164–165) In January 2019, the International Criminal Court in The Hague acquitted Gbagbo for crimes committed during the 5 months he refused to leave office in 2010–2011.[3]

By 2020, Côte d'Ivoire's democracy was more stable.[4] Cocoa remains at the centre of its economy, as it does for neighbouring Ghana, where Cadbury Brothers had turned for a source of cocoa produced by free labour in 1909. Together, the two countries produced 65% of the world's chocolate supply. In July 2019, they agreed to form "a cocoa cartel" and add a fee of $400 per metric ton to improve profits for farmers. The news increased cocoa futures in January 2020 by 15% to $2690 per metric ton. The new tariff went into effect in October 2020,[5] even as the COVID 19 pandemic triggered "ongoing market uncertainty" that led some cocoa farmers to delay investments in their farms, while also disrupting exports abroad.[6] More than a hundred years earlier, in Portuguese colonial São Tomé and Príncipe, the fluctuating price of cocoa and the high costs of unfree labour had threatened the profits of large and small planters alike. Beginning in the late 1890s and continuing for a decade, large planters tried to stabilize profits by buying and storing cocoa beans in Lisbon,

---

[3] Dionne Searcey and Palko Karasz, "Laurent Gbagbo, Former Ivory Coast Leader, Acquitted of Crimes against Humanity," *The New York Times*, January 15, 2019, https://www.nytimes.com/2019/01/15/world/africa/laurent-gbagbo-ivory-coast-icc.html

[4] Fitch Wire, "Cote D'Ivoire President's Re-election Bid Raises Political Risks," *FitchRatings*, August 7, 2020, https://www.fitchratings.com/research/sovereigns/cote-divoire-presidents-reelection-bid-raises-political-risks-07-08-2020

[5] Carol Ryan, "Heard on the Street: The Bittersweet Effect of the New Cocoa Cartel," *Wall Street Journal*, January 17, 2020 (online), January 18, 2020 (US print edition), https://www.wsj.com/articles/cocoa-cartel-is-less-bitter-for-luxury-chocolate-11579256986 and https://global-factiva-com.ezproxy.library.ubc.ca/redir/default.aspx?P=sa&NS=16&AID=9UNI032300&an=DJDN000020200118egli000bp&cat=a&ep=ASI

[6] "Covid-19 Effects on Cocoa Sectors in Ghana and Cote d'Ivoire," *Ministerie van Landbouw, Natuur en Voedselkwaliteit,* The Netherlands, February 25, 2021, https://www.agroberichtenbuitenland.nl/actueel/nieuws/2021/02/25/covid-19-effects-on-cocoa-sectors-in-ghana-and-cote-d%E2%80%99ivoire#:~:text=The%20COVID%2D19%20crisis%20will,the%20sectors%20in%20both%20countries.

where they had an unfortunate tendency to rot. (Higgs 35-36n17) In 2020, the Côte d'Ivoire-Ghana cocoa cartel did encourage some large European firms like Nestlé to purchase cocoa futures at the higher price. Many firms then engaged in "shrinkflation", charging higher prices for smaller portions and passing on the increased price of cocoa to their consumers, not all of whom were tolerant of such tactics.[7] The model for Côte d'Ivoire and Ghana's cocoa cartel was OPEC. Unlike oil, cocoa beans, whether stored in Lisbon, Yamoussoukro or Accra, can spoil, as chocolate manufacturers know. Cocoa remains vulnerable to disease, insects and soil degradation.[8] In Ghana in 2024, swollen shoot disease may have eliminated as much as "500,000 hectares of farmlands".[9] The profit in cocoa is not in the growing but in the processing. Even with periodic declines in the output of cocoa and increases in the cost of cocoa beans on world markets, that profit will still go to largely Western chocolate manufacturers, as it did in 1900 and as it does in 2024.[10]

---

[7] Ryan, "Heard on the Street: The Bittersweet Effect of the New Cocoa Cartel" *Wall Street Journal*, January 17, 2020 (online); "Ghana's 2023/24 Cocoa Output Seen Almost 40% Below Target, Sources Say," *Reuters*, February 22, 2024, https://www.reuters.com/world/africa/ghanas-202324-cocoa-output-expected-be-almost-40-below-target-sources-say-2024-02-22/

[8] Olivier Monnier and Isis Almeida, "Why African cocoa growers are having an OPEC moment," *Bloomberg QuickTake*, July 23, 2019, https://www.bloomberg.com/professional/blog/african-cocoa-growers-opec-moment/

[9] "Ghana's 2023/24 Cocoa Output Seen Almost 40% Below Target, Sources Say," *Reuters*, February 22, 2024.

[10] "Ghana's 2023/24 Cocoa Output Seen Almost 40% Below Target, Sources Say," *Reuters*, February 22, 2024.

# Chapter 3
# 'Beauty, Imagination and Order'; The Flowers of William and May Morris

Helen Mary Elletson

**Abstract** Throughout their lives, William and May Morris were fascinated by the natural world. Native British flora is featured in all their varied artwork, from wallpaper to embroidery, with flowers, plants, trees, leaves, branches, and stems to form the main components. The two Morrises went on to create some of the most iconic Morris & Company designs that are justifiably considered masterpieces and that remain enduringly popular. This chapter will, therefore, focus on the major role that the natural world played in the lives of William and his daughter May and how they revolutionised interior design by taking the power of the flower to transformative new heights.

**Keywords** William Morris · May Morris · Wallpaper · Embroidery · Morris & Company · Arts and crafts

## William Morris

William Morris (24 March 1834–3 October 1896) (Fig. 3.1) was a revolutionary force in Victorian Britain: his work as an artist, designer, craftsman, writer and socialist dramatically changed the fashions and ideologies of the era. Morris would become one of the most significant figures in the arts and crafts movement, a man of far-ranging creativity and knowledge. Morris founded his firm Morris, Marshall, Faulkner & Co. in 1861. The firm fast became highly fashionable and much in demand, and it profoundly influenced interior decoration throughout the Victorian period, with Morris designing tapestries, wallpaper, fabrics, furniture and stained glass windows. In 1875, Morris assumed total control of the company, renamed

H. M. Elletson (✉)
William Morris Society, London, UK

Emery Walker Trust, London, UK
e-mail: researchcurator@williammorrissociety.org; curator@emerywalkertrust.org.uk

© The Author(s), under exclusive license to Springer Nature Switzerland AG 2024
L. M. Mendonça de Carvalho (ed.), *The Victorians: A Botanical Perspective*,
https://doi.org/10.1007/978-3-031-68759-4_3

**Fig. 3.1** William Morris,
photographed by Emery
Walker, c1889, reproduced
from a lantern slide,
William Morris Society

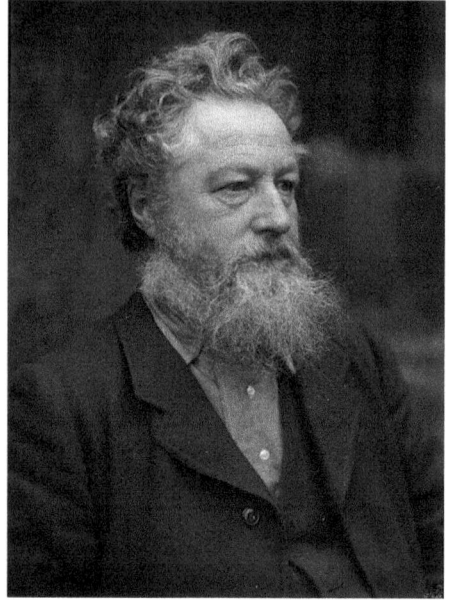

Morris & Co and subsequently traded until 1940, its longevity a testament to the
success of Morris's designs.

## May Morris

May Morris (25 March 1862–17 Oct 1938) (Fig. 3.2) was the second daughter of
William and Jane Morris. Teacher, lecturer, editor, embroiderer, jeweller and
designer, May was accomplished in a wide range of crafts, but it is her work as an
embroiderer that is considered to be her most outstanding achievement. May's
knowledge of needlework, her talent for designing and her brilliance with the needle
led to raising the status of embroidery to fine art. Overshadowed for many years by
her more famous father, May is now beginning to gain the recognition she deserves
as being an incredibly talented craftswoman in her own right and is justifiably
regarded as one of the most significant designers of the Arts and Crafts Movement.

This chapter begins with an exploration of William Morris, referred to as Morris,
whose love of nature resulted from a childhood spent exploring his local Epping
Forest in East London. This ancient woodland with established habitats was threatened with destruction in the Victorian era. Morris campaigned to save the forest,
stressing the importance of its native trees, particularly the hornbeam. The cam-
paigners won, and the site was saved in what was the first major success of the
environmental movement in Europe. However, it was not only the threat to Morris's
beloved forest that cemented his joy in nature, it was also in this woodland that he

**Fig. 3.2** May Morris, unknown photographer, c. 1910, William Morris Society

came across his first floral textile: 'How well I remember as a boy my first acquaintance with a room hung with faded greenery at Queen Elizabeth's Lodge, by Chingford Hatch, in Epping Forest (I wonder what has become of it now), and the impression of romance that it made upon me'.[1]

From a young age, Morris became extremely concerned with the industrialisation and over-production prevalent in Victorian society. It was, therefore, a natural progression that when he was 27 years old, he would combine his natural talent for designing, his love of nature, and his hatred of the Industrial Revolution by forming his own interior design firm in 1861. With the foundation of Morris, Marshall, Faulkner & Company in 1861, Morris strove to create an alternative to the dehumanising industrial systems that produced poor-quality, mass-produced objects, and instead championed the handmade.

The first of the applied arts that Morris mastered was embroidery. The embroidery techniques he employed have been identified as medieval in origin and it is recorded that Morris unpicked antique embroideries to see how they were made. This resulted in his earliest attempt at the craft, with the *If I Can* hanging (1857). This early example expressed Morris's interest in historic furnishings, particularly

---

[1] *The Lesser Arts of Life* (1882), delivered on 21st January 1882 before the Birmingham and Midlands Institute in Birmingham, with the title Some of the Minor Arts of Life and published in the collection Lectures on Art delivered in support for the Society for the Protection of Ancient Buildings London, Macmillan and Co., 1882, pp. 174–232.

the medieval wall hangings he had admired since childhood. If I Can, was Morris's motto, taken from a translation of Jan van Eyck's Als Ich Can. The embroidered panel was designed and stitched by Morris in a repeating pattern of fruit trees with birds in flight. Although this raised work with extremely thick wool embroidered onto a linen ground was primitive in terms of his later work, the piece demonstrated an important learning process, highlighted Morris's attempts to master historical techniques and laid the foundation for Morris's next, more refined embroidery.

The fabulous medieval imagery of a millefleurs tapestry in a fifteenth-century version of an illuminated copy of Jean Froissart's Chronicles, Dance of the Wodehouses, provided the inspiration for Morris's next embroidery, *Daisy* (1860). With its simple outline stitching on blue woollen serge, these embroidered panels were very much in the style of wall hangings from the Middle Ages. Morris and his wife Jane created these hangings for Red House, their first marital home in Bexleyheath, Kent. The daisy motif was utilised in various media manufactured by the Firm, including stained glass and tiles, although it is for wallpaper that the simple daisy is now best-known.

*Daisy* (1864) was the first wallpaper put into production and was Morris's second wallpaper design (Fig. 3.3). This straightforward but beautiful design of repeating white, brown and rose-coloured meadow flowers amongst fields of grass heralded a new era for wallpaper production. However, Morris's designs didn't

**Fig. 3.3** *Daisy* wallpaper, designed by William Morris and manufactured by Jeffrey & Co. for Morris & Co., 1864. Block-printed in distemper colours on paper, William Morris Society

immediately chime with fashionable taste, which in the 1860s generally favoured wallpapers in the super-naturalistic style. Enabled by sophisticated new printing techniques, this look was characterised by complex and unashamedly pretty designs that centred on exuberant flowers and the use of illusory effects such as *trompe l'œil*. As Morris eloquently stated in his lecture, 'Some Hints on Pattern Designing', a successful design should possess a degree of abstraction with a suggestion of the outdoors, rather than 'sham-real boughs and flowers, casting sham-real shadows on your walls.. The alternative fashion was for geometric patterns such as those of A.W.N. Pugin and Owen Jones. Although Morris also focused on plants, like the then-popular floral papers which imitated nature, his work was unprecedented because it celebrated the simple forms he saw in British gardens, fields and hedge-rows rather than exotic, imported blooms. Morris also wrote profusely on unnatural flower propagation and warned against cross-breeding to make showier blooms. Morris's papers celebrate the beauty of repetition and symmetry, and were to become extraordinarily popular amongst his Victorian clientele, starting a floral revolution in interior design.

Much of Morris's design inspiration derived from serious study and research into historic plants. Morris possessed several botanical herbals, and it is known that he owned two copies of John Gerard's famous Herbal, one of them since childhood. John Gerard, an English herbalist and botanist, first published his book on plants and their properties in 1597. The copiously illustrated book contained over 1000 plants and was seen as the best and most exhaustive work of its kind, ensuring its lasting popularity. Gerard's Herbal remained a much-loved reference book through-out Morris's life, frequently referring to it as a source book for plants and their properties. It was through his meticulous study of natural forms that was to ensure Morris's reputation of the leading pattern-designer of the age.

Morris's ability to design stunning repeating patterns is evident when examining his wallpapers and printed cottons. Morris's papers gained popularity in the 1870s and could be found decorating the walls of middle- and upper-class homes as well as two royal residences. The three-dimensionality of the magnificently complex *Jasmine* wallpaper (1872) demonstrates Morris's skill at creating complex patterns with two or three layers of closely interwoven foreground and background (Fig. 3.4).

The luxuriant scrolling foliage, hawthorn leaves and jasmine blossoms give a wonderful impression of a naturalistic garden tangle. Interestingly, the artificial blue and pink colours of the drawing were never reproduced as wallpaper; rather, the greens and blues found in nature were selected for the finished product. Almost all of Morris's wallpapers were printed using hand-cut woodblocks loaded with natu-ral, mineral-based dyes in a process that could take up to 4 weeks.

The original working drawing for *Larkspur* (1874) is unusually monotone in comparison with the multi-coloured end results and demonstrates Morris's working process (Fig. 3.5).

Surprisingly, the drawing features a thistle despite the *Larkspur* title in the upper left-hand corner in Morris's writing. It is only in the completed wallpaper that irregular-shaped petals of the larkspur flower replace the thistle. The larkspur is a cottage garden flower and features in the wallpaper in the blue and pink varieties,

**Fig. 3.4** Design for
*Jasmine* wallpaper,
designed by William
Morris, c. 1872, pencil and
watercolour on paper,
William Morris Society

floating amongst spiralling leaves and curving sprays of foliage. *Larkspur* was first issued as a single-colour paper in 1874, although the multi-coloured version of 1875 produces a more successful pattern with a depth and variety of colour with a dotted background not present in the earlier paper (Fig. 3.6).

The success of Morris's wallpaper designs, such as the iconic *Willow Bough* (1887), relies on Morris's well-practiced and close observation of nature (Fig. 3.7). *Willow Bough* features diagonal willow branches in various shades of green, has been in continuous production since 1887 and is one of Morris's most enduring wallpaper patterns. As Morris stated in his design theory lecture 'Some Hints on Pattern Designing' (1881), he believed all patterns should be made up of 'ornament that reminds us of the outward face of the earth'. In Morris's view, the greatest feature to be evoked was the natural world. Morris strongly believed that patterns should not be meaningless but have a story, writing, 'any decoration is futile … when it does not remind you of something beyond itself, of something of which it is but a visible symbol …I must insist on plenty of meaning in your patterns, I must have unmistakable suggestions of gardens and fields and strange trees, boughs and

**Fig. 3.5** Design for *Larkspur* wallpaper, designed by William Morris, 1874, pencil and watercolour on paper, William Morris Society

tendrils …'.[2] *Willow Bough* soon became one of Morris's timeless masterpieces, proving that it was possible to produce well-designed fabrics using traditional skills and materials on a commercially viable basis.

By the 1880s, Morris's patterns were becoming more sophisticated, incorporating both flora and fauna in his patterns, and *Brother Rabbit* (1882) charmingly fuses the use of animals with birds and fantastic foliage (Fig. 3.8). The name originates from the Uncle Remus stories of Brer Rabbit which Morris read to his young daughters. Like many of Morris's printed cottons, Brother Rabbit was indigo discharged and block printed at Merton Abbey in Surrey, the Morris works premises from 1881. Indigo is a natural dye extracted from the leaves of the indigo plant. It is an ancient dye and had largely gone out of use by the Victorian era and had almost disappeared throughout Europe, replaced with easier to use chemical dyes. Morris consulted his library of old herbals and dyers manuals in order to revive lost skills and partnered with expert Thomas Wardle of Leek in Staffordshire. Together, the two men began resurrecting traditional methods of dyeing cloth. Morris's desperate attempts to revive indigo dyeing, despite its difficult reputation, are clear when he wrote to Georgiana Burne-Jones how he was 'taking in dyeing at every pore'. Telling her,

---

[2] *Some Hints on Pattern-Designing*, 1881.

**Fig. 3.6** *Larkspur* wallpaper, designed by William Morris and manufactured by Jeffrey & Co. for Morris & Co., c 1875. Block-printed in distemper colours on paper, William Morris Society

'The indigo dye vat was 9 feet deep and holds 1,000 gallons, sunk into the earth, it would be a week's work to tell you all the anxieties connected with indigo'.[3] Morris would never give his workers anything to do that he wasn't willing to do himself and famously went around with arms dyed blue for several weeks. Indigo dyeing was a time-consuming and exacting process. First, the undyed cloth was washed and then immersed in the indigo vat for the correct amount of time. When lifted out, the cloth appears green, but turns blue quickly after oxidizing with the air. For the areas not intended to be blue, a bleaching agent was block printed where required. The cloth was washed again, and the blue cleared from the bleached areas, producing a print of dark blue with white.

The fabric was then dried, warmed and prepared for the next colour such as the ancient weld plant for yellow areas and the plant root madder or kermes which gives a strong red in patterns such as *Strawberry Thief* (1883) (Fig. 3.9).

A contemporary of Morris wrote that it was 'as if the cloths were stained through and through with the juices of flowers'.[4] *Strawberry Thief* was the first printed cotton registered using the natural yellow and red dyes added to the basic blue and

---

[3] Norman Kelvin (ed.), *The Collected Letters of William Morris*, vol. 1, p. 476.

[4] May Morris, *William Morris: Artist, Writer, Socialist*, 2 vols (Basil Blackwell, Oxford, 1936) vol. 1, p. 38.

**Fig. 3.7** *Willow Bough* wallpaper, designed by William Morris and manufactured by Jeffrey & Co. for Morris & Co., c. 1877. Block-printed in distemper colours on paper, William Morris Society

**Fig. 3.8** *Brother Rabbit* printed cotton, designed by William Morris and manufactured by Morris & Co., 1884. Indigo-discharged and block-printed, William Morris Society

white ground. The green was achieved by printing yellow on top of blue. Morris was successful in bringing the ancient use of natural plant and vegetable dyes back into use. His reasons for doing so were down to his firm belief that the colours obtained were beautiful and harmonised naturally with each other, and that natural dyes were more permanent than the chemical dyes which faded quickly and unevenly. Morris

**Fig. 3.9** *Strawberry Thief* printed cotton, designed by William Morris and manufactured by Morris & Co., 1883. Indigo-discharged and block-printed, William Morris Society

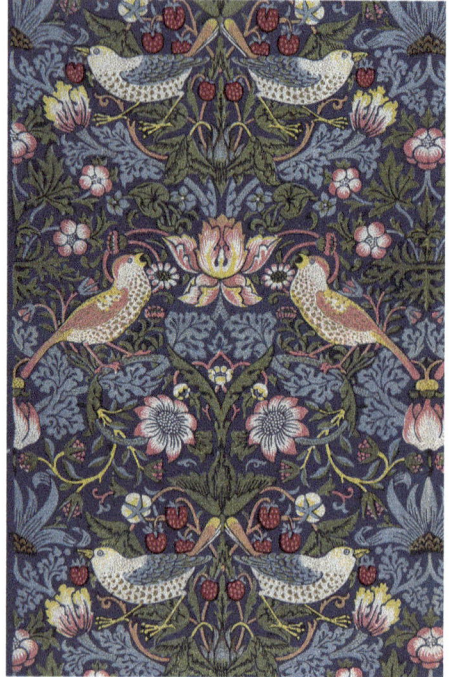

was proved right; *Strawberry Thief* looks as bright and colourful as the day it was printed. The other wonderful feature about this pattern is the story behind it. When Morris saw the thrushes eating the strawberries in his garden, his gardener was about to shoo them away, but Morris told him to stop as the birds had inspired him to create a new design. Featuring pairs of charmingly mischievous birds nestled amongst glorious foliage and flowers, Strawberry Thief became one of Morris's most commercially successful repeating patterns. The inspiration for this print came to Morris whilst at his garden at Kelmscott Manor, his beloved country home in Oxfordshire. Morris came across thrushes stealing fruit from his kitchen garden, providing him with an ideal subject for his next design. This splendidly colourful textile continues to be one of Morris's most popular patterns and is arguably his most recognisable design.

The highly elaborate, large-scale, symmetrical *Pimpernel* (1876) is the first example of Morris using the 'turnover' four-way mirror repeat pattern structure (Fig. 3.10). Like with many of Morris's designs, *Pimpernel* is named after the least prominent flower, the small buds dotted throughout the pattern; the five-petaled wildflowers of the genus Anagallis. The pimpernels are encased amongst delightful large tulip blooms and entwined foliage. *Pimpernel* was one of Morris's favourite wallpapers, used to decorate the dining room of his Georgian riverside home, Kelmscott House in London, now the headquarters of the William Society.

The *Bird* (1878) woven wool is of particular significance to the William Morris Society not only for the fact that the original watercolour is in the collection, but that

**Fig. 3.10** *Pimpernel* wallpaper, designed by William Morris and manufactured by Jeffrey & Co. for Morris & Co., c. 1876. Block-printed in distemper colours on paper, William Morris Society

it was also created by Morris especially for his drawing room at Kelmscott House (Fig. 3.11). This was the first design by Morris to feature birds he drew himself, rather than relying on his colleague Philip Webb.

Birds are also a principal feature of *Rose* (1883), which has clear Eastern influences (Fig. 3.12). The ogee shapes of the tulips are reminiscent of lotus flowers and arabesques are formed with the scrolling branches. Morris championed the rose, and his preference was for the wild rose over all the other roses, believing that it could not be improved upon. He extolled the virtues of his favourite flower in his lectures, when he wrote of the '… exquisite subtlety of form, delicacy of texture, and sweetness of colour, which, blent with the richness which the true garden-rose shares with many other flowers, yet makes it the queen of them all – the flower of flowers'.[5] It therefore seems appropriate that, in 1998, the David Austen rose company introduced the William Morris rose, grown by hand in Shropshire, England. The variety was selected for its apricot/pink flowers, old-fashioned full flower shape with 41 petals, sweet fragrance and hardiness. The William Morris rose grows in the Kelmscott House garden and is widely admired by visitors. Roses feature in numerous Morris patterns, but this glorious *Rose* hand block printed cotton is solely

---

[5] *Some Hints on Pattern-Designing*, 1881.

**Fig. 3.11** *Bird* design for
woven wool, designed by
William Morris, pencil and
watercolour on paper,
1878, William Morris
Society

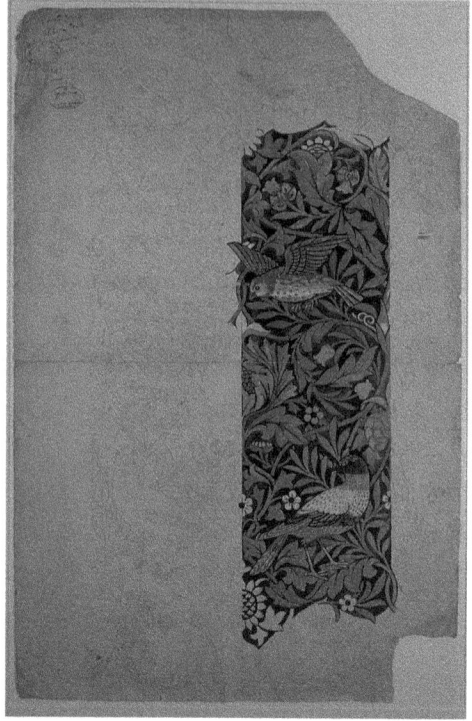

dedicated to Morris's much-loved flower. Morris's remarkable fabrics became incredibly popular and would provide the benchmark of British printed textile achievement.

From the mid-1880s onwards many of Morris's wallpapers showed a structure based on diagonally meandering stems, layering was reduced in favour of backgrounds dotted or covered with tiny stylised flowers and trails.

The *Grafton* wallpaper (1883) is unique as it is the only occasion when Morris used a stencil effect within his designs (Fig. 3.13). It also has a very noticeable use of dots in the background. In his 1889 essay 'Textiles' Morris notes that in printed cloths '…much use can be made of hatching and dotting, which are obviously suitable to the method of block printing'.[6] This was also applicable to Morris's wallpapers. Dotting was achieved by hammering brass pins into the surface of the wood block. Interestingly, there is a note on the watercolour drawing in Morris's handwriting with the instructions to the block cutter, 'leave out pins'. Consequently, the pins were not hammered into the wood block and therefore no dotting is on the *Grafton* finished wallpaper.

---

[6] *Some Hints on Pattern-Designing*, 1881.

**Fig. 3.12** *Rose* printed cotton, designed by William Morris and manufactured by Morris & Co., 1883. Indigo-discharged and block-printed, William Morris Society

**Fig. 3.13** Design for *Grafton* wallpaper, designed by William Morris, 1883. Pencil and watercolour on paper, William Morris Society

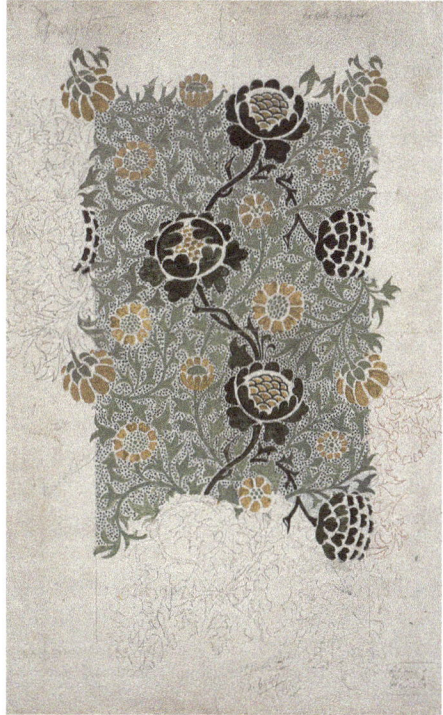

# The Two *Honeysuckles*: A Case Study of the Designs by May and William Morris

*Honeysuckle* is a design that unites father and daughter as it has an identical name. It is also a rare occasion when two designers for Morris & Company created a pattern with the same title. William Morris designed his *Honeysuckle* as a printed cotton and May Morris created hers for wallpaper. Both cottons and wallpapers were hand block printed and this case study will examine the traditional techniques involved in creating these two beautiful designs, as well as exploring *Honeysuckle* in a wider context of other Morris & Company products.

## *Honeysuckle* Designed by May Morris for Wallpaper in 1883

The William Morris Society is fortunate to hold the original watercolour drawing for *Honeysuckle* (1883) and it is the only design for wallpaper by May Morris in the Society's collection (Fig. 3.14). Wallpapers were one of the first products offered by Morris and Company, although Morris's attempts to print his first three repeating wallpaper designs, *Daisy*, *Trellis* and *Fruit*, using oil colours on etched zinc plates were not a success as the results were inconsistent. Morris then had conventional woodblocks hand cut by Barrett & Co., and in 1864 he approached the established wallpaper manufacturer Jeffrey & Co. of Islington to hand print the papers.

The successful relationship between the two companies owed much to Metford Warner and under his lead Jeffrey and company Jeffrey & Co. employed some of the finest designers of the time such as William Burges, Walter Crane, C.F.A. Voysey and A.H. Mackmurdo. Warner was completely sympathetic to Morris's ideals and is quoted as saying that Morris 'allowed nothing to pass until he was quite satisfied it was right in both colour and design'.[7]

Warner set up a special department to handle Morris & Co. wallpaper printing and Morris himself was closely involved in production, choosing colours and overseeing printing. Warner recalled Morris with his hands blue from the dye vat, ordering his workforce '... not to improve my colourings'. Morris did not use synthetic colours, he chose to use natural pigments carried in a water-based medium of size or whiting which produced a thick, chalky effect.

The process of block printing started with the design meticulously engraved onto the surface of a rectangular deal block on which was mounted a layer of fine-grained wood; fruit trees such as cherry and pear offered the right balance between durability and relative ease of carving the design. Those areas not required to print were cut away. Always a perfectionist, if they were not precisely to his requirements Morris would discard a complete set of printing blocks without regard to expense. Multi-coloured patterns required a separately carved block for every pigment in the finished design. The carved surface was inked with paint in a blanket-lined tray and

---

[7] Mackail, 1899, vol. 2, p. 194.

**Fig. 3.14** Design for
*Honeysuckle* wallpaper,
designed by May Morris,
1883. Pencil and
watercolour on paper,
William Morris Society

then lowered face-down onto the paper for printing. 'Pitch' pins on the corners of the blocks helped the printer to line up the design precisely. Each colour was printed separately along the length of the roll, which was then 'festoon' hung up to dry before the next block of a different colour could be applied. The process was laborious but required great skill and precision; a single 30 block roll of paper could take up to 4 weeks to complete.

The Morris & Co. wallpaper book of 1905 promoted the qualities of hand-printed papers over machine-printed:

> HAND-PRINTED PAPERS are produced very slowly, each block used being dipped into pigment and then firmly pressed onto the paper, giving a great body of colour. This process takes place with each separate colour, which is slowly dried before another is applied. The consequence is that in the finished paper there is a considerable mass of solid colour. MACHINE-PRINTED PAPERS are produced at great speed, all the colours being printed at one time and rapidly dried in a heated gallery. In consequence of the speed at which they are printed, there is merely a film of colour deposited on the surface of the paper. FOR PERMANENT USE we strongly recommend the hand-printed papers.

*Honeysuckle* required eight blocks to print and was available in four colourways, including versions with cream, green and blue grounds. Inscribed on the front of the

Society's original design in pencil are notes on the printing blocks and on the verso '*Honeysuckle* Wallpaper' and 'Mr Morris Esq' which has added to the uncertainty whether May or William Morris designed *Honeysuckle*. It also differs to May's other two designs for wallpapers, *Horn Poppy* and *Arcadia*. However, the heavily worked pencil outlines and technique is characteristic of her designs and it was attributed to May Morris for the first time in c1910. It was common practice for designs to come under William Morris's name, to ensure their profitability. It is also stylistically very different to Morris's version of *Honeysuckle* which will be analysed shortly. Confusion over the attribution of the wallpaper design still occurs to this day although, regardless, May's delightful pattern continues to be enduringly popular.

## *Honeysuckle* Designed by William Morris for Printed Cotton in 1876

Having had success with his first wallpaper designs and keen to re-establish the use of natural dyes in textile production Morris began working with Thomas Wardle, the owner of a dyeing works in Leek, Staffordshire. Wardle produced Morris & Company textiles throughout the 1870s, until Morris took the block-printing in-house. His search for a site on which to consolidate all his production workshops ended in 1881 when he found Merton Abbey, an early eighteenth-century calico-printing works. The soft water of the River Wandle was ideal for the natural vegetable dyes which Morris used, preferring these over the harsher synthetic aniline dyes.

The 7-acre site at Merton Abbey already had a dye works and a bleaching field with an area large enough to allow for the necessary washing and drying of lengths of cloth. Block-printing of fabrics was undertaken on long padded tables running the length of the workshop. Designs were carved on a series of fruit-wood blocks, one for each colour used, and the process was similar to the block printing of rolls of wallpaper. The dye pad trays were mounted on trolleys which could be pulled along the length of the table, the inked block pressed down onto the cloth using a lead-weighted mallet. Later the wooden blocks were modified with metal inserts padded with felt to hold the dye.

The Morris & Company dye books demonstrate the complexities of printing and demonstrate Morris's extraordinary knowledge of the technology and chemical elements of the craft. Morris made comprehensive notes on the dying process, including how to maintain the correct atmosphere in the printing sheds, writing, 'some boil a kettle in the shop to keep the air moist; some hang up damp clothes; this last is best'.[8]

Like May Morris's wallpaper, the *Honeysuckle* fabric (1876) also required eight blocks and is a large-scale turn-over repeat (Fig. 3.15). In a lecture on textiles,

---

[8] *The Lesser Arts of Life* (1882).

**Fig. 3.15** *Honeysuckle* printed cotton, designed by William Morris and manufactured by Morris & Co., 1876. Block-printed, William Morris Society

Morris said, 'do not be afraid of large patterns, if properly designed they are more restful to the eye than small ones ... very small rooms, as well as very large ones, look better ornamented with large patterns'.[9]

William Morris said of his *Honeysuckle*, 'it is the most important we have or are likely to have.' May Morris agreed, describing it as 'the most truly Morrisian in character of all his pattern-making ... the most mysterious and poetic – the very symbol of a garden tangle.'[10] *Honeysuckle* was one of Morris's most complex and successful patterns; trial printing started in June of 1876 and took several months to perfect. The Morris and Company registration mark is clear and the Society's large sample beautifully illustrates the block printing process with the irregular end of the printing visible.

The honeysuckle flower was of great significance to Morris & Company, with both William and May Morris using the delicate bloom in a variety of other products, from book covers and embroidery to tapestry and woven wools, demonstrating

---

[9] *Some Hints on Pattern-Designing*, (1881).

[10] May Morris, *William Morris: Artist, Writer, Socialist*, 2 vols (Basil Blackwell, Oxford, 1936) vol. 1, p. 38.

its popularity in Morris & Company products. However, it was in their respective wallpaper and fabric that the Morrises cemented the popularity of this native, fragrant woodland stunner. It seems apt to conclude this case study with a note on the study of flowers and their meanings, which was so popular in Victorian times. Honeysuckle symbolises happiness, which is highly appropriate considering the pleasure these two patterns continue to bring a 140 years after they were designed.

# 'Vividly Picturesque and Uniquely Original': The Gardens of William and May Morris

It would be impossible to do full justice to a study of William and May Morris's flowers without including their gardens. It is possible to gain a thorough understanding of the varied and beautiful gardens of the Morrises homes through contemporary photographs, letters and their own words. This study will begin by looking at descriptions of William Morris's childhood gardens, before exploring the inspiration behind the innovative planting schemes at Red House, and lastly moving on to Morris's particular approach to design in his lectures and how this was reflected in his last two homes; Kelmscott Manor and Kelmscott House.

Morris's birthplace, Elm House in Walthamstow, was sadly destroyed and therefore very little is known of the garden. However, J. W. Mackail, Morris's biographer, tells us that the house had a large lawn surrounded by shrubberies and kitchen gardens, with a large mulberry tree. When William Morris was 6 years old, the family moved to Woodford Hall, which was situated in a 50-acre park on the edge of Epping Forest with the river Lea just 3 miles away. Morris would later create his *Lea* (1885) design, inspired by his childhood memories of the river, one of a series of patterns named after tributaries of the Thames which will be explored later. Morris recalled his second home affectionately, writing; 'To this day when I smell a may tree I think of going to bed [there] by daylight'.[11] He recalled that he and his brothers and sisters had their own gardens and as a little boy he loved the hepatica and herb patch by the kitchen garden.

After the death of Morris's father, the Morris family moved to Water House, Walthamstow, in 1848. The building is now the William Morris Gallery, situated in Lloyd Park, but in Morris's day it had a large lawn leading to a moat 40 feet wide in which Morris and his brothers fished. The moat had an island with aspens, holly, chestnut and hawthorn trees. As mentioned at the start, it was whilst living in Walthamstow that Morris's love of nature originated, particularly with the nearby Epping forest, saying, 'when I was a boy and young man I knew it yard by yard'.[12]

---

[11] Norman Kelvin (ed.), The Collected Letters of William Morris, vol. 2, p. 259.

[12] Letter to the Daily Chronicle supporting a campaign to preserve Epping Forest, 22 April 1895 in Kelvin vol. 4, 1996, p. 268.

Red House, Morris's first married home, in Bexleyheath, Kent, is of great importance to learning of Morris's inspiration when it comes to garden design (Fig. 3.16).

The house was built especially for Morris by Philip Webb in 1859, within an orchard. The building of Red House needed hardly any disturbance to the trees; Webb lists 80 trees on the site, the majority of which were old varieties of fruit trees, including apples, pears, cherries, plums and damson.

Morris had long been inspired by the medieval gardens he had studied in illuminated manuscripts, in literature such as the works of Geoffrey Chaucer, and in religious paintings. Walled and enclosed gardens were particularly favourite features of medieval gardens and Morris admired the practicalities of ancient spaces, with their sensible combination of fruit trees, vegetables, herbs and flowers.

Morris and his new bride Jane moved into their new home in 1860. Jenny and May, their two daughters, were born at Red House and the young family was to spend 5 happy years at the only house they were to own; all the rest were rented, as was common at the time. The garden was planned from the start to be integral to the home. Webb's original plans for the house show that roses, including moss roses, white Jasmine, bergamot, Aaron's Rod and passion flowers were to be planted against the north and west walls. At just over an acre, the garden was a major feature of the house and Mackail described it in detail; 'The garden was planned with the same care and originality as the house; … But in [Morris's] knowledge of gardening he did, and did with reason, pride himself … But of flowers and vegetables and fruit trees he knew all the ways and capabilities. Red House garden, with its long grass walks, its midsummer lilies and autumn sunflowers, its wattled rose trellises inclosing richly flowered square garden plots, was then as unique as the house it surrounded. The building had been planned with such care that hardly a tree had to be cut down; apples fell in at the windows as they stood open on hot autumn nights'.[13]

There are no images of the original garden layout but detailed contemporary descriptions bring it to life. There are also clues to the nature of the garden from the interior decoration of the house, particularly the painted settle in the entrance hall which includes portraits of Morris and his circle situated in a medieval-style garden.

Georgiana Burne-Jones, wife of Edward Burne-Jones and a frequent visitor to Red House, wrote: 'many flowering creepers had been planted against the walls of the house from the earliest possible time, so that there was no look of raw newness about it; and the garden, beautified beforehand by the apple trees, quickly took shape. In front of the house [the garden] was spaced formally into four little square gardens making a big square together; each of the smaller squares had a wattled fence around it with an opening by which one entered, and all over the fence roses grew thickly. Where we sat and talked or looked out into the well-court two sides were formed by the house and the other two by a tall-rose-trellis'.[14]

*Trellis* wallpaper (1862) was one of Morris first wallpaper designs, inspired by the rose trellises at Red House with the birds designed by Webb, the architect of the house (Fig. 3.17). May recalled this wallpaper on the walls of her nursery. May also recollected scented borders which had rosemary and lavender planted.

Aymer Vallance, another of Morris's biographers, noted that the garden, divided into squares, was hedged by sweetbriar or wild rose, saying, 'each enclosure with its own particular show of flowers; on this side a green alley with a bowling green, on that orchard walks amid gnarled old fruit trees; – all struck me as vividly picturesque and uniquely original'.[15] The bowling alley green Vallance mentions was about 75 feet long and 10 feet wide, whereas the square gardens were also quite small, measuring around 30 feet square. Interestingly, a new planting scheme is currently being planned at Red House by the National Trust, which will draw on these sources to reflect Morris's original vision.

Morris moved into Kelmscott Manor, the 'house of his dreams' in 1871 and the garden delighted him as much as the house (Fig. 3.18). Enclosed by high walls and divided by hedges, it conformed to his ideal of a garden 'fenced from the outside world' and he therefore altered it very little. He wrote that the garden had crocuses, aconite, yellow jasmine, scabious, red and white hollyhocks, and cottage annuals such as poppies, china asters, Sweet Sultans and Dianthus. In the summer there were hollyhocks and 'purple-blossom of house-mint and mouse-ears, while autumn

---

[13] Mackail, 1899, vol. 1, p. 147.

[14] Georgiana Burne-Jones, *Memorials of Edward Burne-Jones*, London, 1912, p. 58.

[15] Vallance, Aymer (1897). *William Morris: His Art, His Writings and His Public Life*. London: George Bell and Sons, p. 328.

**Fig. 3.17** *Trellis*
wallpaper, designed by
William Morris and
manufactured by Jeffrey &
Co. for Morris & Co., c.
1862. Block-printed in
distemper colours on
paper, William Morris
Society

brought 'a pale sweetbriar blossom among the scarlet hips'.[16] Also present were roses, including yellow roses in the gable wall of the barn and May Morris recalled the garden 'gay with thousands of tulips'.

May Morris mentioned the 'beautiful wild tulip that my father called the Persian tulip and that he used a great deal in designs, simply runs riot all over the beds'.[17] There were also apple and plum trees, a vegetable garden and a 400-year old mulberry tree, highlighting Morris's preference for native flowers and plants.

Morris described in a February letter, 'Snowdrops are everywhere but mostly double, however, they give one a delightful idea of spring about'. He said in his lectures that a snowdrop was a 'wonder of beauty' and went on to say: 'there are a few violets out here and there and a coloured primrose; and some of the hepatica roots, have flowered, but show no leaves. But how pretty it looks to see the promise of things pushing up through the clean un-sooty soil. I think we shall have a beautiful garden this year'.[18]

---

[16] J.W. Mackail, *The Life of William Morris*, 2 vols (Longmans Green & Co., London, 1899); vol. 1, pp. 236–8.

[17] May Morris, *William Morris: Artist, Writer, Socialist*, 2 vols (Basil Blackwell, Oxford, 1936) vol. 2, p. 281.

[18] Norman Kelvin (ed.), *The Collected Letters of William Morris*, vol. 2, p. 387.

**Fig. 3.18** Kelmscott Manor, unknown photographer, c. 1920, reproduced from a lantern slide, William Morris Society

Although Morris took a keen interest in his garden, it appears that he undertook little practical tasks as Mackail wrote; 'It is doubtful whether he [Morris] was ever seen with a spade in his hands; in later years at Kelmscott his manual work in the garden was almost limited to clipping his yew hedges'. One of these clipped hedges was known as 'Fafnir', after the dragon in Morris's Icelandic saga Sigurd the Volsung.

On another occasion, Morris noticed that his gardener had trellised up the raspberry-canes neatly, 'so that they looked like a medieval garden'. In his 1880 lecture, Making the Best of it, Morris wrote, 'As to colour in gardens. Flowers in masses are mighty strong colour, and if not used with a great deal of caution are very destructive to pleasure in gardening. On the whole, I think the best and safest plan is to mix up your flowers, and rather eschew great masses of colour – in combination I mean'.[19]

---

[19] Making the Best of It, (1880).

Just a year before he died, Morris commented that the garden at Kelmscott Manor looked like part of the house, which he had always felt should be the aim of a garden, writing, 'Many a good house both old and new is marred by the vulgarity and stupidity of its garden, so that one is tormented by having to abstract in one's mind the good building from the nightmare of 'horticulture' which surrounds it'.[20]

Morris approved of native flowers and frequently referred to his love of medieval gardens, which had partitions like exterior rooms, as seen at Red House. He believed that all arts and crafts gardeners should take up the idea that the garden should be in keeping and a reflection of the house, and vernacular architecture should be used as part of the arts and crafts garden.

Flowers were naturalised in the Kelmscott Manor garden and were bordered neatly with Medieval-like low hedges. Morris hated the fashion for carpet bedding, with its strong massing of single bright colours, which he called 'an aberration of the human mind'. Instead, he advocated delicate combinations of scented meadow flowers and the use of native plants.

The last time Morris saw his beloved Kelmscott Manor was 5 months before his death in 1896. He wrote to Georgiana Burne-Jones of the garden, 'I have enjoyed the garden very much, and should never be bored by walking about and in it'.[21]

The frontispiece of Morris's Kelmscott Press edition of *News from Nowhere* illustrates the entrance into the Manor, with the flagged footpath flanked by rose trees (Fig. 3.19). It is interesting to note that the three gardens at the rear of the house were recently replanted to reflect Morris's own scheme and have been populated with the flowers that feature in his designs. The Kelmscott Press published 53 books during its 7-year operation, between 1891 and 1898. The Kelmscott Press is arguably the most famous private press and was the culmination of Morris's successful career as a craftsman and author.

October 1872 saw the Morrises move to Horrington House, in Turnham Green in West London. However, the house has long since been demolished and little is known of it. The only description of the garden is from Morris, who said it was 'a *very* little house with a pretty garden'. After 6 years, the family saw the need to move and their interest turned to Kelmscott House in Hammersmith, also situated in West London (Fig. 3.20).

Morris wrote to Jane, who was on holiday in Italy at the time of his house hunting, about their potential new home and the garden was a strong feature he put to her in his case for buying Kelmscott House; 'The garden is very long and good: it also has a drawback – of being overlooked badly down one half of it, because the wall lowers there: but we might stick up a great high trellis wh: would effectively shut out the onlookers: on the other side there are other gardens & all is quite pretty'.[22] Morris tried to allay Jane's reservations, brought about by Dante Gabriel Rossetti's negative view of the area, by writing, 'The garden is really most beautiful …there is a real green-house down the garden, if you care for that, and capital stabling &

---

[20] Birmingham Guild of Handicraft, (1895).

[21] Norman Kelvin (ed.), *The Collected Letters of William Morris*, vol. 4, p. 458.

[22] Norman Kelvin (ed.), *The Collected Letters of William Morris*, vol. 1, pp. 458–9.

**Fig. 3.19** *News from Nowhere*, William Morris, Kelmscott Press, 1891, William Morris Society

**Fig. 3.20** Kelmscott House, unknown photographer, c.1920, reproduced from a lantern slide, William Morris Society

coach house … The situation is certainly the prettiest in London'.[23] A few days later, he wrote again as Jane was still not convinced; 'There is a good garden and root house, besides the large green house, a tank in the former for watering purposes: there are 2 arbours: there are of big trees 1st a walnut by the stable: 2nd a very fine tulip tree halfway down the lawn (said to be the biggest but one in England). 3rd 2 horse chestnuts at the end of the lawn: beyond that there is a sort of orchard (many good fruit trees in it) with rough grass (gravel walk all around garden): then comes the green-house & beyond that a kitchen: garden with lots of raspberries'.[24]

Other fruits and vegetables we know were in the Hammersmith garden included mulberries, plums, quinces, peas, cherries and pears. The flowers that Morris wrote about in his letters included daffodils, wall-flowers, crocus, peonies, Japanese anemones, china asters, hyacinths, sweet william, roses, foxgloves, orange and white lilies, hollyhocks, poppies and sunflowers. In fact, sunflowers were among Morris's favourite plants, and they appeared in his designs, prose, and lectures. In his famous lecture 'Making the Best of It' (1880) Morris wrote of the Sunflower:

'Though a late comer to our gardens, is by no means to be despised, since it will grow anywhere, and is both interesting and beautiful, with its sharply chiselled yellow florets relieved by the quaintly patterned sad-coloured centre clogged with honey and beset with bees and butterflies'.

The *Sunflower* wallpaper (1879) is a beautiful symmetrical pattern featuring central sunflowers with abundant grapevines and glorious curling leaves (Fig. 3.21). It is flatter and more formalised than *Pimpernel* and *Jasmine*, constructed without the layering technique. The William Morris Society has the original monochrome watercolour design for *Sunflower* in its museum collection, as well as several colourways of the finished hand-block printed wallpaper which range from yellow and red to blue and green, also two glorious foil-embossed wallpapers in gold. This stunning pattern is reminiscent of the Firm's early love of medievalism.

In one of Morris's letters to Jane, he included a hand-drawn plan of the garden, writing that the walls of the kitchen garden were covered with apple, pear and plum trees. He ended his letter to Jane, 'if you could be content to live no nearer London than that, I cannot help thinking that we should do very well there: and certainly the open river and garden at the back are a great advantage'.[25]

He believed that their friends would visit them in Hammersmith 'if only for the sake of the garden and river' and Morris concluded his letter; 'So let us hope we shall all grow younger there, my dear'. Jane was finally persuaded and the family moved in to their riverside home in October 1878.

In his correspondence to Jane and his daughters, Morris often described the garden when they were apart. He wrote to Jane on one occasion; 'I have just come in from the garden, which is looking nice now; the seeds mostly in, & the daffodils almost out in blossom: I am sorry to say though the frost killed almost out all the wall-flowers: poor

---

[23] Kelvin, Letters, vol. 1, p. 459.

[24] Kelvin, Letters, vol. 1, pp. 459–460.

[25] Kelvin, Letters, vol. 1, p. 460.

**Fig. 3.21** Design for
*Sunflower* wallpaper,
designed by William
Morris, 1879, pencil and
watercolour on paper,
William Morris Society

old Matthews is very slow but I don't like sacking him: even on selfish grounds, a new system of horticulture will be more than the garden or I can stand'.[26]

In his lecture, 'Making the Best of It' (1880), Morris wrote 'Before we go inside our house, nay, before we look at its outside, we may consider its garden, chiefly with reference to town gardening, which, indeed I, in common, I suppose, with most others who have tried it, have found uphill work enough … However, uphill work or not, the town garden must not be neglected if we are to be in earnest at making the best of it'. He goes on to state his dislike for the ugly landscape gardening style in imitation of large gardens, using strong words against designed landscapes filling town gardens with formal plants, whereas Morris's plea was for the simplicity of

---

[26] Kelvin, Letters, vol. 1, p. 491.

neat fencing and to fill the spaces with flowers that are 'free and interesting in their growth, leaving Nature to do the desired complexity'.

He wrote to Jenny in April 1882, 'The garden to my cockney eyes is looking pretty well …. I am disappointed with the daffies'. … A friend 'sent us yesterday a lot of peonies, single ones of various kinds very handsome: they are Chinese flowers & look just like the flowers on their embroideries'. In the autumn he wrote again to Jenny, 'The garden here is going the way of all London autumn gardens; but there is still a sort of pale prettiness about it, and there are a good many flowers, in it chiefly Japanese anemones and 'Chaynee oysters".[27] The garden was extensive at nearly an acre long and illustrated Morris's vision that, 'gardens, both private and public, are positive necessities if the citizens are to live reasonable and healthy lives in body and mind'. Sadly, in the 1950s, the A4 road cut through Morris's Hammersmith garden, reducing it by half. However, it remains an interesting plot, the Society owning part which contains a delightful walled garden; a design element which Morris would surely have approved of.

The year 1881 saw the Morris & Company workshops move to Merton Abbey on the banks of the river Wandle, where printed cottons were laid out on banks filled with buttercups to dry naturally in the sun. Morris advocated the necessity of factory gardens, believing in the importance of pleasant surroundings for employees, and immediately planted poplars around the meadow. There was an orchard, a vegetable garden with asparagus and a beautiful flower garden which contained white hawthorn, marsh marigolds, wallflowers, irises, hollyhocks, sunflowers and lilac. We also know from Morris's letters that there was an almond tree and daffodils, and he often used to bring bunches of flowers back home to London with him, including his beloved rose.

Morris died at Kelmscott House on 3rd October 1896. His bedroom was on the ground floor, with a view onto the garden at the front of his riverside home. Contemporary images of the exterior of Kelmscott House perfectly illustrate Morris's views of the garden, 'It should look both orderly and rich. It should be well-fenced from the outside world. It should by no means imitate either the wilfulness or the wildness of Nature, but should look like a thing never to be seen except near a house. It should, in fact, look like a part of the house'.[28]

In 'Making the Best of It' (1880), Morris offered something fresh and different, believing that wallpaper should not slavishly imitate nature but give a beautiful impression of the outdoors:

> Is it not better to be reminded however simply of the close vine trellises which keep out the sun…or of the many-flowered meadows of Picardy…Is not all this better than having to count day after day a few sham-real boughs and flowers, casting sham-real shadows on your walls, with little hint of any-thing beyond Covent Garden in them?[29]

---

[27] Kelvin, Letters, vol. 2, p. 138.

[28] Making the Best of It (1880).

[29] From a lecture, 'Some Hints on Pattern-Designing', delivered 10 December 1881.

# William Morris and His Thames River Patterns

The river Thames held a lifelong fascination for William Morris. His love of the river led Morris to live next to it, take journeys along it, and write about it in his poetry and prose. Most famously though, the river features within his designs for wallpaper and textiles. This section will explore Morris's motivation in creating the series of patterns named after tributaries of the Thames, patterns that reflect his admiration of the great river's flora and fauna.

The influence of the River Thames, its tributaries and its floral landscape flowed through Morris's life and work. Morris came to know the landscape of the Thames intimately and it had a special significance for him, including where he lived. Morris liked to think that the same river which began in the upper reaches near his country home at Kelmscott Manor on the Oxfordshire-Gloucestershire border flowed past his London home at Hammersmith. Morris re-named his Hammersmith residence Kelmscott House to emphasise the link that the river connected his two riverside homes. The Thames provided the setting for a significant part of his leisure time, spent angling and boating with family and friends. In fact, May described her family as 'wet bobs', nearly as much at home on water as on dry land.

It therefore seems no surprise that in August 1880, William Morris embarked with family and friends on an expedition in a boat along the Thames from his home Kelmscott House in Hammersmith on the banks of the Thames. The destination was the family's country residence, Kelmscott Manor, upstream in rural Oxfordshire, and describes the towns visited and people encountered. Morris's account of the trip is fascinating and hilarious, with many incidents along the way. The trip inspired Morris to write 'News from Nowhere' in 1890. This utopian novel focuses on Morris's socialist ideas, and describes a fictional journey along the river, also beginning at Kelmscott House and ending at Kelmscott Manor. The frontispiece features a wonderful depiction of Morris's Oxfordshire home by Charles March Gere, illustrated earlier.

As well as providing inspiration for his designs and novels, the Thames and its surroundings were also the motivation for Morris's poetry, including this extract from his epic poem, 'The Earthly Paradise' (1868–70), where Morris puts into verse his hatred of the industrial city and resulting pollution, and his vision for a transformed future:

Forget six counties overhung with smoke,
Forget the snorting steam and piston stroke,
Forget the spreading of the hideous town;
Think rather of the pack-horse on the down,
And dream of London, small, and white, and clean,
The clear Thames bordered by its gardens green;

Between 1883 and 1885, Morris was motivated to create a series of nine repeating patterns through his love of the Thames. His celebrated Thames textiles are amongst Morris's best-known patterns and were all named after tributaries of the

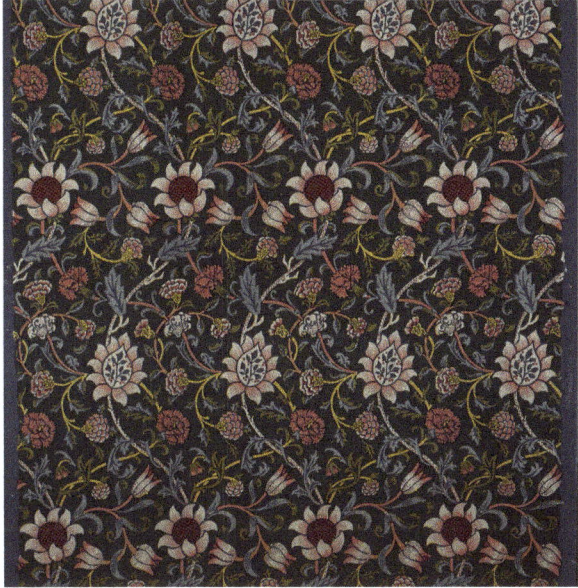

Thames. Throughout his life, Morris was fascinated by textiles, and all the river series are printed cottons. Morris strongly believed that patterns should reflect and be in harmony with the material they were intended for. The flowing nature of the river patterns suits the subtle movement of the fabrics which could be intended for hanging in folds as curtains.

*Evenlode* (1883) was the first in the series and Morris originally planned to call this design colchium or crocus (Fig. 3.22). When the finished design did not include a single crocus, he changed the name to *Evenlode* after a Thames tributary. The renaming may also have been a commercial decision driven by the popularity of spending leisure time on the river. Flowers within this pattern include the sunflower, dianthus, rose, dahlia, pom pom chrysanthemum and wild tulip. The curling, diagonal lines give the impression of flowing water, with offshoots and leaves and stems. The motif of decorative floral elements contained within a larger plant was described by Morris as the 'inhabited leaf'. It is a design component that he borrowed from the historic Persian and Turkish textiles that he had studied in the South Kensington Museum, now the Victoria and Albert Museum. Morris frequently took inspiration from floral features in historic fabrics but, unlike many of his contemporaries, he created his own interpretations rather than merely copying earlier designs. The river Evenlode flows through the Cotswold Hills in Oxfordshire and Gloucestershire, the location of Morris's country residence.

The similar flowing stems of *Evenlode* feature in *Wey* (1883), designed in the same year, and features poppies, pansies, anemones and chrysanthemums set amongst twisting stems (Fig. 3.23). Morris often worked on multiple new designs at the same time, creating four patterns in 1883 for his Thames series. He explained the principles of design in his lectures: 'Remember that a pattern is either right or

**Fig. 3.23** *Wey* printed
cotton, designed by
William Morris and
manufactured by Morris &
Co., 1883. Indigo-
discharged and block-
printed, William Morris
Society

wrong. It cannot be forgiven for blundering, as a picture may be which has otherwise great qualities in it. It is with a pattern as with a fortress, it is no stronger than
its weakest point'.[30]

The wavy diagonal lines of *Kennet* (1883) are reminiscent of flowing water and
are a common feature in many of the Thames Tributary designs (Fig. 3.24). This
feature can clearly be traced back to a seventeenth century Italian fabric that Morris
studied in the South Kensington Museum. The background appears to be bryony
with the distinctive clumps of berries and curling tendrils, and Morris has also
included his distinctive acanthus leaves. The characteristic colours and stylised
plants were key to the success of the Thames patterns. The river Kennet had a long-
standing connection with Morris; his school at Marlborough was flanked by the
Kennet and most of the area surrounding the river is classed as an Area of Outstanding
Natural Beauty. Like the majority of the series, Wey was indigo discharge and block
printed at Morris's workshops at Merton Abbey. *Windrush* (1883) is another of the
Society's original designs by Morris in pencil and watercolour (Fig. 3.25).

When drawing his designs, Morris often completed just enough to show the
woodblock cutter how they would work as a repeating pattern. He made annotations
to record colour notes and experimentation, and other information needed to transfer the design from paper onto wood. Again, Morris depicts vertically meandering
stems with large stylised flower-heads, overlapping with the scrolling foliage and
flower-heads of a far smaller plant. Like *Evenlode*, *Windrush* also features the
inhabited leaf of a floral motif within a flower that Morris derived from Eastern art
and the medieval textiles influenced by it. In a letter to his elder daughter Jenny,
from Kelmscott House on 26th January 1884, Morris wrote that Jane was getting

[30] William Morris, Hopes and Fears for Art, (1882).

**Fig. 3.24** *Kennet* printed cotton, designed by William Morris and manufactured by Morris & Co., 1883. Indigo-discharged and block-printed, William Morris Society

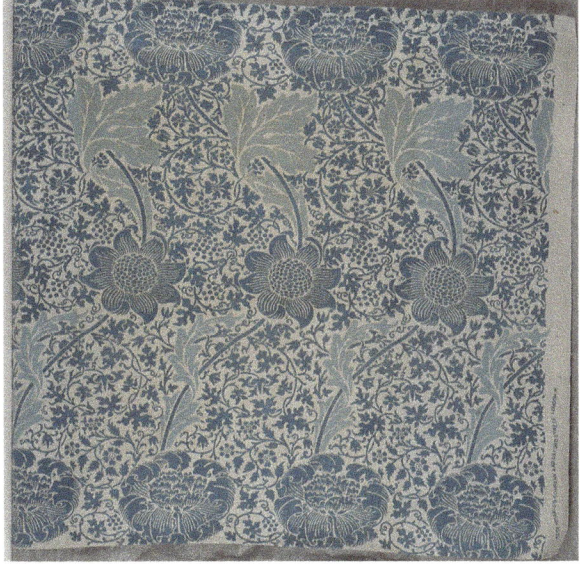

**Fig. 3.25** Design for *Windrush* printed cotton, designed by William Morris, pencil and watercolour, 1883, William Morris Society

**Fig. 3.26** *Lodden* printed cotton, designed by William Morris and manufactured by Morris & Co., 1884. Indigo-discharged and block-printed, William Morris Society

better after a period of ill-health and as a reward for getting well that 'I should hang the drawing room with the blue Windrush as a summer change. I must consent I suppose & salve my conscience on the grounds of its being an advertisement for the goods'. May Morris claims this pattern was 'named in memory of pleasant summer journeys along the Windrush valley'.[31] The river gives its name to the village of Windrush in Gloucestershire.

The subtle beauty of *Lodden* (1884) doesn't feature the strong diagonal of many of the other river patterns and instead has curving vines with flowers and leaves in pink, yellow, green and blue on a more unusual white background (Fig. 3.26). The principle pink flower could be a poppy, other possibilities are cornflower, borage or pimpernel. Also present are some of Morris's favourite fritillarias and dianthus. Morris often designed large-scale patterns, stating in lectures that any room can be improved with the addition of a large design, believing that designers should not be afraid of producing bold patterns. Loddon is a habitat for diverse wildlife and a section of it is a designated Site of Special Scientific Interest due to rare populations of bulbs and pondweed.

A letter written by William Morris to his elder daughter Jenny in September 1883 tells us that one of the cottons he was designing was 'such a big one that if it succeeds I shall call it Wandle: the connection may not seem obvious to you, as the Wet Wandle is not big but small, but you see it will have to be very elaborate and splendid as I want it to honour our helpful stream'.[32] The river Wandle supplied the Morris & Co. factory at Merton Abbey and its soft waters contained the special purity needed for dyeing. The Wandle was also used in the washing of the cloth and to remove excess dye, then the lengths of fabric were dried on the meadow near the

---

[31] Kelvin, Letters, vol. 2, p. 265.

[32] Kelvin, Letters, vol. 2, p. 98.

**Fig. 3.27** *Wandle* printed cotton, designed by William Morris and manufactured by Morris & Co., 1884. Indigo-discharged and block-printed, William Morris Society

workshops in fine weather. Therefore both the river Wandle and the site at Merton Abbey provided ideal conditions for the manufacture of Morris's textiles. *Wandle* (1884) is the largest repeating pattern made by Morris and is certainly a fitting tribute in order to express his gratitude to the river (Fig. 3.27). Flowers appear to be parrot tulips, corn marigolds and pompom chrysanthemums. Additionally, the stems are striped, which textile expert Linda Parry has noted echoes medieval motifs.

*Wandle* was available in both shades of blue and as a multi-coloured version. The former was indigo dyed and discharging and before other colours were printed for the latter. First, the undyed cloth was washed and then immersed in the indigo vat for the correct amount of time. When lifted out, the cloth appears green, but turns blue quickly after oxidizing with the air. For the areas not intended to be blue, a bleaching agent was block printed where required. The cloth was washed again, and the blue cleared from the bleached areas, producing a print of pale and dark blue with white. In order to achieve the multi-coloured print, the fabric was then dried, warmed and prepared for the next colour; the ancient weld plant for the yellow areas and madder root for the strong red.

*Cray* (1884) was the most complex and most expensive of the Thames Tributary designs to print (Fig. 3.28). The more colours that are used for a pattern, the more blocks, time and labour are required. It took 34 printing blocks to produce *Cray*, the largest number of any pattern. *Cray*, along with the other Thames patterns, is complex, with three separate layers of pattern. It has a bold and rich effect, like the multi-coloured splendour of *Wandle*. Morris's designs depict flowers that are stylised, making it difficult to be certain which flowers they are, as Morris believed it was important to study but not to mimic nature. However, the flower with the very pointed petals is likely to be a waterlily and is appropriate in a wetland setting. Water crowfoot could also feature as the small white flowers in the background which actually grow in running water, and the predominant bloom is likely to be chrysanthemum. Morris lectured widely on pattern designing and famously said: 'Ornamental pattern work must possess three qualities: beauty, imagination and

**Fig. 3.28** *Cray* printed cotton, designed by William Morris and manufactured by Morris & Co., 1884. Indigo-discharged and block-printed, William Morris Society

order. Order invents certain beautiful and natural forms which will remind not only of nature but also of much that lies beyond'.[33] The River Cray washes into the Darent, joining the Thames soon after and part of it is a Site of Metropolitan Importance for Nature Conservation.

*Lea* (1885) (Fig. 3.29) is similar in design to Morris's wallpaper *Wallflower* (1890), particularly with the very noticeable use of dots, both as background and in the shading of the leaves (Fig. 3.30). The rosehip detail is present, along with small roses set against the wonderful swirling acanthus leaves, making the pattern seem full of movement. The River Lea is one of the largest rivers running through London.

*Medway* (1885) differs from the other patterns in the Thames series as it is more open and was influenced by Persian sources (Fig. 3.31). It is also the exception as to materials as, unusually, it was also produced as a wallpaper, but renamed *Garden Tulip* (1885). *Medway* also has the distinctive tulips and Morris's favourite fritillaria in the background. *Medway* is the last of Morris's Thames Tributary series. It is named after the River Medway which flows into the Thames Estuary and is the longest of the Thames tributaries. In 1885, Morris stepped back from design ___and manufacturing, handing over substantial control to John Henry Dearle who had progressed through ___Morris & Company, having joined as a shop assistant and then apprentice. Many of Morris & Company's patterns would be created by Dearle from now until his death in 1932.

Morris believed the Thames to be a valuable resource and it was an incredibly rich source of creative inspiration for him. It is no surprise to learn that the Tributary fabrics were hugely popular and commercially successful. Their success was due to Morris's technical and commercial knowledge as well as his understanding of patterns and plants. His designs worked to the strengths and limitations of block printing and natural dyeing, while the choice of Thames tributary names tapped into the fashion for spending leisure time on these rivers. It is certainly true that the Thames and its tributaries had a huge impact on Morris's life and work, leading him to create

---

[33] *Some Hints on Pattern Designing* (1881).

**Fig. 3.29** *Lea* printed cotton, designed by William Morris and manufactured by Morris & Co., 1885. Indigo-discharged and block-printed, William Morris Society

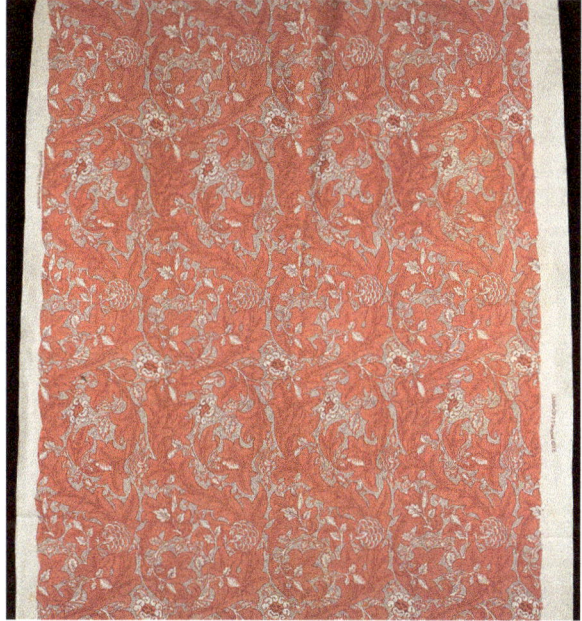

**Fig. 3.30** *Wallflower* wallpaper, designed by William Morris and manufactured by Jeffrey & Co. for Morris & Co., 1890. Block-printed in distemper colours on paper, William Morris Society

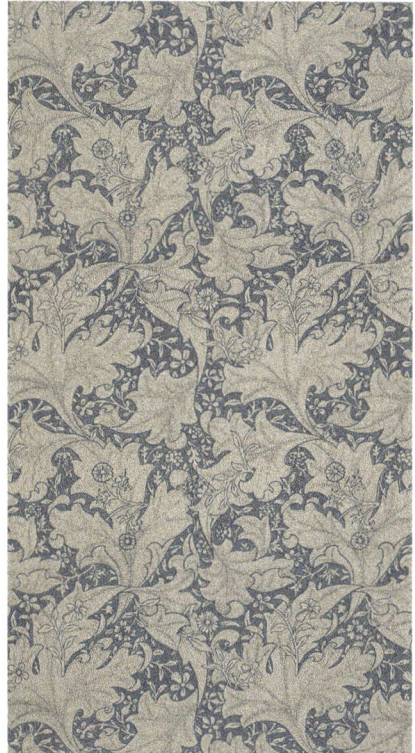

**Fig. 3.31** *Medway* printed cotton, designed by William Morris and manufactured by Morris & Co., 1885. Indigo-discharged and block-printed, William Morris Society

some of his most beautiful and iconic patterns, which are a real celebration of natural British flowers, and that remain enduringly popular to this day.

This study will now focus on May Morris, who inherited a love of nature from her father. May's early study of the natural world was translated into her working practices, especially designing and embroidering. The vast majority of her artworks are floral and perfectly illustrate May's mastery of translating the natural world into stitched pieces.

An embroidery Day Book, covering the years from 1892 to 1896, records her extensive repertoire of floral designs executed in those 4 years. This handwritten ledger by May lists clients orders, designs, time taken to complete, costs and dates of delivery. Items for sale included door hangings, cushion covers, table centres, fire screens and ecclesiastical work such as lectern covers and altar cloths. The William Morris Society has a large collection of May's working drawings and embroideries featuring flora.

*Flowerpot* (c1878) is one of William Morris's best-known embroidery designs and one of the earliest embroidered by May (Fig. 3.32). It was inspired by two panels of seventeenth century French or Italian work that Morris had studied at the South Kensington Museum. Like her father, May travelled extensively both in the UK and abroad, including Italy, Spain, Egypt and Iceland, where she viewed floral embroideries, believing it important to study embroideries of different countries in order to understand the material and effects achieved. She also collected a variety of embroidered textiles, similar in style to *Flowerpot*, ranging from sixteenth century Italian to nineteenth century Turkish. The original design features a decorative vase containing a combination of swirling leaf forms, pale blue peonies and yellow chrysanthemum-like flowers.

Designed by May, *Westward Ho!* (c1880) was named after the Devon village where the Morris's close friend Cormell Price was headmaster (Fig. 3.33). Stylised

**Fig. 3.32** Design for
*Flowerpot* embroidery,
designed by William
Morris and embroidered by
May Morris, c. 1878,
William Morris Society

**Fig. 3.33** *Westward Ho!* embroidery, designed by May Morris and embroidered by Jane Morris, c. 1880, William Morris Society

poppies are enclosed by swirling ochre acanthus leaves on a delicate blue back-ground. The finished embroidery, worked by Jane, May's mother, in darning and satin stitch is encased within a green and yellow chequered border. It is a beautiful collaboration between mother and daughter, showcasing Jane's outstanding ability

**Fig. 3.34** *Bunch of Grapes* embroidery, designed and embroidered by May Morris, c. 1885, William Morris Society

of translating May's design. It remained within the Morris family until relatively recently and retains its exhibition label, which states, 'not for sale'.

In her teachings, May advised the embroiderer to go to nature and study the flowers and animals found there. Her love of traditional countryside hedgerows is clear in her own carefully observed sketches, such as those found in her own sketchbook. May's preliminary drawings offer a wealth of charmingly illustrated English flora and fauna, including rose buds and hawthorn blossoms.

Even the smaller-sized examples of May embroideries, such as *Bunch of Grapes* (c.1885), display her stitching skill, fine colour sense and creative originality (Fig. 3.34). May encouraged the embroiderer to express their individuality with subtle shading and freehand stitching as demonstrated in *Bunch of Grapes*, with its delightful heart-shaped vine enclosing a bunch of variously coloured grapes with hues of pale yellow, cream and green.

Each piece is individually designed and crafted, and like the majority of the William Morris Society's collection of embroideries by May, it has a background surface that is entirely stitched. *Grapevine* (c1885) features a swirling vine laden with rich purple fruit surrounded by light and dark green veined leaves (Fig. 3.35). The size and subject matter suggest that *Grapevine* may have been designed as a small altar frontal.

A glorious example of May's love of stitching onto the Morris and Company *Oak* silk damask is demonstrated in the *Vine* (c1887) embroidered panel designed by John Henry Dearle, originally intended as one panel of a three-part screen and on permanent display at Kelmscott House, alongside the working drawing (Fig. 3.36). It is interesting to note the changes from the design to the finished piece, illustrating

**Fig. 3.35** *Grapevine* embroidery, designed and embroidered by May Morris, c. 1885, William Morris Society

**Fig. 3.36** *Vine* embroidery, designed by John Henry Dearle and embroidered by May Morris, c. 1887, William Morris Society

the freedom that May had in altering another artist's design. Most noticeably, she has changed the flowers at the bottom of the design from harebells and peonies to cornflowers and tulips. The rest of the embroidery remains the same as the original, showing vines interspersed with dense foliage overlaid with scrolling tracery of various stems as to make the background and foreground completely interwoven.

*Orange Tree* (c1885) is a beautifully preserved example of May's delicately balanced design and bold colouring, worked in long and short stitch, stem stitch and French knots for the centres of the oranges (Fig. 3.37). Amid the radiant oranges are elegant sprays of orange blossom, complemented by a vibrant blue background. May was known for her use of colour, she wrote in the arts and crafts essay in 1903, 'colour cannot be too bright in itself (but) …. It should not disturb the eye'. Her favourite colour was blue, writing that it was 'one of the (most) pleasant to have constantly under one's eye'. As Orange Tree does not appear in the Day Book, it is believed to be a unique design and probably a special commission.

The Emery Walker Trust's beautiful poppy silk cushion (c1895) was stitched by Dorothy, Emery Walker's daughter, from a design by her close friend May (Fig. 3.38). Dorothy, who was a skilled embroiderer herself has clearly adopted May's belief that 'Clear and beautiful colouring, sometimes complex, sometimes simple, is one of the principal features of fine embroidery'. The Poppy is another popular flower, featuring in a range of Morris & Company products, including the *Pink and Poppy* wallpaper, designed by William Morris in 1880 (Fig. 3.39).

In this pattern, the fragile, curling petals of the poppy flowers droop downwards, while the marigolds, with their frilly petals, and the pinks, seen in profile, lift the pattern upwards. The 'pink' of the title is a type of flower, part of the Dianthus genus. Veining and dotting, printed by pins hammered into the blocks, give movement to this gorgeous, eye-catching pattern. The Society holds the original watercolour for *Pink and Poppy*, as well as several colourways of the printed wallpaper.

Much of Morris & Company's success in this period must be attributed to May's designing, embroidering and supervising the department at Hammersmith Terrace.

**Fig. 3.38** *Poppy* embroidered cushion, designed by May Morris and embroidered by Dorothy Walker, c. 1895, The Emery Walker Trust

**Fig. 3.39** Design for *Pink & Poppy* wallpaper, designed by William Morris, 1880. Pencil and watercolour on paper, William Morris Society

**Fig. 3.40** *Millefleurs*,
tapestry cushion designed
and woven by May Morris,
c. 1910, The Emery Walker
Trust

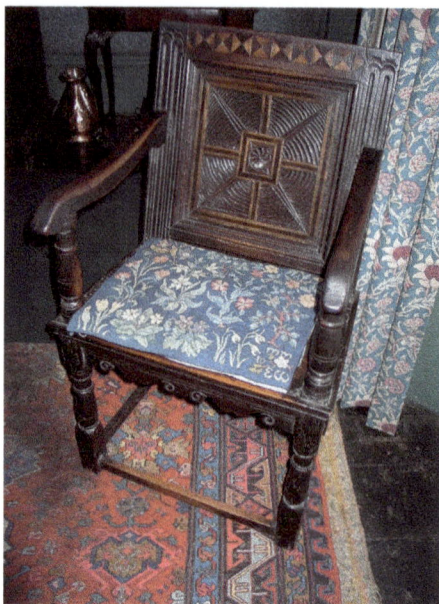

By now, May was established as an expert needle worker and authority, although like her multi-talented father, May was also a true polymath and skilled in a number of crafts, including designing for wallpaper. May also applied her talents to spinning, weaving and tapestry. An example of May's delightful tapestries is on display at Emery Walker's House in Hammersmith. May's next-door neighbours were the Walkers at 7 Hammersmith Terrace. Emery Walker was a typographer, printer, engraver and co-founder of the Doves Press. In his lovingly preserved arts and crafts home can be seen William Morris's seventeenth century library chair from Kelmscott House. It was given by Jane Morris to Emery Walker after her husband's death. May's woven tapestry cushion cover encapsulates the close friendship between the Morrises and the Walkers, with its dedication 'MM to EW'.

It incorporates another medieval-inspired floral design, millefleurs, with English spring and midsummer flowers of crocus, violets, primrose and snowdrops (Fig. 3.40). May created it especially for her late father's chair and it again has May's distinctive organic, stylised foliage and flowers.

May also stitched pieces for herself. The delightful *June* (1909–10) (Fig. 3.41) wool on linen embroidery depicts her family home of Kelmscott Manor with lines from her father's epic poem, The Earthly Paradise; 'This little stream whose hamlets scarce have names, This far off lonely mother of the Thames'. May was talented at designing lettering and several of her embroideries have calligraphy incorporated, a design element which was only found in her father's tapestries. May's *June* embroidered frieze was designed for her sitting room at Hammersmith Terrace and offered a lovely connection between her London home and Kelmscott Manor, her country home; both houses having the Thames flowing past. Scattered around the four

**Fig. 3.41** *June* embroidery (detail), designed and embroidered by May Morris, c. 1909–10, reproduced from a lantern slide used by May Morris, William Morris Society

**Fig. 3.42** *Bedcover*, designed and embroidered by May Morris, c. 1919, The Emery Walker Trust

slender trees are various field and hedgerow flowers that were found around Kelmscott Manor.

A fascinating example of May's exceptional design skills and use of a great variety of floral motifs is demonstrated in the beautiful, stitched coverlet on the bed at Emery Walker's House, which was embroidered by May in wool on a felted wool ground (Fig. 3.42).

The floral designs were inspired by the flowers found in the gardens and fields surrounding Kelmscott Manor and are alternately contained within a knot work structure, a device found in historic examples. Flowers included are those particularly admired by May and her father, such as daffodil, daisy, forget me not, horned poppy, rose and tulip. May created this cover for Mary Grace, Emery Walker's wife, at the end of her life when she was bedridden and it has extremely poignant associations. Designed and entirely stitched by May, it encapsulates her talents and knowledge, use of colour and design, and demonstrates her extensive knowledge of British flora. May and her skills were certainly appreciated by the Walkers; this embroidery meant so much to the whole family that it was used as the pall cover for each of their coffins in turn, on Mary Grace's in 1920, on Emery's in 1933, on Dorothy's in 1963 and finally on Elizabeth de Haas's, the custodian of the house, in 1999. It is now one of the Trust's treasures. Although May's active participation in Morris & Company ceased in the early twentieth century, she continued to produce stunning floral embroideries, acted as a consultant and was still taking on commissions up until the last few months of her life.

## Conclusion

This study has sought to prove that the natural world, and flowers in particular, had an enormous influence on the life and works of William and May Morris. By examining the floral inspiration behind their patterns, it is clear that the Morrises celebrated the splendour of buds, blooms, stems and leaves, to create some of the most iconic patterns of the nineteenth century. May is finally becoming known as an incredibly talented designer in her own right, and rightly taking her place at the centre of the Arts and Crafts Movement. Morris is also enjoying a resurgence in popularity, and it is no understatement to say that he is one of the most influential designers of all time, revolutionising interior design and pattern making by utilising the natural beauty of the flower. Both May and Morris's products are now thought of as quintessentially English due to their innovative design featuring flora and fauna, use of natural materials and insistence on quality. Their brilliance in transforming the home with natural ornament offers a true celebration of the natural British flower and it is no surprise that the floral patterns of this famous father and daughter duo remain as enduringly popular today as they were in their own time over 150 years ago.

## The William Morris Society

The William Morris Society aims to perpetuate the memory of one of the greatest men of the Victorian, or any, age. Established in 1955, the Society's roots go back to 1918 when May, William Morris's younger daughter, founded the Kelmscott Fellowship, which later merged with the William Morris Society. The life, work and ideas of William Morris are as important today as they were in his lifetime and the Society exists to make them as widely known as possible.

The variety of Morris's ideas and activities bring together those who are interested in him as a designer, craftsman, poet and socialist, who admire his robust and generous personality, his creative energy and his courage. His ideas on how we live and how we might live, on creative work, leisure and machinery, on ecology and conservation, on politics and the place of arts in our lives remain as stimulating now as they were over a century ago.

The Society's office and museum are in the basement and Coach House of Kelmscott House, Hammersmith, Morris's London home for the last 18 years of his life. During this time, he ran his manufacturing company, Morris and Company, he founded the Kelmscott Press and he held Socialist League (later the Hammersmith Socialist Society) meetings in the Coach House.

Membership of the Society is open to everyone, and members receive the Journal, Magazine and commentaries on all aspects of Morris's work. The Society has an exciting variety of exhibitions and events throughout the year, as well as running an inspiring educational programme of both formal and informal learning.

## The Emery Walker Trust

The Emery Walker Trust was established in 1999 in order to preserve and protect Emery Walker's House at 7 Hammersmith Terrace as the best example of an Arts and Crafts domestic interior in London. Its aims are to conserve and interpret the house, its collection and its unique riverside garden to wider and more diverse audiences to foster a greater appreciation of the arts and crafts movement. The collection at Emery Walker's House reflects the occupancy, between 1903 and 1999, of the last four interrelated owners of the house: Emery Walker (1851–1933, who lived there from 1903 to 1933), his wife Mary Grace Walker (1849–1920, who was an occasional resident from 1903 to 1920), their daughter Dorothy Walker (1878–1963, who lived there from 1903 to 1963) and her companion Elizabeth de Haas (1918–1999, who lived there from 1948 to 1999); and the attempts of the latter to preserve the house as it was during the residence of the Walker family.

Emery Walker was an important figure at the centre of the Arts and Crafts Movement and the private press revival. He became close friends with some of the most influential minds of the time, including William Morris, Walter Crane, Philip Webb and George Bernard Shaw. These friendships reflect Walker's interests and, as a result, his collecting. His key role in printing and artistic circles led to the formation of wide-ranging friendships, particularly with some of the pioneers of the Arts and Crafts Movement. A highly significant collection of objects relating to these key personalities are reflected in the collections.

However, it is not solely the extraordinary Arts and Crafts collections that make this historic house so important. It is the eclectic nature of the collections that make it a truly authentic Arts and Crafts home. From eighteenth and nineteenth-century family Irish table linen and English furniture to Middle Eastern carpets and Chinese porcelain, the house is an accurate reflection of a genuine Arts and Crafts collection. Emery Walker's House is widely considered to be the only surviving true Arts and Crafts interior in Britain and its significance cannot be overstated.

# Bibliography

## *Publications and Lectures by William and May Morris*

Morris M (1903) Arts and crafts essays. Arts and Crafts Exhibition Society, Longmans, Green and Company, London

Morris M (1893) Decorative needlework. Joseph Hughes, London

Morris M (1910–15) The introductions to the collected works of William Morris, 24 vols. Longmans, London

Morris M (1936) William Morris: artist, writer, socialist, 2 vols. Basil Blackwell, Oxford

Morris W (1882a) Hopes and fears for art, published in1919 in five lectures by William Morris. Longmans, Green and Company, London

Morris W (1880) Making the best of it, 1880, published in 1919 in five lectures by William Morris. Longmans, Green and Company, London

Morris W (1890) News from Nowhere or, an epoch of rest, originally published in the commonweal, January—October 1890; reprinted by Reeves and Turner, 1891

Morris W (1881) Some hints on pattern designing, delivered *10 December 1881* before the Working Men's College at the College, Queen's Square, Bloomsbury, published in 1899 by the Chiswick Press for Longmans & Co., London

Morris W (1889) Textiles, from arts and crafts essays. Arts and Crafts Exhibition Society, Longmans, Green and Company, London

Morris W (1868) The earthly paradise. F.S. Ellis, London

Morris W (1882b) The lesser arts of life, delivered on 21st January 1882 before the Birmingham and midlands Institute in Birmingham, with the title some of the minor arts of life and published in the collection lectures on art delivered in support for the Society for the Protection of ancient buildings London, Macmillan and Co.

## *Other Publications*

Burne-Jones G (1912) Memorials of Edward Burne-Jones. Macmillan, London

Fairclough O, Leary E (1981) Textiles by William Morris and Morris & Company 1861–1940. Thames and Hudson, London

Fiell C, Fiell M (1999) William Morris. Taschen, London

MacCarthy F (1994) William Morris: a life for our time. Faber & Faber, London

Mackail JW (1899) The life of William Morris, 2 vols. Longmans Green & Co., London

Kelvin N (ed) (1984–96) The collected letters of William Morris, 4 vols. Princeton University Press, Princeton, NJ

Parry L (1996) William Morris. Philip Wilson, London

Parry L (1983) William Morris textiles. Weidenfeld and Nicolson, London

Rodgers D (1996) William Morris at home. Ebury Press, London

Thompson EP (1955) William Morris: romantic to revolutionary. Lawrence & Wishart, London

Vallance A (1897) William Morris: His art, his writings and his public life. George Bell and Sons, London

Waggoner D, Kirkham P (2003) The beauty of life: William Morris and the art of design. Thames and Hudson, London

Watkinson R (1990) William Morris as designer. Trefoil Books, London

# Chapter 4
# Charles Darwin, Victorian Botany, and Victorian Culture

Jonathan Smith

**Abstract** Despite not regarding himself as a botanist, Charles Darwin conducted extensive botanical research, particularly in the decades following the publication of *Origin of Species*, and he authored numerous botanical articles and books. His work on orchids, insectivorous species, plant movement, and cross-fertilization had dramatic impact on botanical science, blurring the boundary between plants and animals and deploying natural selection to solve several botanical puzzles. Darwin's botanical studies were also widely disseminated and popularized in Victorian culture, incorporated into debates about breeding, sexuality, and marriage between cousins; fueling discussions about the origins and role of beauty; and inspiring early science fiction tales and comic 'skits' about man-eating plants and monstrous vegetables.

**Keywords** Charles Darwin · Insectivorous plants · Orchids · Evolutionary aesthetics · Cross-fertilization · Plant movement · Cousin marriage

## Introduction

Despite growing attention in recent years, Charles Darwin's botanical writings remain perhaps the most overlooked aspect of his career. Even those familiar with Darwin are often surprised to learn that plants were his major experimental interest in the decades following the publication of *On the Origin of Species* in 1859. Indeed, his first book after *Origin* was a volume entitled *On the Various Contrivances by Which British and Foreign Orchids Are Fertilised by Insects* (1862b). By 1880, just 18 months before his death, Darwin had published an additional five botanical books: on climbing plants, insectivorous plants, plant fertilization, the different forms of flowers in plants of the same species, and plant movement. Several

J. Smith (✉)
University of Michigan-Dearborn, Dearborn, MI, USA
e-mail: jonsmith@umich.edu

© The Author(s), under exclusive license to Springer Nature Switzerland AG 2024     93
L. M. Mendonça de Carvalho (ed.), *The Victorians: A Botanical Perspective*,
https://doi.org/10.1007/978-3-031-68759-4_4

appeared in second editions. If these works were not as widely read and did not provoke the same level of controversy as *Origin of Species* or *The Descent of Man* (1871), botanists hailed them as having revolutionized their field, and they were widely reviewed and commented upon in both the scientific and popular press. Darwin's studies of orchid fertilization, climbing plants, and insectivorous species in particular captured the Victorian public's imagination. In blurring the boundaries between plants and animals, Darwin unsettled one of the natural world's most fundamental divisions, and his work on the relationships between flowers and their insect pollinators contributed to the development and dissemination of an evolutionary account of beauty that challenged centuries of aesthetic theory.

Plants were a keen interest of Darwin's long before *Origin of Species*. In his autobiographical recollections he noted that by the time he was eight he had 'tried to make out the names of plants' (Darwin 1958, p. 22). More fancifully, he told another little boy that he 'could produce variously coloured Polyanthuses and Primroses by watering them with certain coloured fluids' (Darwin 1958, p. 23). As an undergraduate at Cambridge he attended the botany lectures of John Stevens Henslow, who soon became his chief mentor. Darwin accompanied Henslow on his long daily walks, dined frequently with his family, and took part in the botanical excursions he organized. It was through Henslow that Darwin received the offer to sail on the *H.M.S. Beagle* (Browne 1995; Darwin 1958; Walters and Stowe 2001). Darwin lamented in his published *Journal of Researches* (the book we now call *The Voyage of the Beagle*) his 'ignorance' of botany, saying he 'collected more blindly in this department of natural history than in any other' (Darwin 1839, p. 629), but he was attentive to plants, collecting them throughout the journey and sending them to Henslow for identification and analysis. While Darwin produced no *Botany* to accompany his *Geology* (Darwin 1842–1846) and *Zoology* (Darwin 1838–1843) of the voyage, the seeds of many of his later botanical interests were planted in his travels. The forests of Brazil left an especially vivid impression on him, helping spur his later interests in orchids and climbing plants and the role of insects in plant fertilization. In 1838–39, after his return to England, his speculations about species led him to investigate the cross-fertilization of flowers, particularly native orchids, by insects, a subject to which he gave some attention in virtually every subsequent summer. Another important botanical element from the voyage in the development of Darwin's thinking about species came from the plants of the Galapagos. In the first edition of his *Journal* in 1839, Darwin related, in one of his Addenda, his surprise at learning from Henslow, who had only just begun to examine Darwin's Galapagos specimens, that in several cases distinct species of the same genus appeared on separate islands (Darwin 1839, p. 629). Darwin's Galapagos plants then passed to the young botanist Joseph Hooker, Darwin's good friend (and Henslow's future son-in-law) who had recently returned from his own voyage to the southern hemisphere. Hooker (1847) completed the analysis, confirming Henslow's preliminary conclusions. In 1845, in the second edition of the *Journal*, Darwin's much-revised chapter on the Galapagos laid out the case for the unusual characteristics of the archipelago's flora and fauna. By then, he was able to see that the plants were as distinctive as the finches, mockingbirds, and tortoises. 'The botany of this group is fully as interesting as the zoology' (Darwin 1845, p. 392), Darwin declared.

During the decade from 1845 to 1855 Darwin directed most of his attention to barnacles, but as he turned to writing his 'big book' on species, Darwin maintained and expanded his interest in plants. His study filled with potted plants and trays of germinating seeds. He visited nurseries and corresponded with nurserymen, read and published in *The Gardeners' Chronicle*. His friend Hooker supplied the steady stream of botanical advice and exotic specimens from the Royal Botanic Gardens at Kew (where Hooker was Assistant Director under his father, William Jackson Hooker, and succeeded his father as Director in 1865) that Darwin increasingly demanded. By the first edition of *Origin of Species*, Darwin had collected so much information on crossing and fertilization that he felt comfortable declaring as 'a general law of nature' his belief that 'no organic being self-fertilises itself for an eternity of generations' (Darwin 1859, p. 97).

Despite his frequent disclaimers that he was no botanist, Darwin in the years after *Origin of Species* worked and published extensively on plants. His paper on 'the remarkable sexual relations' of two forms of *Primula*—the first installment of what would become *The Different Forms of Flowers on Plants of the Same Species*— was read at the Linnean Society in November 1861 and published in the Society's *Proceedings* the following March (Darwin 1862a). *Orchids* appeared 2 months later, in May 1862 (Darwin 1862b). In the winter of 1862–63, Darwin had a small hothouse constructed at his home near the Kentish village of Down to facilitate his experiments on tropical plants ('Darwin's hothouse' n.d.). A greenhouse had been built at Down in 1855–56 in part to accommodate his botanical work, but during the intervening years he had had to use the hothouses of a neighbor or rely on observations by the botanists at Kew on many orchid species. In 1864 he added another large greenhouse comprised of two halves kept at different temperatures. This small complex, with its four separate areas, facilitated and consolidated Darwin's botanical experiments over the remainder of his life. In early 1865 he delivered his long paper on climbing plants to the Linnean; it was published in the Society's *Journal* later that year as well as separately by Longmans, though it did not sell well until the second edition of 1875 (Darwin 1865, 1875b). In 1875 he also brought to fruition 15 years' study of insect-eating species like the sundew (*Drosera rotundifolia*) and Venus fly-trap (*Dionaea muscipula*) with the publication of *Insectivorous Plants* (Darwin 1875a). In 1876 he published a comprehensive study of fertilization in *The Effects of Cross and Self-Fertilisation in the Vegetable Kingdom* (Darwin 1876). An expanded edition of *Orchids* appeared the following year, as did the issue in book form of the Linnean Society papers of the previous decade on primroses, cowslips, oxslips, flax, and loosestrife. In 1880 came *The Power of Movement in Plants*, which vastly expanded his earlier study of climbing plants, and a new edition of *The Different Forms of Flowers*. Much of this botanical work began to find its way into new editions of *Origin of Species*, especially in the heavily revised fourth edition in 1866. While *The Variation of Animals and Plants Under Domestication* (1868), a book chock-full of the evidence he had been forced to omit from *Origin*, dealt primarily with animals, two chapters were devoted to cereal plants, vegetables, fruit, ornamental trees, and cultivated flowers. Despite being forever and memorably linked to animals, Darwin was nearly as well known to colleagues and to his Victorian contemporaries for his plants (Allen 1977; Ayres 2008; Kohn 2008).

Darwin frequently downplayed the significance of his botanical work or acknowledged his botanical ignorance. Opening his introduction to *The Different Forms of Flowers*, he apologized that his subject 'ought to have been treated by a professed botanist, to which distinction I can lay no claim' (1877, p. 1). Yet many botanists of the day regarded Darwin's plant research as significant, even as having revolutionized the field. Natural selection, as it had in zoology, provided a different approach to botanical classification and new explanations of geographical distribution and morphology. In seconding the nomination of Darwin for the 1864 Copley Medal, the Royal Society's highest honor, Hugh Falconer cited *Orchids* and the recently published papers that would become *Different Forms of Flowers* as 'of the highest order of importance. They open a new mine of observation, upon a field which had been barely struck upon before' (Falconer 1864). Darwin's work on fertilization indeed opened the botanical floodgates to a host of similar studies. Hooker complained in 1860 that Darwin's work on the importance of cross-fertilization was 'much underrated' (Hooker 1860, p. x), and he used the platform of his Presidential Address to the British Association for the Advancement of Science in 1868 to call Darwin's fertilization studies 'the greatest botanical discoveries made during the last ten years' (Hooker 1869, p. lxvi). By 1878, though, William Ogle, the editor of the *British and Foreign Medico-Chirurgical Review*, was looking back upon Darwin's *Orchids* as a 'classical work' that disclosed 'a wide and unexplored region to the research of physiological botanists' (1878, p. vii). Just a few years later, the Darwinian botanist Hermann Müller (1883) was warning that recent investigators had endorsed and extended Darwin's emphasis on the benefits of cross-fertilization too uncritically, raising it to the level of dogma. Darwinians of course had a vested interest in constructing contemporary botany as undergoing fundamental changes in response to Darwin's views, but they were sufficiently successful that even those opposed to Darwinian interpretations of botanical phenomena usually acknowledged the influence of Darwin's work.

Darwin's botany had several layers of significance for evolutionary theory and its applications. At a general level, it broke down rigid distinctions between plants and animals. Insectivorous species captured and digested insect 'prey.' Climbing plants were capable of what seemed almost voluntary movement. Indeed, Darwin came to argue, movement in response to various environmental stimuli was a phenomenon common to all plant species, and thus plants, while lacking the nervous system of animals, should be seen as operating in closely analogous ways. '[I]t is impossible not to be struck with the resemblance between the … movements of plants and many of the actions performed unconsciously by the lower animals' (1880, p. 571), he wrote in the conclusion to *The Power of Movement in Plants*. Indeed, he went on, it is 'hardly an exaggeration to say that the tip of the radicle … acts like the brain of one of the lower animals' (1880, p. 573). Ever the boundary-blurrer, Darwin confessed in his autobiographical reflections that it 'always pleased me to exalt plants in the scale of organised beings' (Darwin 1958, p. 135). Darwin's allies and supporters worked to publicize this exaltation of plants in the periodical press, while opponents resisted it. In 'On the Border Territory Between the Animal and the Vegetable Kingdoms' for *Macmillan's Magazine*, T. H. Huxley (1876) declared that the old distinctions between these two kingdoms had been completely broken down, with the differences between plants and animals a matter of degree rather than kind. In the *Cornhill Magazine*, Andrew Wilson also answered with a negative the question of his title, 'Can we separate Animals from Plants?' (Wilson 1878). Reviewing Darwin's *Power of Movement in Plants* for the *Edinburgh Review*, however, the popular science writer R. J. Mann complained that analogies between plant movement and animal locomotion, 'especially dear to the hearts of evolutionists' (1881, p. 506), should not be allowed to obscure the crucial distinction that the latter is volitional while the former is not. Mann understood the importance to Darwin and his followers of blurring lines between supposedly unbridgeable divides in the organic world.

Darwin's botanical books also built the case for natural selection. His first book after *Origin of Species* was not the promised work filled with the supporting evidence speed had compelled him to excise from *Origin* but a detailed examination of orchid fertilization. This struck many contemporaries as odd, but *Orchids* offered an empirical study in which Darwin invoked natural selection to account for the elaborate markings, fanciful structures, and gorgeous color of orchids. He did not employ the term 'natural selection' until more than halfway through the book, and most of the limited number of explicit uses of it came in the final chapter, but Darwin made clear that orchids buttressed his faith in the explanatory power of his theory. 'The more I study nature,' he wrote in the closing pages:

> the more I become impressed with ever-increasing force with the conclusion, that the contrivances and beautiful adaptations slowly acquired through each part occasionally varying in a slight degree in many ways, with the preservation or natural selection of those variations which are beneficial to the organism … , transcend in an incomparable degree the contrivances and adaptations which the most fertile imagination of the most imaginative man could suggest with unlimited time at his disposal (Darwin 1862b, p. 351).

Natural selection generates more beneficial adaptations from the myriad of slight variations in every organism than could be dreamed up by the most imaginative person. Darwin's interest in orchids stemmed from the fact that they seemed to pose a problem for his theory. Perhaps more than any other group of organisms, orchids displayed numerous 'perfect contrivances' (Darwin 1862b, p. 355)—elaborate structures involved in pollination that seemed to be explicable only as the work of a benevolent and all-powerful, but also whimsical, Creator. George Levine (2010, p. 109) has called Darwin's nature 'a Rube Goldberg sort of world,' and nowhere is that a more accurate description than with orchids. Their jury-rigged contraptions could not have arisen naturally, many naturalists argued; they had to have been designed by a mind. The beauty of orchids, moreover, seemed to pose another problem for natural selection. How could an entity putatively concerned only with beneficial adaptations produce such un-utilitarian colors and markings? Surely natural selection could not explain orchids any more than the peacock's tail? Orchids seemed to show that the divine Creator took pleasure in his creative work and in the natural world, and that he wanted humans to share in that pleasure. No, answered Darwin, orchids were the product of natural selection, of the evolutionary dance between flowers and insects. Their beauty and bizarre structures had nothing to do with us, or with our pleasure or God's, but instead served the purpose of attracting certain insects in a manner that assured cross-fertilization rather than self-fertilization. Darwin had been criticized for asserting in *Origin of Species* that the need for cross-fertilization is a universal law in nature; with *Orchids* he began to show that natural selection had brought this about. The book, he explained at the outset, 'affords me ... an opportunity of attempting to show that the study of organic beings may be as interesting to an observer who is fully convinced that the structure of each is due to secondary laws, as to one who views every trifling detail of struc-ture as the result of the direct interposition of the Creator' (1862b, p. 2). The main 'secondary law' Darwin had in mind was natural selection, and he clearly thought natural selection offered a true explanation, not merely a more 'interesting' explana-tion, of orchids. Typically understated and unassuming, Darwin was nonetheless bold: he turned a group that his critics were using against him into support for his theory.

Darwin obviously could not make his case for the universal importance of cross-fertilization with a book on a single group. *The Different Forms of Flowers* extended the results of *Orchids* on the role of insects in effecting cross-fertilization, and it did so with another set of familiar but odd species. Why would plants like the oxlip (*Primula elatior*), cowslip (*Primula veris*), and loosestrife (*Lythrum salicaria*) have two or even three distinctly different sexual forms, their stamens and pistils set at different heights? Much to his surprise, Darwin found that the different sexual forms functioned almost as separate sexes. A flower with long stamens and short pistils achieved full fertility only when pollinated by a flower with short stamens rather than with pollen from its own long stamens or even a separate long-stamened indi-vidual. Flowers with short stamens similarly produced more seeds and more vigor-ous offspring when fertilized with pollen from flowers with long stamens. An insect visiting a short-stamened flower would be dusted with pollen in a position on its body from which the pollen was easily transferred only to the pistils of a flower of the other form, and vice versa. These different sexual forms thus functioned not only to minimize self-fertilization but also to facilitate crosses with a different sex-ual form. By the time he published his comprehensive book on *Cross and Self-Fertilisation* across the vegetable kingdom, Darwin could point both to almost four decades of his own observations on the subject and to numerous works by other naturalists since the appearance of *Orchids* to buttress his earlier claims about plant fertilization and the evolutionary advantages of crosses.

The three books that questioned the plant–animal boundary—*Climbing Plants*, *Insectivorous Plants*, and *The Power of Movement in Plants*—also showed how both typical and unusual forms of plant behavior could be accounted for by natural selection. In *Climbing Plants* Darwin rebutted the notion that these species pos-sessed an innate tendency to grow in an upward spiral, arguing instead that the many 'beautiful adaptations' of climbing plants 'illustrate in a striking manner the prin-ciple of the gradual evolution of species' rather than the action of divine design

(1875b, p. vi). As *Cross and Self-Fertilisation* showed that the benefits of cross-fertilization were not limited to orchids and cowslips, so *The Power of Movement* demonstrated that plant movement was not limited to climbing plants—heliotropism, the 'sleep' and sensitive responses of leaves in various species, even the inexorably downward motion of the root tip were all for Darwin modifications by natural selection of a common property of circular movement in plants that he called 'circumnutation.' Far more surprising, even shocking, was his demonstration in *Insectivorous Plants* that some species like the sundew (*Drosera rotundifolia*) and butterwort (*Pinguicula vulgaris*) could trap or ensnare insects and then dissolve and digest them. Mechanisms ranging from the simplistic to the elaborate had evolved not to facilitate cross-fertilization but to provide a dietary supplement for species in nutrient-poor soils. Darwin was particularly astonished by the sundew leaf's sensitivity to minute quantities of substances like ammonia or to the faint pressure of a bit of hair, which could cause sudden movement of the leaf or its tentacles. This sensitivity, he marveled, 'exceed[s] that of the most delicate part of the human body' (1875a, p. 272). Impulses were somehow being transmitted from one part of the leaf to another without the benefit of a nervous system. Darwin did not initially trust the results of his experiments on the sundew and delayed publication for a year to run the experiments again. Convinced that he was right, Darwin took to calling the sundew a 'disguised animal' or even a 'sagacious animal' in his letters to the botanists Hooker and Gray (Darwin 1860, 1863b). Although he made such comments with tongue firmly in cheek, Darwin delighted in the fact that these plants behaved almost as if they were animals, and he delighted in the fact that natural selection could explain these strange and unusual species. Darwin's comment to his publisher about *Orchids*, that 'I think this little volume will do good to the Origin, as it will show that I have worked hard at details, & it will, *perhaps*, serve [to] illustrate how natural History may be worked under the belief of the modification of Species' (Darwin 1861, original emphasis), proved to be accurate in some degree for all his botanical works.

Those botanical works were important culturally as well as scientifically. *Orchids* appeared in the midst of an orchid mania, when collectors were combing tropical rainforests around the globe for new and exotic specimens. Explorer-naturalists like Henry Walter Bates and Alfred Russel Wallace financed their expeditions and researches through the collection and sale of exotic plants and animals, including orchids. Aristocrats cultivated them in the conservatories of their stately homes. Nurserymen vied to offer the latest types. Botanical and horticultural publications featured the new discoveries. Darwin's network of correspondents included figures from all these areas of orchid interest. Darwin feared his book would only appeal to those with 'a strong taste for Natural History' (1862b, p. 2), and it did not sell well, but as the next book by the author of *On the Origin of Species*, it received attention and reviews, and writers of popular works of botany made its results known over the ensuing decades. Indeed, as the century progressed, orchids became a common feature in fiction as markers of exoticism and luxury, and even figured prominently in a sub-genre of tales of deadly plants (Endersby 2016b; Miller 2012).

*Orchids* also figured in cultural debates about 'consanguineous marriage' in humans—chiefly, marriage between first cousins. Darwin's work on plant fertilization and especially on the advantages of cross-fertilization offered a range of potential applications to humans, but from the beginning it was cousin marriage on which Darwin focused. In the final paragraph of *Orchids*, ruminating on the 'astonishing fact' that such elaborate mechanisms for transporting pollen had evolved in preference to simpler, more efficient self-fertilization, Darwin closed with a sentence about the significance of this result for humans: 'may we not further infer as probable, ... that marriage between near relations is likewise in some way injurious,— that some unknown great good is derived from the union of individuals which have been kept distinct for many generations?' (1862b, p. 360). This was an issue that cut close to the bone for Darwin. His wife, Emma Wedgwood, granddaughter of the pottery founder, was also his first cousin, and several other cousin marriages had taken place between the Darwins and the Wedgwoods. The families had lived not far from each other, and their children grew up in similar social and environmental conditions. Darwin worried incessantly that his own chronic ill health and that of his children (three of whom had died in childhood or infancy) was the result of too much interbreeding, and his subsequent fertilization experiments increased his anxieties (Browne 2002, pp. 279–282). Darwin's botany thus entered an ongoing debate, primarily involving agriculturalists and breeders, but increasingly concerned with humans, about the effects of in-breeding. The evidence was anecdotal and conflicting—writers could draw on accounts of monstrosities and sterility as well as those of higher yields and more robust offspring, and while many breeders pointed to the ill-effects of in-breeding, others were vociferous about its virtues and denied that it was injurious in any inherent way. Applying this ambiguous evidence to humans, and especially to the case of cousin-marriage, was even more problematic. In England, marriage between first cousins was prohibited until the sixteenth century but became legal under Henry VIII. It grew to be fairly common over the ensuing centuries among propertied families, serving as a means to consolidate wealth and reinforce kinship networks. Yet most discussions of the subject in the 1860s and 1870s began by noting a strongly held popular presumption *against* cousin-marriage, the offspring of which were felt more likely to be physically or mentally deformed, more susceptible to disease, less fertile, and shorter lived (Anderson 1986; Trumbach 1978; Wolfram 1987).

Darwin's *Orchids* was almost immediately absorbed into public discussions about whether or not cousin marriages, though legal, should be outlawed on the grounds that the children of such unions were disproportionately inferior both physically and mentally (Adam 1865; Child 1863). Darwin found his results challenged on the grounds that they did not justify the sweeping proclamation of a natural law against perpetual self-fertilization. Such reactions seemed to spur Darwin into more direct research on the subject. In 1863 he was asking Asa Gray for information on American laws about cousin marriage (Darwin 1863a). In 1870 Darwin's friend, neighbor, and local MP John Lubbock—a polymathic figure who was himself an authority on both the role of social insects in the fertilization of plants and the marriage practices in other cultures—failed in his attempt during the debate over the

Census Act to insert into the Census a question about cousin marriage for which Darwin provided the wording and which Lubbock described to Darwin as 'your amendment' (Lubbock 1870). In the middle of the decade Darwin's son George began to publish the results of his independent researches on consanguineous marriage. George concluded that the available evidence did not support claims for the deleterious effects of cousin-marriage, although knowledge of his father's as yet unpublished experiments for *Cross and Self-Fertilisation* induced him to allow that the children of such unions might be at a disadvantage if they lived in considerable poverty (G. Darwin 1873 and 1875). Charles incorporated George's results into subsequent editions of his books, and he removed from the second edition of *Orchids* his closing speculation that marriage between near relations may be injurious. The Darwins' work remained part of these debates through the 1870s and well beyond Charles's death in 1882.

These Victorian debates over cousin-marriage were part of changing cultural understandings of incest and shaped by the early anthropological writings of figures like Lubbock about marriage in 'primitive' societies. For Darwin as for most of his contemporaries, however, the issue mainly concerned the health of offspring and of the family (whether writ small or, as social class or nation, large). Did cousin-marriage constitute in-breeding that, as with the self-fertilization of plants, led to fewer, less vigorous, and less fertile offspring? Was some cousin-marriage neutral or even advantageous, and if so, how much? Could exogamous marriage counteract any ill effects of too much cousin-marriage? Did the fact that cousin-marriage was common among 'savages' mean its practice in Victorian Britain was 'natural' or a sign of degeneration? Victorian fiction also reflected these uncertainties over cousin-marriage, which was surprisingly ubiquitous in novels of the period and represented the most common fictional form of what scholars call 'familiar' or 'companionate' marriage as opposed to romantic marriage. Despite our modern affection for Elizabeth Bennet and Mr. Darcy, Jane Eyre and Mr. Rochester, romantic love frequently leads in Victorian fiction to unhappiness or misery. Passion for a virtual stranger may have been exciting, but it was often depicted as risky. At the same time, familiar marriages were often presented as fulfilling and meaningful rather than characterized by the lack of desire, and this was certainly how Darwin spoke of his own marriage (Corbett 2008; Schaffer 2016).

Perhaps more surprising was the dissemination of Darwin's work from *Different Forms of Flowers* on *Primula* and *Lythrum* in particular. The case of the cowslip, with its two different sexual forms and thus four different reproductive combinations, was fairly simple. But the case of loosestrife, with *three* different sexual forms, was far more complex. Since each of the three forms came in two different variations, Darwin discovered that an astonishing 18 different reproductive combinations were possible. 'In their manner of fertilisation,' he remarked, 'these plants offer a more remarkable case than can be found in any other plant or animal' (Darwin 1877, p. 137). Loosestrife sexuality, he summarized, consisted of 'a triple union between three hermaphrodites—each hermaphrodite being in its female organ quite distinct from the other two hermaphrodites and partially distinct in its male organs, and each furnished with two sets of males' (Darwin 1877, p. 138). This was, arguably, the apex of queerness, in both the Victorian and modern senses of that term, in all of nineteenth-century natural history. The 18 different sexual combinations, moreover, yielded a spectrum of fertility. Only the six crosses involving comparable sexual elements of different sexual forms consistently resulted in full fertility. The 12 crosses involving sexual elements of different size, on the other hand, ranged from nearly full fertility to complete sterility. Darwin's text revels in the complexity and multiple possibilities of loosestrife reproduction, and his descriptions and conclusions about them circulated widely in Victorian culture. Reviewers often discussed them in detail, and authors of popular botany books frequently included extended accounts of Darwin's experiments. Whether or not they accepted Darwin's evolutionary gloss of these phenomena, virtually every commentator joined in marveling at the 'remarkable sexual relations' of *Primula* and the 'triple union between three hermaphrodites' of *Lythrum*.

The evolution of Darwin's terminology for these species also reflects his desire to capture their sexual queerness. When he began working on *Primula*, with their two different sexual forms, Darwin coined the term 'heteromorphic' to describe fertilization between flowers of different forms, which were on average more fertile, and 'homomorphic' to characterize the fertilization, almost invariably less productive, between flowers of the same form. While Darwin was apparently not exposed to the contemporary Continental discussions of human sexuality that were constructing the categories of 'homosexual' and 'heterosexual,' with homosexuality positioned as a deviation from the 'natural,' instinctual drive toward procreative sexual intercourse within marriage, his initial description of plant sexuality obviously bears significant resemblance to that discourse. When he discovered the *three* different forms of *Lythrum*, however, Darwin abandoned his hetero/homo dichotomy in favor of 'dimorphic' and 'trimorphic.' While he later accepted the German botanist Friedrich Hildebrand's term 'heterostyled' to describe any species with multiple sexual forms, Darwin nonetheless retained 'dimorphic' and 'trimorphic,' complaining that 'the term "heterostyled" does not express all the differences between the forms' (Darwin 1877, p. 2n).

At the same time as he chafed over terminological constraints that failed to capture sexual difference, however, Darwin was deploying a linguistic framework that relied on cultural and specifically Anglican notions of sexuality and marriage. He had already spoken in *Orchids* of moths that 'performed their office of marriage-priests' (1862b, p. 41). In *Different Forms of Flowers*, his acceptance of 'heterostyled,' his own initial adoption of 'heteromorphic' and 'homomorphic,' and his definition of heteromorphic crosses as those of full fertility and homomorphic crosses as those of lesser fertility or sterility all marked a willingness to force his queer flowers into traditional sexual binaries. In Darwin's view, these species had evolved away from a reliance on homomorphic crosses, which were now at best a kind of sexual backup system. Natural selection favored sexual dimorphism throughout the animal and vegetable kingdoms, and hermaphroditic plants with flowers of different forms were evolving to function as if they were sexually dimorphic. Indeed, Darwin registered the superior fertility of heteromorphic crosses by labeling them 'legitimate,' while homomorphic crosses were 'illegitimate.' This decision is especially striking in the case of *Lythrum*, in which the 12 'illegitimate' crosses displayed a wide range of fertility. Only the most fertile crosses were 'legitimate,' and the most fertile crosses were between different forms. The 'triple union of three hermaphrodites,' moreover, was a case in which 'nature has ordained a most complex marriage-arrangement' (Darwin 1877, p. 138). While it is significant that 'nature' does the 'ordaining,' the language is that of an Anglican wedding, and the triple union of three hermaphrodites becomes a 'marriage.' Where only fully pro-creative marriages are deemed to be 'legitimate,' it is impossible, too, not to hear the morality of the divine injunction in Genesis to 'be fruitful, and multiply.'

As Gowan Dawson (2007) has shown, Darwin's work on mate choice and repro-duction among animals (both human and otherwise) in *The Descent of Man* was both risky and risqué. Darwin's critics often yoked *Descent* and its theory of sexual selection to other culturally unsettling movements associated with sexuality, whether that was the eroticism of avant-garde literary and artistic movements like Pre-Raphaelitism, Aestheticism, and Decadence, or social and political movements in favor of women's rights, free love, and open marriage. Darwin's allies labored mightily to maintain the respectability of Darwin and his theories, and to distance both from these figures and associations. Darwin's work stirred up enough contro-versy without the addition of incendiary affiliations on a topic as fraught for the Victorians as sex. Darwin's public persona as a figure of moral probity—a serious, honorable, bourgeois family man—was essential for insuring his scientific theories a fair hearing. Plant fertilization was a safer topic than animal mating, and, as with the breeding of domesticated animals, could be discussed in relation to longstand-ing discourses of agriculturalists and nurserymen. Nonetheless, the anthropomor-phic language that Darwin, like his contemporaries, frequently employed when talking about plant fertilization—marriage, parents and children, unions, legitimate and illegitimate—could seem to invite applications to Victorian society in a host of politically or morally controversial ways. As Darwin harnessed the evidence of *Orchids* to address a question about human sexuality and marriage—cousin-marriage—that did not generate moral panic, so he constrained his sexually and reproductively diverse flowers, consciously or not, within the linguistic framework of his society's dominant and culturally sanctioned model for procreation.

Arguably the most attention-getting portion of Darwin's botanical writings, how-ever, was his book on *Insectivorous Plants*. As I have mentioned, this book played a key role in Darwin's blurring of the boundary between plants and animals. Coupled with his studies of climbing plants, *Insectivorous Plants* helped show that plants could move and act much like animals did. They seemed able to set traps, to capture and digest insect prey, to move in response to stimuli. When the great Victorian art and social critic John Ruskin wrote his own book about flowers, his disparaging references to contemporary botany were thinly disguised attacks on Darwin's botanical writings and Darwin's work on insectivorous species and plant fertiliza-tion in particular. Ruskin lamented 'the recent phrenzy for the investigation of digestive and reproductive operations in plants' and declared himself 'amazed and saddened' by efforts to explain 'every possible spur, spike, jag, sting, rent, blotch, flaw, freckle, filth, or venom, which can be detected in the construction, or distilled from the dissolution, of vegetable organism' (Ruskin 1906, pp. 390–391). Describing an unpublished pamphlet sent to him by an unnamed author, on 'every sort of plant that looked or behaved like an animal, and every sort of animal that looked or behaved like a plant,' Ruskin makes clear his sense of the pointlessness of such an exercise: the unnamed author 'gave descriptions of walking trees, and rooted beasts; of flesh-eating flowers, and mud-eating worms; of sensitive leaves, and insensitive persons; and concludes triumphantly, that nobody could say either what a plant was, or what a person was' (1906, pp. 507–508). Like other opponents of Darwinism, Ruskin understood the value to evolutionists in breaking down absolute distinctions between plants and animals, and his sense of the 'phrenzy' for 'flesh-eating flowers' points to the particular popularity of Darwin's insectivorous species. The Victorian Little Shop of Horrors opened by Darwin's researches on the sundew (*Drosera rotundifolia*), a denizen of British fens, marshes, and bogs, exerted a dark and sometimes darkly comic fascination on his contemporaries. *Insectivorous Plants* was widely reviewed and popularized, and it provided fodder for *Punch* and other comic magazines.

Darwin's insectivorous plants were frequently discussed in the popular press in relation both to Gothic tales of horror and suspense, of imprisonment, torture, and death, and to the modern, updated narratives of 'sensation,' wildly popular in the 1860s and 1870s, that set Gothic plots in middle-class Victorian homes rather than medieval castles, drawing on recent real-life murders and scandals. In her review of Darwin's book, Ellice Hopkins noted that in the case of the Venus fly-trap, 'the old haunting story of the prisoner who finds the walls of his cell gradually closing in upon him comes true, not in gloomy human dungeons, but down among the starry moss, and windy lights, and lovely glancing things, and all the wide peacefulness of upland nature' (1879, p. 44). When Darwin recruited J. Burdon Sanderson, Professor of Physiology at University College London, to attach electrodes to the leaves of insectivorous species and induce movement in them with pulses of current, experiments Sanderson then described to a popular audience at the Royal Institution in 1874, the connections to Mary Shelley's *Frankenstein* (and Luigi Galvani's late-eighteenth-century frog experiments that helped inspire it) were inevitable. J. E. Taylor's reaction in his popular *The Sagacity and Morality of Plants* to the 'cunning story of temptation, deceit, and final ruin' offered by pitcher plants, whose bright colors and aromatic nectar lure insects down a literally slippery slope into a pool of liquid that drowns and digests them, acknowledged the connection to contemporary thrillers: '[f]or "pitchers" and "flies" we have only to substitute the various dissipations of "fast life," and the characters who indulge in them—and both fact and moral hold equally true!' (Taylor 1884, p. 273). References to 'carnivorous' plants upped the ante, sensationalizing Darwin's 'insectivorous' even though critics of Darwin complained that 'insectivorous' was already too sensational and misleading. And in the decades that followed, writers of 'scientific romances' and speculative fictions like H. G. Wells and Grant Allen made not just 'carnivorous' but man-eating plants a minor staple of their own literary diets. Fascinatingly, as Jim Endersby (2016a) has shown, these stories frequently featured man-eating orchids in a mash-up of two of Darwin's favorite botanical subjects. As the century turned,

Samuel Butler issued a revised edition of his satiric novel *Erewhon* that included a chapter on 'The Rights of Vegetables' clearly indebted to Darwin's botanical work. The original edition, heavily influenced by Butler's reading of *Origin of Species*, had been published in 1872, before the appearance of *Insectivorous Plants* and *The Power of Movement*, but the revised edition of 1901 spoke of an Erewhonian botanist who argued that 'animals and plants were cousins, and would have been seen to be so, all along, if people had not made an arbitrary and unreasonable division between what they chose to call the animal and vegetable kingdoms' (Butler 1901, p. 289), citing as prominent examples various insectivorous species.

If *Insectivorous Plants* had an impact on Victorian popular culture, Darwin's movement and fertilization studies helped fuel high-minded debates over theories of beauty that also played out in popular writings on science and aesthetics. Darwin's evolutionary account of beauty was developed most fully in *The Descent of Man, and Selection in Relation to Sex*. While *Descent* was the book in which Darwin finally spelled out the implications of natural selection for the development of *Homo sapiens*, those implications were obvious in *Origin of Species* and had been directly addressed by others in the intervening period, most prominently by Darwin's friend and ally T. H. Huxley in *Man's Place in Nature*. More ambitious and original in *Descent* were Darwin's efforts to demonstrate that the supposedly unique, non-physical features separating us from animals—our mental powers, moral sense, and aesthetic sense—were different only in degree from those of animals, and had been inherited from them. The bulk of the book, in fact, was devoted to Darwin's theory of sexual selection (Richards 2017). Sexual selection addressed mate choice, and in particular sought to explain the presence of 'secondary sexual characters'—the peacock's tail, the stag's antlers, the male mandrill's blue face and red nose—found only in one sex, usually the male. While antlers and spurs and pincers in many species give the male an advantage in battles with other males for the right to select a mate, or even the ability to grasp and hold a female against her will, many secondary sexual characters are used by males to attract or compete for the notice of the females, who become the selectors of the most pleasing males. Some of these secondary sexual characters, like the peacock's tail, seemed to pose a problem for natural selection, for why would natural selection generate a feature that made a bird conspicuous to predators and unable to fly? The answer, for Darwin, lay in the aesthetic preferences of countless generations of peahens selecting males with longer, more beautifully colored, and more beautifully ornamented tails. So, too, with the male mandrill's blue face and red nose, the song of the male nightingale, the rasping music of male crickets, and even the red-striped, bright yellow abdomen of the male green huntsman spider. Since evolution depended on the passing down of beneficial variations, mating and reproduction were critical. Darwin thus surveyed cases throughout the animal kingdom, in the process finding the origin of virtually every form of art and beauty, from color, ornament, and scent to vocal and instrumental music, dance, and architecture (Lightman and Zon 2014).

This was a new way of looking at beauty. By the late 1870s, it was being referred to as 'evolutionary aesthetics' or 'physiological aesthetics.' Like the Aesthetic or 'art for art's sake' movement emerging at around the same time, evolutionary aesthetics rejected the more traditional view that beauty in nature was an expression of the Creator's delight in the creation and concern for human pleasure. Since in the creation story in Genesis humans had been given dominion over nature, the existence of beauty in the natural world was a divine gift to humans, a sign of God's omnipotence and benevolence. Darwin had acknowledged in *Origin of Species* that if structures 'have been created for the sake of beauty, to delight man or the Creator,' it was 'absolutely fatal' to his theory (1872, pp. 159–160). Evolutionary aesthetics, on the other hand, made beauty about sex and reproduction. It had nothing to do with a God's pleasure or ours. Nor was beauty connected to the good, ugliness to evil. Sexual selection was a utilitarian theory of beauty, but one that focused on the reproductive imperatives of organisms. The human aesthetic sense, like the human moral sense, was not unique to humans and had evolved.

Sexual selection was controversial, rejected by the vast majority of Darwin's contemporaries, including most evolutionists. Darwin's work on plant fertilization, on the other hand, as we have seen, did become widely accepted in his own day. Color, scent, and some floral structures attracted insects in the right way and at the right time to facilitate cross-fertilization, and cross-fertilization led to the production of more and more vigorous offspring. *Orchids* demonstrated this particularly well, but all of Darwin's books on plant fertilization made the case that cross-fertilization conferred an adaptive advantage. Darwin's botany books were quieter and more modest works, aesthetically speaking, than *Descent of Man*, but they added to the evidence of aesthetic preferences even among insects. Darwin did not dwell on the aesthetic implications of insects' attraction to flowers; nor did he draw attention to the fact that botanical beauty had nothing to do with us and everything to do with the plants themselves. His contemporaries, however, whether allies or opponents, had no trouble seeing the implications of Darwin's ideas and appreciating that sexual and natural selection were being used to provide a natural, material, evolutionary account of the beautiful (Smith 2006, pp. 160–178).

The work of promoting the aesthetic dimensions of Darwin's fertilization studies was largely carried out by others, and overwhelmingly in periodical articles and books aimed at popular audiences. By the mid-1870s Darwin's botanical work was helping to fuel a new wave of botanical popularization, and descriptions of Darwin's results often comprised the bulk or served as the basis of these efforts. Some writers endorsed and even extended Darwin's use of natural selection and his challenge to traditional understandings of floral beauty, while others challenged or simply remained silent about these implications. J. E. Taylor endorsed the 'new philosophy of flowers,' asserting that flowers were not created for the delight and use of humans but had evolved for their own survival. Nonetheless, he argued, Darwin's work could be easily incorporated into, and would in fact strengthen, a theological understanding of nature. 'The teachings of modern botanists and naturalists concerning flowers and the insects,' Taylor wrote, 'are of a higher and more spiritual kind than those they have replaced' and will 'excite more reverence and admiration of Creatorial wisdom than the older teachings could educe' (1878, p. 3). In the final paragraph of *The Sagacity and Morality of Plants*, Taylor offered a revision of the *Origin*'s conclusion, arguing that 'it does not seem possible to contemplate this tangle of Vegetable Life and its conditions … without feeling that underneath and behind all are the Untiring purposes of Divine Wisdom and Love!' (1884, p. 303). M. C. Cooke, on the other hand, kept natural selection at arm's length in his *Freaks and Marvels of Plant Life* (1882), published by the Society for Promoting Christian Knowledge. Although 14 of the book's 20 chapters are explicit summaries of Darwin's work, Cooke contends that he is not concerned with arguing the merits of natural selection but with honoring Darwin as a patient observer and collector of facts, and he repeatedly frames Darwin's work in terms of Divine design.

By far the most prolific popularizer of Darwin's botany, however, was Grant Allen. Keenly interested in natural science, Allen as a young man sought to earn his living by popularizing and extending the evolutionary theories of Darwin, Spencer, Wallace, and Huxley (Clodd 1900; Lightman 2007, pp. 219–294; Morton 2005; Smith 2004). His first book, *Physiological Aesthetics* (1877), synthesized Darwin's work on mate-selection in animals and the evolutionary relationship between flowers and insects with physiological psychology and the physiology of sight and hearing. Allen argued that aesthetic feelings have a physical basis, are the product of natural and sexual selection, and thus are not unique to humans. His follow-up book, *The Colour Sense* (1879), developed the more specific claim that the human color-sense had in fact been inherited from fruit-eating ancestors whose ability to discriminate color had been developed in the same way and for the same reason as that of birds and insects. These books led to Allen's career as a writer of popular science essays for a wide range of middle-class periodicals, and as a writer of stories and novels, many with scientific themes. Allen's science essays were often collected and published in book form, including *The Evolutionist at Large* (1881), *The Colours of Flowers* (1882), and *Flowers and Their Pedigrees* (1883). Wallace (1882, p. 381), reviewing one of Allen's books in this period, declared that Allen 'stands at the head of living writers as a popular exponent of the evolution theory.' His popularization efforts promoted the aesthetic implications of Darwin's fertilization studies, often drawing conclusions from them that Darwin had eschewed, at least in print. In one of his works, Allen made clear that Darwin posed a fundamental challenge to traditional thinking about beauty in nature: 'The old school of thinking imagined that beauty was given to flowers and insects for the sake of man alone: it would not, perhaps, be too much to say that, if the new school be right, the beauty is not in the flowers and insects themselves at all, but is read into them by the fancy of the human race' (Allen 1881, p. 46). Moreover, according to Allen it is not fanciful to speak of insects as having particular aesthetic 'tastes.' While the ability to distinguish colors 'is a mere question of the presence or absence of nerve-centres,' the association of food with color ultimately associates pleasure with color, even for insects: 'creatures which pass all their lives in the search for bright flowers must almost inevitably come to feel pleasure in the perception of brilliant colours' (Allen 1884, pp. 452, 465). Colored flowers have not been divinely created for human benefit and pleasure but have been generated by the reproductive needs of plants and the nutritional needs of insects.

Evolutionary aesthetics did not displace the traditional approaches to beauty that it challenged. As Benjamin Morgan (2017) has shown, however, it was one of the most significant examples of a surprisingly wide range of materialist approaches to aesthetics in the second half of the nineteenth century and on into the twentieth. These approaches deployed the latest scientific thought to understand aesthetic experience as a physical and embodied process not unique to humans. The roots of these approaches were firmly planted, moreover, in Darwin's theory of sexual selection and his botanical work on plant fertilization.

## Conclusion

Charles Darwin's botanical investigations and writings were extensive and influential. They were his major experimental interest in the decades following the publication of *Origin of Species*, and they played a significant role in providing empirical and theoretical support to his claims in *Origin* and *Descent of Man*. His work on plant fertilization, plant movement, and insectivorous species were regarded by contemporaries as revolutionary, and allies and opponents alike appreciated the implications of his destabilizing of the boundary between plants and animals. Darwin's botany was widely reviewed and popularized as well, becoming well known to the public. His research on orchids and fertilization generally was absorbed into debates over in-breeding and cousin marriage. His studies of insect-eating plants like the common sundew were sources of fascination and horror to the public, who shivered at seeing in the flowers of English meadows and marshes the same kind of macabre tales of murder and incest to be found in the pages of the *Times* and in the latest fictional thriller. Novelists in turn appropriated Darwin's plants for late-century fictions about botanical monsters run amuck and alluring but deadly femmes fatales of the vegetable world. Darwin's botany also played a key role in the development of an evolutionary aesthetics that both countered traditional understandings of the source and role of beauty in nature and offered an alternative to the anti-utilitarianism of the 'art for art's sake' movement of the late century. From the realms of popular culture to philosophical debates about beauty, the significance of Darwin's plants was nearly as well known to the Victorians as his views on apes and his work on finches and pigeons.

## References

Adam W (1865) Consanguinity in marriage. Fortnightly Rev 2:710-730 and 3:74-88
Allen G (1877) Physiological aesthetics. Appleton, New York
Allen G (1879) The colour-sense: its origin and development. Trübner, London
Allen G (1881) The evolutionist at large. Fitzgerald, New York
Allen G (1882) The colours of flowers as illustrated in the British flora. Macmillan, London
Allen G (1883) Flowers and their pedigrees. Longmans, Green, London
Allen G (1884) Our debt to insects. Gentlemen's Mag 256:452–469
Allen M (1977) Darwin and his flowers. Taplinger, New York

Anderson NF (1986) Cousin marriage in Victorian England. J of Family History 11:285–301

Ayres P (2008) The aliveness of plants: the Darwins at the dawn of plant science. Pickering & Chatto, London

Browne J (1995) Charles Darwin: voyaging. Knopf, New York

Browne J (2002) Charles Darwin: the power of place. Knopf, New York

Butler S (1901) Erewhon, or over the range, Revised edition. Grant Richards, London

Child GW (1863) Marriages of consanguinity. Westminster Rev 80:88–109

Clodd E (1900) Grant Allen; a memoir. Grant Richards, London

Cooke MC (1882) Freaks and marvels of plant life. Society for Promoting Christian Knowledge, London

Corbett MJ (2008) Family likeness: sex, marriage, and incest from Jane Austen to Virginia Woolf. Cornell University Press, Ithaca

Darwin C (ed) (1838–1843) The zoology of the voyage of HMS Beagle. Colburn, London

Darwin C (1839) Journal of researches into the geology and natural history of the various countries visited by HMS Beagle. Colburn, London

Darwin C (1842–1846) The geology of the voyage of the HMS Beagle. Smith, Elder, London

Darwin C (1845) Journal of researches into the natural history and geology of the various countries visited during the voyage of HMS Beagle round the world. Murray, London

Darwin C (1859) On the origin of species. Murray, London

Darwin C (1860) Letter 3008. In: Darwin correspondence project. https://www.darwinproject.ac.uk/letter/DCP-LETT-3008.xml. Accessed 4 Mar 2024

Darwin C (1861) Letter 3264. In: Darwin correspondence project. https://www.darwinproject.ac.uk/letter/DCP-LETT-3264.xml. Accessed 5 Mar 2024

Darwin C (1862a) On the two forms, or dimorphic condition, in the species of *Primula*, and on their remarkable sexual relations. Pro Linnean Soc (Botany) 6:77–96

Darwin C (1862b) On the various contrivances by which British and foreign orchids are fertilised by insects, and on the good effects of intercrossing. Murray, London

Darwin C (1863a) Letter 4110. In: Darwin correspondence project. https://www.darwinproject.ac.uk/letter/DCP-LETT-4110.xml. Accessed 5 Mar 2024

Darwin C (1863b) Letter 4262. In: Darwin correspondence project. https://www.darwinproject.ac.uk/letter/DCP-LETT-4262.xml. Accessed 5 Mar 2024

Darwin C (1865) On the movements and habits of climbing plants. J Linnean Soc (Botany) 9:1–118

Darwin C (1868) The variation of animals and plants under domestication. Murray, London

Darwin C (1871) The descent of man, and selection in relation to sex. Murray, London

Darwin C (1872) The origin of species, 6th edn. Murray, London

Darwin C (1875a) Insectivorous plants. Murray, London

Darwin C (1875b) On the movements and habits of climbing plants. Murray, London

Darwin C (1876) The effects of cross and self-fertilisation in the vegetable kingdom. Murray, London

Darwin C (1877) The different forms of flowers on plants of the same species. Murray, London

Darwin C (1880) The power of movement in plants. Murray, London

Darwin C (1958) The autobiography of Charles Darwin 1809–1882. Collins, London

Darwin G (1873) On beneficial restrictions to liberty of marriage. Contemp Rev 22:412–426

Darwin G (1875) Marriages between first cousins in England and their effects. Fortnight Rev 18:22–41

Darwin's hothouse and lists of hothouse plants (n.d.) In: Darwin correspondence project. https://www.darwinproject.ac.uk/commentary/life-sciences/darwin-and-down/darwin-s-hothouse-and-lists-hothouse-plants. Accessed 5 Mar 2024

Dawson G (2007) Darwin, literature and Victorian respectability. Cambridge University Press, Cambridge

Endersby J (2016a) Deceived by orchids: sex, science, fiction and Darwin. Brit J History Sci 49:205–229

Endersby J (2016b) Orchid: a cultural history. University of Chicago Press, Chicago

Falconer H (1864) Letter 4464. In: Darwin correspondence Project. https://www.darwinproject.ac.uk/letter/?docId=letters/DCP-LETT-4644.xml. Accessed 5 Mar 2024

Hooker JD (1847) An enumeration of the plants of the Galapagos archipelago. Trans Linnean Soc 20(2):163–223

Hooker JD (1860) The botany of the antarctic voyage of H.M. discovery ships Erebus and Terror, Part III: Flora Tasmaniae. Lovell, Reeve, London

Hooker JD (1869) Address by the president. In: Report of the thirty-eighth meeting of the British Association for the Advancement of science. Murray, London, pp lviii–lxxv

Hopkins E (1879) Carnivorous plants. Contemporary Rev 35:37–50

Huxley TH (1876) On the border territory between the animal and the vegetable kingdoms. Macmillan's Mag 33:373–384

Kohn D (2008) Darwin's garden: an evolutionary adventure. New York Botanical Garden, New York

Levine G (2010) Darwin the writer. Oxford University Press, Oxford

Lightman B (2007) Victorian popularizers of science: designing nature for new audiences. University of Chicago Press, Chicago

Lightman B, Zon B (eds) (2014) Evolution and Victorian culture. Cambridge University Press, Cambridge

Lubbock J (1870) Letter 7288. In: Darwin correspondence project. https://www.darwinproject.ac.uk/letter/?docId=letters/DCP-LETT-7288.xml. Accessed 5 Mar 2024

Mann R (1881) Darwin on the movements of plants. Edinburgh Rev 153:497–514

Miller TS (2012) Lives of the monster plants: the revenge of the vegetable in the age of animal studies. J Fantastic Arts 23:460–479

Morgan B (2017) The outward mind: materialist aesthetics in Victorian science and literature. University of Chicago Press, Chicago

Morton P (2005) 'The busiest man in England': Grant Allen and the writing trade, 1875–1900. Palgrave Macmillan, New York

Müller H (1883) The fertilisation of flowers. Macmillan, London

Ogle W (ed) (1878) Flowers and their unbidden guests. Kegan Paul, London

Richards E (2017) Darwin and the making of sexual selection. University of Chicago Press, Chicago

Ruskin J (1906) Proserpina. In: Cook ET, Wedderburn A (eds) The works of John Ruskin, vol 25, library edition. George Allen, London

Schaffer T (2016) Romance's rival: familiar marriage in Victorian fiction. Oxford University Press, Oxford

Smith J (2004) Grant Allen, physiological aesthetics, and the dissemination of Darwin's botany. In: Cantor G, Shuttleworth S (eds) Science serialized: representations of the sciences in nineteenth-century periodicals. MIT Press, Cambridge, pp 285–306

Smith J (2006) Charles Darwin and Victorian visual culture. Cambridge University Press, Cambridge

Taylor JE (1878) Flowers: their origin, shapes, perfumes, and colours. Hardwicke & Bogue, London

Taylor JE (1884) The sagacity and morality of plants. Chatto & Windus, London

Trumbach R (1978) The rise of the egalitarian family: aristocratic kinship and domestic relations in eighteenth-century England. Academic Press, New York

Wallace AR (1882) Review of vignettes from nature, by Grant Allen. Nature 25:381–382

Walters SM, Stowe EA (2001) Darwin's mentor: John Stevens Henslow, 1796–1861. Cambridge University Press, Cambridge

Wilson A (1878) Can we separate animals from plants? Cornhill Mag 37:336–350

Wolfram S (1987) In-laws and outlaws: kinship and marriage in England. St. Martin's, New York

# Chapter 5
# Moving Plants in the Victorian Era: Glass, Transplants, and the Wardian Case

Luke Keogh and Angela Kreutz

**Abstract** This chapter introduces the Wardian case as a Victorian technology used for both presenting plants in the home and allowing plants to move beyond their home range. Looking closely at the Wardian case, we see the varying relationships of people with plants during the Victorian era. On the one hand, we witness the emerging middle class and passion for plants, in particular showcasing these rare and unique plants in their homes and greenhouses. On the other hand, we witness the journey that many plants took from the far reaches of the globe to European centers and for agricultural interests. This chapter introduces the Wardian case and the journey of plants in the Victorian era and touches upon some of the implications of these global transplants.

**Keywords** Wardian case · Nathaniel Bagshaw Ward · Banana · Cinchona · Crystal Palace · Glass · Gardening · Transplanting plants

## Introduction: Inventing the Wardian Case

In 1829, the doctor and amateur naturalist Nathaniel Bagshaw Ward (1842a) placed the chrysalis of a moth, soil, and dried leaves inside a glass bottle, screwed the cap tightly, and placed it on the windowsill. He wanted to observe the moth hatch. As he waited and watched, something else sprouted inside the bottle—a fern and meadow grass. The moth hatched and was allowed to go. However, the plants that grew inside the bottle captured Ward's attention. A keen naturalist, Ward had tried for many years to grow the fern in his garden, but the polluted city air that surrounded his central London home prevented it. Inside the bottle, the fern thrived (Keogh 2020).

Ward discovered a new method to keep plants alive. Under glass, in the presence of sunlight and moisture, plants can survive in this microenvironment for long

L. Keogh (✉) · A. Kreutz
Deakin University, Geelong, Australia
e-mail: luke.keogh@deakin.edu.au

**Fig. 5.1** The contest for the Bouquet: The Family of Robert Gordon in their New York Dining-Room, 1866. Notice Wardian case on windowsill (left). By Seymour Joseph Guy, oil on canvas. Collection of the Metropolitan Museum of Art, New York

periods without water. Ward suggested that they could survive as long as 8 months without watering. Over the coming years, Ward experimented further with plants under glass. His experiments inside his house led to the discovery of the Wardian case. Ward's discovery has two widespread implications. First, it brought plants inside the Victorian home (Fig. 5.1), exemplified by Ward's immediate use of all range of glass cases in his London house at Wellclose Square. The wider implications of Ward's invention were for the long-distance transport of live plants. For much of the eighteenth and early nineteenth centuries, transporting plants between countries was a great challenge. In 1819, John Livingstone, a keen botanist and surgeon posted to Macao for the East India Company, estimated that only 1 in 1000 plants survived the journey from China to Britain (Livingstone 1822).

In 1833, Ward tested his invention by transporting two traveling style cases on the longest journey then known—to Australia. The plants survived the 6-month journey to the Royal Botanic Gardens, Sydney. The cases were repacked and sent back to Ward in London. Again, they survived. When Ward and his friend George Loddiges, a well-known nursery owner, went aboard the ship in London to inspect the plants, he wrote, "I shall not readily forget the delight expressed by Mr.

**Fig. 5.2** Travelling style Wardian case, as prescribed by N.B. Ward. From N.B. Ward, *On the Growth of Plants in Closely Glazed Cases* (London: John Van Voorst, 1852)

G. Loddiges, who accompanied me on board, at the beautiful appearance of the fronds of *Gleichenia microphylla*, a plant now for the first time seen alive in this country" (Ward 1842a, 47). The experiment was successful.

A traveling Wardian case looked in some ways like a miniature greenhouse made with a rectangular timber base and a sloping roof that held glass inserts (see Fig. 5.2). Planting a Wardian case was quite simple. Inside the case, first a layer of rocks or broken bricks was placed, followed by a layer of sphagnum moss or leaf litter, and then a layer of soil. Once plants were in the soil, battens were often placed across the inside of the box to hold everything in place. The plants were well watered and often allowed to settle for a few days or even weeks. Often, there was a hole in the base of the case that allowed water to drain before a plug was placed in the hole. Overwatering plants was one of the most common mistakes. The case was then closed and not opened until the plants arrived at their destination.

In the Victorian era, the Wardian case was the primary means for transplanting live plants around the globe. After witnessing the first successful transport, Ward's friend George Loddiges, of the famous Loddiges & Sons nursery in Hackney, put into use more than 500 Wardian cases for his international shipments (Loddiges 1842). It was used for more than a century by nurseries, botanical gardens, plant explorers, and agriculturalists. The Royal Botanic Gardens, Kew, would use the case extensively to move plants, particularly for their plant hunters and for sending plants to their network of botanical gardens in the colonies (Desmond 2007). To complete the cycle of transport, many of Kew's cases were returned with unique foreign plants that botanists at Kew would name and describe in their publications.

Nathaniel Bagshaw Ward was not only a keen experimenter but was also well connected with nineteenth-century London and European naturalists. Among his correspondents were Charles Darwin and Michael Faraday. His friends included Kew's directors William Hooker and Joseph Hooker, as well as the famous nursery-man George Loddiges, the Harvard botanist Asa Gray and the accomplished Irish botanist William Henry Harvey. Ward was well connected and happy to host a plant lover at his home. All of these connections greatly helped in promoting the case and led to its wide usage. It was used by institutions and companies all over the world, from Britain to Germany and from Australia to Belgium, and the case travelled all over the world from Calcutta to Cameroon and from Melbourne to Peking.

Even in the early Victorian era, the informal networks that spread throughout the connecting world helped to spread the Wardian case (Brockway 1979; Endersby 2008). This is most evident in how the news of the Wardian case reached botanists in Paris. George Loddiges was very good friends with Nathaniel Ward; in fact, plants from Loddiges' nursery filled the first traveling Wardian case sent to Sydney in 1833. After the first successful trip, Loddiges actively used the Wardian case to send exotic plants to his customers and collectors in distant locations around the globe (Loddiges 1842). In 1836, one of the recipients of these cases was Nathaniel Wallich, a botanist and surgeon for the East India Company and superintendent at the East India Botanic Gardens in Calcutta. Immediately upon seeing the cases, Wallich realized their usefulness and filled Loddiges' case with Indian plants and sent it to colleagues at the Jardin des Plantes in Paris (Mirbel 1838). The botanists in Paris were equally impressed and had copies made. They returned the case to Wallich but sent the copied cases to the famous Cel nursery in southern France. The Cel's were also impressed with the case and began to use it in their business of moving plants around the world. The spread of the Wardian case throughout these early Victorian networks established it as the key technology in live plant transport for the remainder of the nineteenth century.

The most famous plants that were moved in Wardian cases were successful economic plants such as Cavendish banana, cinchona, coffee, tea, and rubber (these will be discussed further below). These plants were distributed beyond their home range and would go on to have a significant impact on agricultural economies in distant locations. There were also many ornamentals that were sent, including fuchsia, ixora, and many varieties of roses. A single case being sent could be filled with up to 100 plants of many plant varieties.

In the nineteenth century, a Wardian case filled with ferns became a feature of many middle- to upper-class Victorian homes. It has been suggested that before Ward, the Scottish botanist Allan MacConichie (1839) developed an airtight system for preserving plants. MacConichie did not promote his invention widely and nor did he have extensive networks like Ward and remaining details of his invention are limited (Allen 1969; Whittingham 2012). The use of similar inventions is not surprising. Looking back to the late eighteenth century, we also see that there was a long list of experiments on moving live plants that paved the way for the success of the Wardian case (Ellis 1770; Lindley 1824; Lindley and Loudon 1835).

The Wardian case has captured a wide audience of scholars interested in the Victorian era, and while many have looked closely at ornamental Wardian cases, the connections between the indoor botanical style and the traveling Wardian cases are an interesting feature of plants in the Victorian era. The earlier work of David Allen (1969) alerted us to the passions and destruction associated with the fern craze, which have been widely taken up by scholars for many years. In recent years, the Wardian case has received attention for its artificial manipulation of nature (Darby 2007). Or as a counterpoint to the coal smoke and factory emissions of the mid-nineteenth century that allowed Victorians "to conceptualize their increasingly fraught relationship to the natural world" (Wells 2018, 159). The fern-dominated mid-nineteenth century style of the Wardian case has been said to bring us closer to the Carboniferous landscape; indeed, it might be a "complex symbolic carrier of the pre-industrial world" (Yuval-Naeh 2019). However, in Whittingham's (2012) *Fern Fever*, which covered both the fern craze and indoor gardening, she clearly noted that there was a connection between the indoor gardening crazes, the tastes of the Victorian era and long-distance plant transport. As Whittingham (2012, 19) suggested, "The important legacy of the Wardian case was thus in greatly increasing the survival rates of plants being transported across the world." To bring plants inside and to spread the colonial ambitions of the era required connections across a variety of locations. The Wardian case allows us to bring in a logistical dimension to the study of plants in the Victoria era (Keogh 2020).

Very few objects illuminate the passion for plants in the Victorian era as the Wardian case does. Successfully sending live plants around the world in the nineteenth and early twentieth centuries in Wardian cases helped to transform our global environment. This technology reached well beyond the geographic limits of Victorian England. Not only did the Wardian case allow for an increased number and variety of plants that could be moved, but it also unleashed a whole range of other ecological impacts that are still with us today (Keogh 2019). The remainder of this chapter proceeds in two parts. First, we briefly examine the types of Wardian cases used in the nineteenth century and examine the implications of gardening under glass. Then, we look briefly at the journeys of three well-known plants—bananas, tea, and cinchona—to show the journeys of plants and the impact they had.

## Designs with Glass

Nathaniel Bagshaw Ward never patented his invention, and throughout his life, the Wardian case involved many designs and styles. However, we can identify two distinct styles of Wardian cases. The first type, often most reminiscent of the Victorian era, is the ornamental Wardian case (Allen 1969; Whittingham 2012). These were beautifully constructed and ornate glass cases that adorned the inside of your home (Fig. 5.1). Inside the cases were anything from ferns and greenery or rare plants that preferred a humid atmosphere inside the glass cases. In the mid-nineteenth century, ferns were highly prized and were a common variety to keep under glass. Indoor gardening and the fern craze have been covered extensively by David Allen (1969) and more recently by Sarah Whittingham (2012) in the beautifully illustrated *Fern Fever*. By the end of the century, other varieties, such as orchids or rhododendrons, were also highly prized. The ornamental Wardian case was a precursor to today's terrariums that have made a resurgence in recent years; this style dates back to the early nineteenth century.

The other type was traveling Wardian cases, which were sturdy and made to withstand a long sea journey (Fig. 5.3). It was made of strong timber—some of the first ones used by Loddiges' nursery were built with teak from old East India Company ships (Loddiges 1842). The sloping roof of the case held glass inserts. The top of the glass was often covered with wire gauze or timber battens to protect the glass while still allowing light to enter. Inside the case one could send plants in pots or in soil directly placed in the bottom of the case. Either way, the battens were fastened internally to keep the plants and soil secure. Inside these cases were anything from ornamental plants traveling with colonists to rare plants returning with plant hunters or economic plants for plantations.

The implications of the Wardian case also include a growing appetite for glass houses and gardening under glass; the glass houses at Chatsworth House offer an example here. The sixth Duke of Devonshire, William Cavendish, and his head gardener Joseph Paxton wanted to build a new hothouse at the Duke's stately home Chatsworth House in Derbyshire in the heart of England (Colquhoun 2006). Taken

**Fig. 5.3** Wardian case containing plants (late nineteenth century) used to Kew Botanic Gardens. Dorling Kindersley/Alamy

with the fever for exotics of the time and encouraged by the vision and talent of the head gardener Joseph Paxton, they set about building a great glasshouse that they envisaged would be unrivalled in Europe. Paxton wanted to use a wood structure and curvilinear glass to make a palm house that was original in design and grand in proportion. They were very much aware of the recent experiment of the Wardian case. For advice, they took their designs, including a miniature model, to both George Loddiges at Hackney and later to John Lindley at the Horticultural Society.

Confident that it would meet their expectations, Cavendish and Paxton then put their minds to filling the glasshouse with plants. What they most wanted were rare plants that would match the grandeur of the architecture. Paxton and the Duke selected their young Chatsworth gardener John Gibson to travel to Calcutta, India, who would use the Wardian case to collect a wide variety of plants to help fill their new glasshouse. The young gardener set out in 1835 with the primary purpose of collecting *Amherstia nobilis*, the pride of Burma, a newly discovered tropical tree regarded and sought for its extravagant large, drooping blood red collection of flowers. To fill the glasshouse at Chatsworth, Gibson collected orchids and other rare ornamentals for the Duke of Devonshire (Lemmon 1968). By 1837, Gibson's journey, although filled with many ups and downs, was largely successful. He arrived at Chatsworth with more than 40 boxes of plants that were moved from India to Chatsworth; not all the boxes were Wardian cases, but a high proportion of these were.

Over a decade later, Joseph Paxton became even more prominent in gardening circles and was designing the Crystal Palace that would host the Great Exhibition in Hyde Park. The Great Exhibition of the Works of Industry of All Nations would showcase the great industrial and economic exploits of nations from around the globe. In autumn 1850, Ward welcomed one of Paxton's gardeners to see his Wardian cases (Ward 1850). The gardener was calling to question Ward about his plant boxes, both the original construction and the newer types of cases he was now using. The Great Exhibition was an important moment for the Wardian case. First, it featured the Wardian case as one of many interesting industrial exploits. Second, with the large construction of the Crystal Palace, glass became inexpensive on a commercial scale. The construction of the Crystal Palace was an architectural achievement of the time. Covering almost 18 acres, the main building was more than 1800 feet (563 m) long and 400 feet (124 m) wide, and at its greatest height, it was 108 feet (33 m) tall (V&A 2019). It was largely made of glass. In total, 300,000 glass sheets (4 ft. 1 inch by 10 inches/1.3 m × 25.3 cm) were used to cover the large structure. Such a construction would not have been possible without the repeal of the glass tax years earlier. At their height, the Chance Bros. who held the glass contract, made 63,000 sheets in just 14 days during January 1851. With the construction of the Crystal Palace the mass production of glass became possible. Readily available and affordable glass would influence its widespread use in new Wardian cases.

Opened in May 1851 and concluded in October, the Great Exhibition saw more than 6 million people through the gates. Inside this glass structure were not only the largest collection of manufactured products from around the world but also thriving products from nature, including a row of large elm trees that already existed in Hyde

Park and were covered by the building. The historian David Allen has said that the palace made "the whole country extremely aware of 'foliage inside glass'" (Allen 1969). However, on a large scale, it was still representative of what was possible in a greenhouse. The Wardian case appeared in various parts of the exhibition. People could view economic and ornamental plants brought from far-off regions. They could also view curious objects such as Ward's glass bottle with a plant in it that had apparently not been watered for 18 years (Barber 1980; Hershey 1996). There were many different varieties of Wardian cases throughout the exhibition. While many of the well-adorned parlor cases were quite curious, in the East gallery there were also traveling Wardian cases (Illustrated London News, 2 Aug., 1851).

According to a journalist for the *Illustrated London News* the Wardian case, particularly these traveling cases, was an important part of the British industry and worldly connections worthy to be displayed alongside other industrial items such as steam engines or vulcanized rubber. By 1851 not a week passed when ships did not arrive in Britain with cases full of plants from "the remotest habitable regions" around the globe. In these cases, the journalist went on, "have thus conferred upon us a power of procuring exotic vegetable productions" that could not have otherwise been obtained if it was not for the Wardian case. Many ornamental plants were found throughout the exhibition growing in Wardian cases. Some of the plants shown inside cases were the porcelain flowers (*Hoya carnosa*) and closely associated *Hoya bella* flowers. Positioned in the northern transept of the Great Exhibition, one could find a case that housed both ferns and cacti (Ward 1851). The Wardian case was an object of great accessibility. Not only was it worthy of being shown at the Great Exhibition, it was also something that could adorn your home. The journalist for the *Illustrated London News* left the final comment for bringing nature indoors. "The cultivation of plants is an occupation delightful in itself, and one that is calculated to afford intense pleasure to those who follow the amusement." Every child in London, proposed the journalist, needed a Wardian case so that a love of cultivation would grow with them as they aged. The Wardian case at the Great Exhibition was a major milestone in its uptake and acceptance. Probably more so for ornamental reasons and for its use by amateurs as a parlor centerpiece. However, even as it was shown for its aesthetics, it was not lost on many commentators how important this technology was in bringing plants across the ocean. Even "uninteresting" cases, such as those carried on ships, were shown at the Great Exhibition.

Another aspect of Nathaniel Ward's work was to show the significance of gardening under glass for a wider public. Together with a number of scientists and medical practitioners, Ward played an important role in abolishing the glass tax. From the middle of the eighteenth century, glass was taxed on its weight. This practice continued for nearly a century. In the 1840s, a wide section of the community, from scientists to medical practitioners, argued against the tax. Considering that many of its readers had their own greenhouses, it was not surprising that the *Gardeners' Chronicle*, led by John Lindley, was particularly vocal about abolishing the tax. Throughout 1844 and 1845 the newspaper ran long articles calling for the end of the tax.

The Commission of Inquiry into the State of Large Towns and Populous Districts was one of the government avenues where opinions on the glass issue could be heard. Nathaniel Ward noted, when he was called upon to give evidence, that if the tax could be removed, then all people, rich and poor, could have Wardian cases in their home. Pushed on the issue, Ward (1844, 45) even told the commission, "but there is light enough in the most dirty parts of London to grow plants of the most delicate kind." Following the recommendations of a range of experts, Robert Peel's government abolished the glass tax. On 22 February 1845, soon after Peel's speech to the House of Commons, the *Gardeners' Chronicle* led with an editorial from Lindley on the issue. "Whatever the opinion of politicians may be respecting Sir Robert Peel's financial measures, there can be no doubt that the Right Hon. Gentleman deserves the hearty thanks of all persons having gardens." Glass could now become cheap and durable. To be used in anything from greenhouses to Wardian cases. It was an important technical turning point that led to much wider use of glass in homes and in long-distance plant transports.

## Transplanting Bananas

At home in London, the Wardian case had broad public appeal. From glass houses to the Crystal Palace and from homes to hospitals, the use and visibility of the case was widespread. By its very nature, the Wardian case was also a cosmopolitan traveler. It was a significant transporter for many plants. As a box for moving plants, the Wardian case took part in two important processes that were both in tension and mutually beneficial. First, the Wardian case allowed us to move important economic plants for such things as colonial agriculture and plantations that had many ongoing and long-term effects. As this process was taking place in the mid-nineteenth century, botanists were still coming to terms with the wide diversity of plant life on the planet. Second, the Wardian case allowed many varieties of plants to be moved from remote regions of the world to scientific centers so that they could be described and classified. The remainder of this chapter highlights these two processes by examining three important crops moved in Wardian cases: bananas, ornamental Chinese plants and cinchona. First, we turn to the banana journey and its spread throughout the Pacific.

On 11 April 1839, the missionary Reverend John Williams left London for Samoa. In the previous two decades, he had traveled widely in the Pacific, spreading God's word and trying to convert native populations. In the 1830s, he returned to Britain and, after many publications and public lectures, garnered wide fame. While in Britain, he translated the New Testament into the Roratongan language of the native people of the Cook Islands. Soon after completing his translation, he again set sail for the South Pacific. So popular had Williams' mission become he was able to purchase his own ship. Traveling with him on his new ship the *Camden* were two Wardian cases. Inside one case was the plant

*Musa cavendishii*, a Cavendish banana plant. Originally domesticated in tropical Asia and New Guinea by native peoples over thousands of years, the Cavendish banana first arrived in Britain from Mauritius in 1829. The original owner had two plants and when he died they were put up for sale. One went to the Continent and the other found its way into the Duke of Devonshire's glass house. Joseph Paxton cultivated it, and by 1838, he reported that he had 100 plants to distribute. The missionary John Williams heard about the value of this variety of banana and asked his friend Nathaniel Bagshaw Ward if the *Musa* would travel in a Wardian case to the Pacific. Ward encouraged the transplant (Ward 1842b). Paxton gave Williams a plant to take to the South Pacific and they were packed in Wardian cases. Six months later, the banana was carried to shore at the small South Pacific island of Upolu, which is part of Samoa. When Williams unpacked the Wardian case, he thought the plant almost dead and did not give it a chance of survival. William Mills, the missionary based at Upolu, was able to get one plant to grow and planted it in their small garden (Murray 1876; Prout 1846). It struck, and by 1840 some reports said there were more than 300 bananas protruding from the introduced tree and that 30 suckers could be transplanted (Ward 1849).

The banana is nutritious and flavorsome and easily cultivated by cloning, all attributes that worked neatly with the missionary's project. Starchy varieties of plantains, however, had traveled with Polynesian people as they spread eastward thousands of years earlier. The missionaries argued that the native varieties of banana were too tall and were often blown over in heavy storms. With such vulnerability, they were not a stable food supply. As the missionaries traveled, so too did the banana. It spread quickly. One missionary reflected years later, "Our teachers have taken the plant, whose value is now so well known, wherever they have gone" (Murray 1876, 271). With their help, the Cavendish banana was now found throughout the Pacific, from Tahiti to the Torres Strait and even as far as Hawaii. Along the way, by spreading only this variety of plant, the wide genetic variety of many different types of plantains used in the Pacific began to decline with the reliance on only one variety of banana. As missionaries like to tell it, the arrival of the missionaries to the Pacific brought many fruits both civilizing and economic. The Wardian case traveled on this civilizing mission and carried a valuable plant. Colonial botany had many flow-on effects beyond single transplants. Nathaniel Ward liked to tell the missionaries version of the story; it also helped to promote the value of his glazed cases (Ward 1842b). For Ward, the utility and necessity of understanding the natural products of a location were not only for transporting plants and understanding the botanical products of a location but also as a way to colonize (Ward 1840). The movement of plants in the Victorian era occurred in many directions and facilitated a wide range of activities. In the following decade, the Wardian case was again used in a number of important transplants, this time carrying plants from China.

## Chinese Plants in Wardian Cases

Since the eighteenth century, there was a thirst for exotic Chinese products in Europe. This went hand-in-hand with China becoming a flourishing market for British products. Traders, surgeons and ship captains stationed in Canton or Macao fed this growing market with products from the east. Chinese plants were popular for gardeners. John Reeves, stationed in Macao in the late eighteenth century, worked for the East India Company. He used early style plant boxes, very reminiscent of a Wardian case, to send many beautiful plants, such as chrysanthemums and roses, to the Royal Horticultural Society in London (Bailey 2019). Of all the products from China, tea was the most desired.

Following the Treaty of Nanking (1842) there was far greater access to China for foreigners. The Royal Horticultural Society was one of the first to notice the opportunities and quickly organized to send a plant collector to China. They chose Robert Fortune. A knowledgeable young gardener, Fortune, learned about his trade in various estates in Scotland and was one of the first to graduate with a Certificate of Horticulture from the Royal Horticultural Society. After serving at the Royal Botanic Gardens of Edinburgh, he returned to London to take up a post at the Society's Chiswick garden. Here, Fortune was noticed by the head of the garden John Lindley. Fortune was organized to travel to China to collect ornamental plants that would be of interest to British gardeners.

Fortune would go on to be one of the most well-known plant hunters of the nineteenth century. His travels and exploits are described not only in his own published works but also by historians (Fortune 1847a, 1852, 1857, 1863; Watt 2017; Fromer 2008). The focus of this section is on Fortune's use of the Wardian case and its usefulness in moving Chinese plants. Fortune was an accomplished botanist and robust traveler and had a talent for retelling (and embellishing) the stories of his journeys. His four books were popular and widely circulated. From the first journey, he was an avid promoter of the Wardian case. Like no other plant explorer before him, describing the Wardian case was an integral part of showing the distance he traveled and the challenges facing plant hunters. Following his final journey in the 1860s, he still told readers that travel in remote places was not the hard part, "the difficulty—the great difficulty—was to transport living plants" (Fortune 1868, 608). Fortune learned about the intricacies of the Wardian case on his first expedition to China.

As Fortune was preparing to leave for China, he gained intimate knowledge of the box from George Loddiges and John Lindley, who were both on the committee overseeing the journey. When he was readying to leave, he was given three Wardian cases filled with fruit trees and ornamental plants. The plants inside were to be used as gifts to contacts in China and he was also to record how the plant boxes held up on the journey, and the results would make up an important first result of the journey (Fortune 1847b). One letter was important for Fortune to get his plants home. It was written by the wealthy ship owner Joseph Somes and addressed to ten ship captains and "the commanders or many other of my ships that may be in China." The plants in Wardian cases, said Somes, were to come under special watch from the captains.

"It is my particular wish that you receive on board yr. ships any cases which Mr. F. may wish to send to London as Freight and that you will pay the most particular attention to their preservation from injury during the voyage" (Somes 1843). Captains were even expected to carry boxes of seeds in their cabins.

Fortune left by the *Emu* arriving in Hong Kong in July 1843. The cases he had with him were "strongly and coarsely made" (Markham 1862). The thick glass was fixed firmly to the box, and each piece of sheet glass was medium sized. On the outside, large iron bars extended across the box to protect the glass. At the top there was a canvas covering that could be unrolled to screen the plants from the sun. There was 8 inches of soil that were kept down by cross battens. When he arrived in Hong Kong most of the plants were healthy, although a few were quite "exhausted" (Fletcher 1969). One of his earliest collections from around the Amoy region was placed in Wardian cases as he traveled by river through the region. However, when navigating the Formosa Channel, his boat was struck by a severe storm and the boxes and plants were all destroyed (Fortune 1847a). In general, most of the Fortune plants collected were in and around the recently opened ports of Guangzhou (Canton), Xiamen (Amoy), Fuzhou (Foochow), Ningbo (Ningpo), and Shanghai. Halfway through his time in China, he sent Wardian cases to the Horticultural Society, but few plants survived. In late 1844, he sent more plants with varying results. Finally, after nearly 3 years of traveling through the coastal ports of China, Fortune boarded the *John Cooper* in Hong Kong with 18 Wardian cases ready to return to London.

Fortune gave a full report on the use of the Wardian case that was later published in the Society's journal. The transport of plants was so important that Fortune included the report as an appendix to the second edition of his popular travel book, *Three Years' Wanderings in the Northern Provinces of China* (1847). Fortune understood that many readers had friends in distant countries who may wish to collect plants. Tending to a case of plants on a long journey not only afforded "amusement" but also enabled one to "enrich one country with the productions of another." (Fortune 1847a, 411) After detailing how he used the cases he concluded by reflecting on Livingstone's estimate to the Society in 1819 that only one in one thousand plants survived the journey from China, "we have made some improvement in the introduction of Chinese plants since the days of Mr. Livingstone." (Fortune 1847a, 420) Fortune's collections were prized when they came home; in just this collection, there were 130 different tree peonies of various colors. On this first journey, Fortune became well practiced in using the Wardian case for transporting plants long distances. This experience would be used repeatedly for the next two decades on his further travels.

The journey to China was successful for both Fortune and the Royal Horticultural Society. Many ornamental plants were introduced, and society was able to distribute many Chinese plants to its members and subscribers (Fig. 5.4). Fortune was also a good observer, and many of his descriptions of botanical practices were valued by curious British people wishing to further exploit China. Fortune's observations of tea manufacturing were significant for his future career and for the future of such an industry within the borders of the British Empire (Fortune 1847a; Watt 2017). There

**Fig. 5.4** *Daphne fortuni,* "the harbinger of spring", one of the many ornamental plants Robert Fortune moved in Wardian cases. This one arrived at the Chelsea Physic Garden in 1844 and flowered in England for the first time in 1846. From *Journal of Horticultural Society,* vol.2, 1847, p.35

had been much confusion in the botanical literature over green and black tea. No westerner had witnessed the process in detail. By collecting herbarium specimens of tea plants from various locations, Fortune showed them to be the same species.

Following his return to London, Fortune was employed as a curator at the Chelsea Physic Garden, which was the botanic garden of the Worshipful Society of Apothecaries and one of the oldest gardens in Britain. At Chelsea, he became friends with Nathaniel Ward who was a member of the garden committee. Ward and Fortune would remain friends for the next two decades (Fortune 1868; Keogh 2020). While Fortune was working in Chelsea, the prospects of the tea industry in India were gathering pace. Already there were auctions of Assam grown tea in London. And now the British East India Company were looking to develop the industry on a large scale with Chinese plants. After only 18 months working at the Chelsea garden, the Company contracted Fortune to go to China to collect tea for them so that they could further develop the industry in India.

We see with Fortune's first journey to China, the Wardian case became a key technology for moving plants in the Victoria era. Additionally, London horticultural figures such as George Loddiges and John Lindley were heavily involved in the transplant of plants and the use of Wardian cases. While much has been made of Fortune's later journeys in China in search of tea, it is in the first journey that many of the practices for the later journeys were formed. In Fortune's later journeys, he transported hundreds of Wardian cases full of tea seedlings to India. He would also transport Wardian cases full of tea to the United States. With regard to tea, particularly in India, it was not so much the plants that Fortune transplanted that were significant but the accompanying Chinese workers, as well as their equipment, who understood how to manufacture tea, which was an important ingredient for the emergence of the Indian tea industry (Keogh 2020). Indeed, with all the hype around Fortune's tea theft, it was discovered that India's native variety of tea (*Camelia sinensis* var. *assam*) was better suited for cultivation in India.

## Finding Cinchona

In 1853, on his third trip to China, Robert Fortune was given a special consignment of six cinchona plants to take with him (*Bulletin of Miscellaneous Information* 1931). Fortune packed these into Wardian cases and took them overland to the Calcutta Botanic Gardens. He left the plants in Calcutta and went on to China to gather more tea. At the time, cinchona was much more valuable than tea. Native to South America, it was widely known that the bark of the plant helped to prevent fevers and was an effective drug for fighting malaria. Those six plants that Fortune took in Wardian cases were originally sent to gardens in Britain by the French naturalist Hugh Weddell. He had also sent plants to Paris. From Paris one plant was sent to Dutch botanists in Java. The British and Dutch were very keen to appropriate cinchona from South America. The Wardian case was a key tool in their efforts. The work of Fortune in moving tea from China had shown imperialists that moving large quantities of valuable plants was possible. A few years later, Clements Markham, who led the British efforts to transplant cinchona, sat down with Fortune to learn the best ways to use Wardian cases.

The British and Dutch governments decided to introduce cinchona from its native South American locations to plantations in their colonies. It is one of the most infamous acts of biological espionage in world history (Crawford 2016; Honigsbaum 2002; Brockway 1979; Hobhouse 1985). There were a whole cast of gardeners and plant hunters enlisted to find plants and deliver them to the Asian colonies. It was moved from South America as both seeds and live plants. For the latter, the Wardian case was the prime mover. In particular, the case was used by the two leading figures employed in the transplant: Justus Karl Hasskarl for the Dutch and Clements Markham for the British. In the 1860s, soon after the imperial powers transplanted cinchona, South American nations banned the removal of the plant from their native locations. However, it was too late, and the seeds were in the soil in Java and India.

Hasskarl was a German-born botanist working for the Dutch. Following his training, he worked at the famous Buitenzorg Botanical Garden, where he introduced the systematic arrangement and authored their first catalog (*Chemist and Druggist* 1894). Following a disagreement with the director, he left Java and worked in Germany as a journalist and translator. In 1852, he was contracted by the Dutch government to travel to South America to collect cinchona seeds and plants for the purpose of starting plantations in the Dutch Indies. Hasskarl set out for Bolivia traveling under an assumed name to avoid suspicion and made collections. He packaged 400 cinchona plants into 21 Wardian cases. In late 1854, he set sail for Java with cases and a large quantity of seeds (Taylor 1945). Only 40 plants survived the journey. Of the cinchona plants that survived they were found to include a number of known species, including *C. succirubra* and a new species of cinchona, *Cinchona pahudiana*. Unsurprisingly, the new species was named after the Governor General of the Dutch East Indies, Charles Ferdinand Pahud. The new plants were planted by the enthusiastic colonial naturalist Franz Junghuhn, known as the Humboldt of Java. Junghuhn directed the efforts to set up cinchona plantations. He was familiar with the Wardian case having traveled to Java in 1855, with four cases full of cinchona. Despite having a number of cinchona plants to choose from, Junghuhn preferred *C. pahudiana*. By 1860, there were nearly 1 million plants of *C. pahudiana* in plantations on Java. After the plantings had taken hold, the bark of this tree was tested, and it was soon realized that it contained little quinine. Such an embarrassment was the whole project that the failure was debated in Dutch Parliament in the early 1860s. By 1865 the Dutch were close to abandoning their cinchona plantations.

Five years later, Clements Markham thought that the British had the knowledge that Hasskarl did not. The British effort to acquire cinchona was spread over three locations in an attempt to collect different varieties of the plant (Headrick 2010). With this diversity, once back in India, they would see which ones were best suited to cultivation. Markham, who coordinated the expedition, traveled to Bolivia with his wife Minna. He collected with the help (and significant labor) of Robert Weir, a nurseryman. They collected *C. calisaya* specimens. Richard Spruce was a former Kew gardener who, along with Robert Cross, was sent to collect in Ecuador. They were responsible for collecting *C. officinalis* and *C. succirubra* plants and seeds. G.J. Pritchett was given a wide license to collect any varieties he could in Peru. While plants were being collected in South America, a site for a plantation was selected and cleared near Ootacamund (today Udagamandalam), Madras, in southern India. At the same time, a new propagating house at the Royal Botanic Gardens, Kew, was erected to serve as the halfway point for the many plants on the journey from South America to India.

The logistics of the project were extensive and the Wardian case was an important technology for the collectors. Thirty Wardian cases were constructed at Kew, packed as flat-packs, and sent around Cape Horn to ports in South America ready for collectors. They were 3 feet 2 inches long, 1 foot 10 inches wide, and 3 feet 2 inches high. When full of soil and plants each weighed 336 lb. (152 kg). Fifteen went to Ecuador for Spruce's collection and 15 went to southern Peru for Markham's plants. Pritchett was originally sent to gather seeds, but six cases were made by a carpenter in Lima to take his plants to India. Together with the gardener Weir, Markham traveled from Islay over the Andes to the cinchona regions around La Paz, Bolivia, where plants and seeds were collected. They arrived in Islay (today Mollendo, Peru) with plants to fill the 15 cases. At the port of Islay, Markham used the garden of an Irishman who ran a blacksmiths shop. The blacksmith also got them the soil for their Wardian cases. The cases were packed with 458 *C. calisaya* plants. Before setting off, the plants appeared to be in good health. However, Markham suggested that once on board the steamer, locals tried to sabotage the operation by pouring boiling water in the cases potentially to avoid important botanical products being exported from their home country. The sabotage was discovered and avoided and the cases left on the steamer on their way through Panama and a very hot journey through the Red Sea.

Markham (1862) described the adventures of collecting cinchona in popular travel accounts when he returned to Britain. In these aggrandized stories, the Wardian case appeared prominently. Moving plants was the purpose of the journey; therefore the technology for moving them earned an important place in the story. However, for all the adventures that Markham described, it was Robert Weir, the gardener who did most of the labor. In addition, it was Weir that would come under fire when the plants arrived in southern India in bad condition. The plants were sickly and almost lost. On the other hand, the seeds and plants collected by Robert Cross were well packed and had cool weather for their journey. Cross also accompanied the plants for their journey. They went via Kew for a short stop and then sent on to India and arrived at Ootacumund. Four hundred and sixty-three plants survived and they grew well in the propagating houses. By December 1861, the first plants had become 5200 plants. These formed the basis of the plantations in India.

It was neither Dutch nor British efforts to move cinchona in Wardian cases led to commercialization. Many years later, a different variety of cinchona—*Cinchona ledgeriana*—was collected by a local Peruvian Manuel Incra Mamani that proved successful. These seeds were later purchased by the Dutch and created the basis for their commercial plantations in Indonesia in the late nineteenth century. Although high in quinine content (the alkaloid used from the plant), *C. ledgeriana* was not well suited to plantations. Therefore, the Dutch inventively grafted scions of *C. ledgeriana* onto the hardy root stock of *C. succirubra* originally collected by Cross and widely cultivated in India (Fig. 5.5). While much debate carried on whether the high yield of the scion would be drained by lower yielding understock, it was ultimately shown that *C. ledgeriana* and all its quinine-containing bark thrived in these hybrid conditions. These hybrid seedlings were distributed throughout Java and led to the establishment of Dutch cinchona plantations in Java.

Cinchona was a plant that required years of trial and error to make it a commercially viable agricultural crop. As plant hunters scoured South America for the best plants to take back to the Dutch and British colonies, there was a flood of work in just trying to understand the genus. In the literature, there were as many as 330 different names given to various cinchona plants since the eighteenth century. Botanists and chemists grappled with diverse species that could be found throughout much of South America and were trying to determine what was the best variety to cultivate. Along the way, there were many failures. Early cultivation by the Dutch and the British were two examples of this failure. Today, after much lumping by botanists, there are 25 different plants that make up the cinchona genus, but only three have enough alkaloids in their bark to warrant commercial cultivation—*C. ledgeriana*, *C. officinalis*, and *C. succirubra* (Cowan 1929; Andersson 1998). However, even in

**Fig. 5.5**   Grafting *C.ledgeriana* scions on *C.succirubra* root stock at Tjinjiroean plantation, West Java. Collection of the Wellcome Library

this list, the first variety contains three times as much quinine as the last variety and is far superior. This was not widely known until well into the twentieth century. However, it was the variety *C. succirubra* moved in Wardian cases in the 1860s that formed an important root stock for Dutch cinchona plantations.

## Conclusion: Complexities of Moving Plants

Often, when thinking about our relationship with plants, we might think of those valuable plants that touch our lives every day. Bananas, tea, and cinchona are three well known examples of plants that were acclimatized beyond their home range in the Victorian era and had an immediate impact. All three had large economic impacts on the structure of agricultural economies for native and distant regions. Colonists and capitalists benefited, while native peoples often lost valuable products. In these three examples, there is great complexity in how we move plants. By moving plants in cases, we decreased the diversity of local crops, such as bananas. By moving plants in Wardian cases, we also began to understand the value of local varieties of plants, such as the native tea that grew in India. By moving plants in cases, we also hybridized nature to the needs of people and economies, such as varieties of cinchona in Indonesia. The Wardian case allowed live plants to travel a great distance in the Victorian era.

The Wardian case has largely been preserved in much of the historical literature as occupying a significant place in the natural history crazes of Victorian England (Allen 1969, 1994, 1996, 2001; Barber 1980; Gould 1998). By 1851, the Wardian case full of plants was exhibited at the Great Exhibition. Inside the Crystal Palace, people could view live ornamental ferns brought from distant regions; they could also view Ward's glass bottle with a plant in it that had not been watered for 18 years. As Darby aptly concluded about the Wardian case in the Victorian era, it was both "an extreme and characteristic example of the Victorians' artificial manipulation of nature" (Darby 2007, 647). It was certainly an artificial manipulation of nature but casting our eyes wider than the context of Victorian England; the scale of movement that the case facilitated was significant (Keogh 2020; Rigby 1998; Klemun 2012).

It is also important to reflect upon what Frank Turner (1997, 288) noted: "The world of Victorian science in practice within the empire involved a whole host of people about whom we know little." Looking closely at plants, we see that there were many hands that worked the soil to allow plants to travel. Focusing on the movement of plants allows us to see the extent to which the Wardian case was used as a unified transport technology in a time when the world was becoming increasingly connected. We also must not forget that the Wardian case moved not only plants but also a range of unwanted species, including plant pathogens and diseases that have caused wide-scale environmental problems (Keogh 2019, 2020). To bring plants inside and to spread the colonial ambitions of the era required connections across a variety of locations, fields, and distances. Indeed, bringing a logistical dimension to the study of plants in the Victorian era allows us to see the Wardian case and the movement of plants far more holistically.

# References

Allen DE (1969) The Victorian Fern craze: a history of Pteridomania. Hutchinson, London

Allen DE (1994) The naturalist in Britain. A social history. Princeton University Press, Princeton

Allen DE (2001) Naturalists and society: the culture of natural history in Britain, 1700–1900. Ashgate, Aldershot

Allen DE (1996) Tastes and crazes. In: Jardine N, Secord JA, Spray EC (eds) Cultures of natural history. Cambridge University Press, Cambridge, pp 394–407

Andersson L (1998) A revision of the genus cinchona (Rubiaceae-Cinchoneae). Mem NY Bot Gard 80:1–75

Bailey K (2019) John reeves: pioneering collector of Chinese plants and botanical art. Royal Horticultural Society, London

Barber L (1980) The heyday of natural history, 1820–1870. Doubleday, New York

Brockway L (1979) Science and colonial expansion: the role of the Royal Botanic Gardens. Academic Press, New York

Bulletin of Miscellaneous Information (1931) Introduction of cinchona to India. Bull Misc Inf 3:113–117

Chemist and Druggist (1894) 'Obituary: Hasskarl', 20 Jan., 73–74

Colquhoun K (2006) "The busiest man in England": a life of Joseph Paxton, gardener, architect and Victorian visionary. Godine, Boston

Cowan JM (1929) Cinchona in the empire. Progress and prospects of its cultivation. Empire Forestry J 8(1):45–53

Crawford MJ (2016) Between bureaucrats and bark collectors: Spain's Royal Reserve of Quina and the limits of European botany in the late eighteenth-century Spanish Atlantic world. In: Manning P, Rood D (eds) Global scientific practice in an age of revolutions, 1750–1850. University of Pittsburgh Press, Pittsburgh, pp 21–37

Darby MF (2007) Unnatural history: Ward's glass cases. Vic Lit Cult 35(2):635–647

Desmond R (2007) The history of the Royal Botanic Gardens Kew. Kew Publishing, Kew

Ellis J (1770) Directions for bringing over seeds and plants from the east-indies and other distant countries in a state of vegetation. Davis, London

Endersby J (2008) Imperial nature: Joseph hooker and the practices of Victorian science. University of Chicago Press, Chicago

Fletcher H (1969) The story of the Royal Horticultural Society, 1804–1968. Oxford University Press, Oxford

Fortune R (1847a) Three years' wanderings in the northern provinces of China. John Murray, London

Fortune R (1847b) Experience in the transmission of living plants to and from distant countries by sea. J Hort Soc London 2:115–121

Fortune R (1852) A journey to the tea countries of China. John Murray, London

Fortune R (1857) A residence among the Chinese: inland, on the coast, and at sea. John Murray, London

Fortune R (1863) Yedo and Peking: a narrative of a journey to the capitals of Japan and China. John Murray, London

Fortune R (1868) Ward's plant cases. *Gardeners' Chronicle* 27 Jun., p.608

Fromer J (2008) A necessary luxury: tea in Victorian England. Ohio University Press, Athens, GA

Gould SJ (1998) Seeing eye to eye, through a glass clearly. In: Leonardo's mountain of clams and the diet of Worms: essays on natural history. Harmony Books, New York, pp 57–73

Headrick DR (2010) Power over peoples: technology, environments, and Western imperialism, 1400 to the present. Princeton University Press, Princeton

Hershey D (1996) Doctor Ward's accidental terrarium. Am Biol Teach 58(5):276–281

Hobhouse H (1985) Seeds of change: five plants that transformed mankind. Sidgwick & Jackson, London

Honigsbaum M (2002) The fever trail: the hunt for the cure for malaria. Farrar, Straus and Giroux, New York

Keogh L (2019) The Wardian case: environmental histories of a box for moving plants. Environ Hist 25:219–244

Keogh L (2020) The Wardian case: how a simple box moved plants and changed the world. University of Chicago Press, Chicago

Klemun M (2012) Live plants on the way: ship, Island, botanical garden, paradise and container as systemic flexible connected spaces in between. J Hist Sci Technol 5. Available: https://www.johost.eu/vol5_spring_2012/marianne_klemun_2.htm

Lemmon K (1968) The Golden age of plant hunters. Barnes, New York

Lindley J (1824) Instructions for packing living plants in foreign countries. Trans Hort Soc London 5:192–200

Lindley J, Loudon JC (1835) Implements of gardening. In: An Encyclopedia of gardening comprising the theory and practice of horticulture, floriculture, arboriculture, and landscape-gardening. Longman, London, pp 585–586

Livingstone J (1822) Observations on the difficulties which have existed in the transportation of plants from China to England, and suggestions for obviating them. Trans Hort Soc London 3:421–429

Loddiges G (1842) Copy of a letter from G. Loddiges, Esq., to the author. In: Ward NB (ed) On the growth of plants in closely glazed cases. John van Voorst, London, pp 86–87

Markham C (1862) Travels in Peru and India while superintending the collection of Chinchona plants and seeds in South America, and their introduction into India. John Murray, London

MacConochie A (1839) On the use of glass cases for rearing plants similar to those recommended by N.B. Ward, Esq. Annual Report and Proceedings of the Botanical Society of Edinburgh 3: 96–97

Mirbel C (1838) Method of transporting exotic plants. Madras J Lit Sci 7:191–192

Murray AW (1876) Forty years Mission work in Polynesia and New Guinea, from 1835–1875. James Nisbet, London

Prout E (1846) Memoirs of the life of the Reverend John Williams. John Snow, London

Rigby N (1998) The Politics and Pragmatics of Seaborne Plant Transportation, 1769–1805. In: Lincoln M (ed) Science and exploration in the Pacific. Boydell Press, Suffolk, pp 81–100

Somes J (1843) Letter to ship captains, 22 Feb, folder 1, Fortune Papers, Royal Horticultural Society London

Taylor N (1945) Cinchona in Java: the story of quinine. Greenberg, New York

Turner F (1997) Practicing science: an introduction. In: Lightman B (ed) Victorian science in context. University Chicago Press, Chicago, pp 283–289

Victoria and Albert Museum (2019) Building the museum. Victoria and Albert Museum https://wwwvamacuk/articles/building-the-museum#slideshow=31131014&slide=0 Accessed 28 August 2022

Ward NB (1840) Letter to Asa Gray 6 Mar. Harvard Botanical Libraries, Biodiversity Heritage Library https://wwwbiodiversitylibraryorg/page/53546432 Accessed 28 Aug 2022

Ward NB (1842a) On the growth of plants in closely glazed cases. John van Voorst, London

Ward NB (1842b) Letter to WJ Hooker, 19 December, directors correspondence, Vol 18B. Folio 225. Archives of the Royal Botanic Gardens, Kew

Ward NB (1844) Nathaniel Bagshaw Ward, Esq, surgeon examined. In: First report of the commissioners for the inquiry into the state of large towns and populous districts, vol 1. Clowers and Sons, London, pp 41–45

Ward NB (1849) Letter to the Linnaen society, read 1 November. Proc Linn Soc Lond 1:157

Ward NB (1850), Letter to Asa Gray autumn 1850. Biodiversity heritage library. https://www.biodiversitylibrary.org/page/53546432#page/29/mode/1up Accessed 28 August 2022

Ward NB (1851) Letter to the editor. Floricultural Cabinet and Florist's Magazine, Oct, p 260

Watt A (2017) Robert Fortune: a plant hunter in the orient. Kew Publishing, Kew

Wells L (2018) Close encounters of the Wardian kind: terrariums and pollution in the Victorian parlor. Vic Stud 60(2):158–170

Whittingham S (2012) Fern fever: the story of Pteridomania. Francis Lincoln, London

Yuval-Naeh N (2019) Cultivating the carboniferous: coal as a botanical curiosity in Victorian culture. Vic Stud 61(3):419–445

# Chapter 6
# Circulation and Civility: Mid-Victorian Botany and Microscopical Method

Meegan Kennedy

**Abstract** This chapter examines important work on method by two German botanists, Matthias Jakob Schleiden and Hermann Schacht, and how this work was (or was not) recirculated in British books on microscopical method during the 1850s. Schleiden, concerned with scientific error, argued that microscopists must train themselves in a methodical practice. He laid out an inductively grounded system of practice in an essay that was translated into English in 1849 and widely reviewed. His student Schacht published a book on microscopical method, translated for British readers in 1853, that followed Schleiden's precepts. These texts were amongst the earliest to instruct English-speaking readers on Continental approaches to microscopy. In the 1850s, British microscopists were concerned about German and French supremacy in microscopical research and sought to establish a more methodical microscopical practice amongst their readers. Although these British writers largely echo Schleiden's precepts, they all ignore Schleiden's essay, citing similar work by his student Schacht or other writers instead. This is unexpected given that Schleiden was a figure of great repute for his earlier work on cell theory; he was one of the first to write intensively about method; and few other Continental microscopists had published on method in English at this time. Given British expectations of scientific civility and concerns about fostering a larger microscopical community in Britain, Schleiden's combativeness in print, which was widely censured as ungentlemanly, may have contributed to the poor uptake in Britain of his work on method.

**Keywords** British · German · Botany · Translation · Microscope · Error · Method · Instruction · Practice · Patience · Observation · Education · Inductive · Perspective · Intervention · Reagent · Citation · Civility · Community · Nationalism

M. Kennedy (✉)
Florida State University, Tallahassee, FL, USA
e-mail: meegan.kennedy@fsu.edu

© The Author(s), under exclusive license to Springer Nature Switzerland AG 2024
L. M. Mendonça de Carvalho (ed.), *The Victorians: A Botanical Perspective*,
https://doi.org/10.1007/978-3-031-68759-4_6

## Chapter Introduction

This chapter examines important work on method by two German botanists, Matthias Jakob Schleiden and Hermann Schacht, and how this work was (or was not) recirculated in British books on microscopical method during the 1850s. Schleiden, acutely concerned with scientific error, argued that microscopists must train themselves in a methodical practice. In a methodological essay that was translated into English in 1849 and widely reviewed, he proposed an inductively grounded system of practice. His student Schacht then published a book on microscopical method, translated for British readers in 1853, that largely followed Schleiden's precepts. These texts were amongst the earliest to instruct English-speaking readers on Continental approaches to microscopy. This essay examines Schleiden's and Schacht's recommendations on method and their relation to the work of British microscopists, who in the 1850s were concerned about German and French supremacy in microscopical research and sought to establish a more methodical microscopical practice amongst their readers. Although these British writers—mostly medical men—largely echo Schleiden's precepts, they all ignore Schleiden's essay, citing similar work by Schacht or others instead. This is unexpected given that Schleiden was a figure of great repute for his earlier work on cell theory; he was one of the first to write intensively about method; and few other Continental microscopists had published on method in English at this time. Given British expectations of scientific civility and concerns about fostering a larger microscopical community in Britain, Schleiden's combativeness in print, which was widely censured as ungentlemanly, may have influenced British microscopists' reluctance to quote Schleiden on method, even as they echoed his precepts.

In this chapter, I differentiate "methods," individual techniques or practices, from "method" as a systemic approach to observation, designed to minimize error. Many nineteenth-century writers discuss "methods"; they present instructions on specific practices—on how to use various elements of the microscopical apparatus or how to execute some particular task. Such passages usually arise in the course of a discussion organized by apparatus or object. Some writers, however, also present rules or principles of practical "method" (in the singular) as an overall organizing principle of microscopical work. In some cases "method" acts also as a major element in a text that argues for care and systematic planning in microscopical practice generally; a text may even be organized by elements of method (preparing, seeing, interpreting, recording). Schleiden was amongst the earliest to publish on "method" in this way.

# Chapter

Method emerges in nineteenth-century microscopical texts as a response to long-standing concerns about errors of observation. Microscopists in the German lands enjoyed access to courses on practical microscopy and published texts systematizing error in the 1840s, earlier than most British writers. The discourse around error was particularly consequential at this time, when the microscope was still of uncertain status as a scientific tool and, as L. S. Jacyna says, "the basic verbal and pictorial technologies of the field [of histology] were still in the process of formation" (Jacyna 2003, 81). Of the five genres of microscopical text that Jutta Schickore identifies from this period, two particularly contribute to this concern with error: "methodological sections in handbooks of anatomy and physiology of plants and animals" and "instruction manuals ... [for] the medical student and practitioner" (Schickore 2007, 222). Schleiden and Schacht offer advice on practical method as a means of meeting and preventing many causes of microscopical error. While they were not the only Continental microscopists concerned about practical method, their work was among the most accessible to Victorian readers. They benefited from training and working in the German lands, which were known as a center of empirical research on microscopical structures. Other investigators working here during this period include Christian Gottfried Ehrenberg, Johannes Müller, Jakob Henle, Robert Remak, Rudolf Virchow, and Albert von Kölliker, but these microscopists did not publish much on practical method. Among those who did write on microscopical method were Hugo von Mohl, a German botanist (*Mikrographie, oder Anleitung zur Kenntniss und zum Gebrauche des Mikroskops*, 1846); Julius Vogel, a German pathologist (*Anleitung zum Gebrauch des Mikroskopes*, 1841); and Jan Evangelista Purkyně (Purkinje), a Czech physiologist ("Mikroskop: Anwendung und Gebrauch bei physiologischen Untersuchungen," 1844). However, these works were not translated into English. (Vogel's great *Pathological Anatomy*, which was published in English in 1847, includes an introduction elaborating general investigative method but not microscopical method).

Another center of microscopical work in the 1830s and 1840s was Paris, where physician-researchers gathered from across the Continent to work out questions of histology, pathology, and cell theory in a mecca for clinical medicine. Among these were Gabriel Andral, Alfred Donné, Gottlieb Gluge, David Gruby, Hermann Lebert, François Magendie, Louis Mandl, and Charles-Philippe Robin, along with the optician Charles Chevalier. None of them, however, wrote books on method in a form accessible to English-speaking readers. The optician Chevalier's *Des Microscopes et de leur Usage* (1839) offers an important early example of practical instruction with passages on careful work, perseverance, and method. Mandl's *Traité pratique du microscope* (1839) includes a section on how to use the microscope and a chapter dedicated to causes of error, and Lebert's *Physiologie pathologique* (1845), like Schleiden's *Grundzüge*, primarily makes arguments about structure but includes a section on microscopical practice and error. Robin's *Du microscope et des injections* (1848) includes practical instruction as well. However, none of these books enjoyed an English translation.

Scientific investigators in Britain who were Anglican clergy or gentlemen of leisure might know enough German or French to read such books in the original, but language was evidently enough of a barrier for general readers that the editors of the *Quarterly Journal of Microscopical Science*, at its founding in 1853, included as a recurring feature the translation of important articles into English. The English translations of Schleiden and Schacht would have been especially valuable during the 1840s and early 1850s for their role in enabling Continental approaches to circulate through a wider English-speaking readership.

Schleiden was a German lawyer who, after struggling with depression, had given up law for the study of medicine and botany. In 1838 he published an influential article building on Robert Brown's discovery of the cell nucleus to argue that the cell is the basic structure of all plants. Theodor Schwann, a physician and physiologist, and a friend of Schleiden, referenced Schleiden's work in supporting his own argument that the cell is the basis of animal form and growth as well. Indeed, the English translation of Schwann's important work included Schleiden's article (Schleiden 1847). Schleiden was not the only investigator promoting a cell theory—Purkyně, for example, had written on the topic—and he had misinterpreted the role of the cell nucleus, considering it to be a site for the crystallization of new cells. Indeed, his thoughts on cell genesis were entirely wrong. However, his work was influential, helping turn botanists' attention toward the structure and life cycle of plants, in particular the role of cells in these matters. He was instrumental in making botany both a more inductive science and a more microscopical one; in his advocacy for microscopical method and (later) Zeiss lenses, he contributed to microscopy in general as well.

Both Schleiden (1849) and Schacht (1853) wrote treatises on botany that were noted for their grounding in microscopy. While these German botanists conducted important research on plant structures, their books were also instrumental in elaborating on practical method. They offered systematic guidelines for making and recording observations at a time when British microscopists were becoming increasingly interested in such models. Indeed, as Soraya de Chadarevian notes, due to Schleiden "the microscope came to dominate the work of the botanist" (1993, 534). Translated texts by Schleiden and Schacht, with John Quekett's *Practical Treatise on the Microscope* (1848), immediately preceded a surge of British texts on microscopical method. These texts present what amounts to a structured course of training in specific, favored physical and cognitive actions that would enable observers to prepare specimens, manipulate the apparatus, observe the image, record information, and share the resulting microscopical knowledge. Much of the British advice on microscopical practice at mid-century was published by medical men: the physicians John Hughes Bennett (*On the Employment of the Microscope in Medical Studies*, 1841 and *An Introduction to Clinical Medicine*, 1850), Lionel Beale (*The Microscope and Its Application to Practical Medicine*, 1854, and *How to Work with the Microscope*, 1857), and William Benjamin Carpenter (*The Microscope: And its Revelations*, 1856). Jabez Hogg (*The Microscope: Its History, Construction, and*

*Application*, 1854), like Quekett, was a surgeon. Additionally, amongst less-frequently cited books, the Danish pathologist Adolphe Hannover published *Om Mikroskopets Bygning og dets Brug* (1847), which John Goodsir translated into English as *On the Construction and Use of the Microscope* in 1853. The advice in these books adopts and reframes an interest in rules of practice that had previously been detailed by Schleiden, Schacht, and other German microscopists working in botany. This German interest in method was made accessible to the larger micro-scopical community in Britain through English translations.

## Practical Method: Schleiden

Schleiden, a professor of botany at the University of Jena, published *Grundzüge der wissenschaftlichen Botanik* in 1842, hoping to champion a more active, investiga-tive practice, and to oust a type of botany that he dismissed as mere species-collecting and species-naming. In the *Grundzüge* he argues for the natural system of botanical classification and the need to examine the whole life history of a plant. However, the book was not available in English until 1849, when Edwin Lankester published a translation of the second (1845) edition of Schleiden's treatise as *Principles of Scientific Botany; or, Botany as an Inductive Science*. Lankester, a surgeon and public health reformer, was an accomplished microscopist; he was an active member of the Microscopical Society of London (the elite British organiza-tion for microscopical work), where he was elected President in 1858–59, and he was a founding editor of the *Quarterly Journal of Microscopical Science*. His trans-lation of Schleiden's treatise appeared at the price of 21 shillings with the publishers Longman, Brown, Green, and Longmans, a large old London firm known for pub-lishing medical and scientific titles, textbooks, and reference works as well as litera-ture. Longman had published major medical and scientific texts on general scientific methodology, such as Herschel's A *Preliminary Discourse on the Study of Natural Philosophy* (1830), and specific practice, such as Henry Hughes' *Clinical Introduction to the Practice of Auscultation* (1845). The *Grundzüge* was a research text focused on Schleiden's botanical investigations, but in the second edition he added an extensive methodological introduction and a subtitle claiming botany as an inductive science. Although, as de Chadarevian points out, the *Grundzüge* "did not carry the word *Lehrbuch* [textbook] in its title," Schleiden used this book to argue that instruction in general principles of practical method would help secure the quality of microscopical work. This was the element of his text "best received" by reviewers, even some who disagreed with his research conclusions (de Chadarevian 1993, 535–6).

When Lankester published his translation of Schleiden's *Principles*, only one major English work devoted to microscopical practice had been published—Quekett's *Practical Treatise on the Microscope*. British interest in such subjects was not as well-established as in Continental circles. Perhaps for this reason, Lankester chose to condense the methodological introduction significantly. He wrote, in the

"Translator's Preface," that the "second German edition" of Schleiden's treatise "was accompanied with a Methodological Introduction," but that "[a]s the discussion of these general principles occupies a considerable space in the original, and it was deemed desirable not to increase the bulk of the present Work, this Introduction has been omitted." (In a footnote he directs readers to Herschel's and Whewell's general guides to inductive science instead). Lankester largely deleted Schleiden's philosophical remarks on vision and botany in general. He explains, "In [their] place I have, however, inserted a short summary of these observations" (Lankester 1849, iii). This summary provides the first four paragraphs of the book, taken from an abstract authored by Schleiden himself. Lankester also abridged Schleiden's extended instructions on specific aspects of microscopical method and moved them to an Appendix. Schleiden's original "Methodologische Einleitung" takes up 166 pages and does appear in the 1842 (first) edition; the abbreviated version in the translation ("Appendix D. On the Use of the Microscope in Botanical Investigations") requires only 16.5 pages. Concerned about cost—the treatise already ran to some 600 pages, with plates—Lankester anxiously excuses his inclusion of this Appendix, since "the observations of the Author seem … judicious, and we have none of precisely the same kind in our own language, and especially as this book may fall into the hands of students who are not acquainted with the powers or the manner of using this instrument" (iii). These defensive remarks suggest that in addition to the ever-present concern about cost, he feared criticism from British microscopists dismissive of such attention to method, or who might take offense at the suggestion that their work required correction. In fact, British reviews of the translation were more likely to criticize Lankester for abridging Schleiden's advice on method.

Schleiden writes in reaction to what he considered unbridled speculation, complaining that "whims of the imagination have taken the place of earnest and accurate scientific investigation" (Schleiden 1849, 313). He believed that method could protect scientific work from going astray. He argues, "In an inductive inquiry into natural objects, before all things we must strive after the discovery of leading maxims, which should be securely founded, and by which we may decide upon the admissibility of hypotheses, and through which alone science can be made free from fiction" (470). Schleiden provides these "leading maxims" in his Methodological Introduction. In the included passages, Schleiden argues overall that observers need to consider "*the method of microscopical investigation*" (583, his emphasis). "Method" means, in part, that the observer must not work hastily or carelessly; he declares, "Patience and perseverance will always have to be applied as the most important of all means of success" (415). Schleiden's most important advice for observers is that they should strive to adopt habits of patient, methodical work.

Schleiden outlines two main practices that, he believes, underlie methodical work at the microscope. First, the investigator must recognize that the human using the microscope is responsible for many of the errors attributed to the instrument, and must train the eye to avoid these errors. Schleiden rejects the idea "that microscopical researches can never be depended upon" since "the microscope is frequently very deceptive." To this he replies that "the microscope is perfectly innocent of every thing of which it is accused"; that "errors … are only committed by the erring judgment" (Schleiden 1849, 583). He also rebuts the notion that "nothing more is requisite for microscopical investigation than a good instrument and an object, and that it is only necessary to keep the eye over the eye-piece, in order to be *au fait.*" Rather, he says, "we must . . . gradually learn to see through the medium of the microscope," especially since it is "a much more difficult instrument than our eye." (584)

This notion that the eye needs training is not original to Schleiden, but he frames it for microscopical vision specifically, pointing out how ordinary vision enjoys several advantages over microscopical vision. For example, in everyday life, "our eye is moveable, and we may wander about with it among the objects." The movements of our bodies and eyes allow observers to accumulate diverse "impressions" of an object. These impressions are further diversified by our natural binocular vision. And finally, "we are able to move ourselves or the objects," multiplying "different perceptions" of an object to provide better grounding for our inferences about it. These "different views" include three-dimensional perspective and physical touch (Schleiden 1849, 584). In contrast, he says, observers using compound microscopes use monocular vision (a binocular microscope was proposed by John Riddell, a professor of chemistry at Tulane University, in 1854, but such instruments did not come into common usage for some years afterward). Additionally, says Schleiden, microscopists "see the object always in an isolated condition," and perceive only a narrow plane of surface, rather than the whole object (585). Finally, observers cannot move themselves or the microscopic objects, or touch them, as we do with ordinary objects, rendering the observation more susceptible to error. He laments microscopists' lack of attention to "fundamental principles" of observation, lamenting, "We find, alas! no information at all respecting these things" in scientific texts, "because no one has as yet occupied himself with the theory of microscopical observation" (586).

Second, Schleiden lays out several steps that trained observers can follow to artificially construct *multiple views* on an object, providing a diversity of impressions somewhat like what we naturally enjoy with ordinary objects. Observers should "examin[e] thoroughly every aspect of the ... object" and "remov[e] from it everything which does not belong to it," such as chromatic aberration and other optical phenomena. (Schleiden 1849, 585). Observers should "distinguish" extraneous elements in the image, such as dust and bubbles of air or oil, as well as misleading phenomena such as Brownian motion or currents of mixed fluids (587–88).

Most important, Schleiden suggests a "fundamental rule": microscopical observers "must observe frequently and with the most profound attention," a strategy he calls "many-sided comprehension" (*Vielseitigen Auffassung*) (Schleiden 1849, 591, 589; 1842, 151). By "frequently" he appears to mean taking multiple (different) views. Schleiden did not invent this concept; for example, the British instrument-maker Andrew Ross, in an 1839 article on "Microscope" for the *Penny Cyclopaedia of the Society for the Diffusion of Useful Knowledge* explains,

> In investigating any new or unknown specimen, it should be viewed in turns by every description of light, direct and oblique, as a transparent object and as an opaque object, with strong and with faint light, with large angular pencils and with small angular pencils thrown

in all possible directions. Every change will probably develop some new fact in reference of the structure of the object, which should itself be varied in the mode of mounting in every possible way. It should be seen both wet and dry, and immersed in fluids of various qualities and densities, such as water, alcohol, oil, and Canada balsam (Ross 1839, 181).

But this is the extent of Ross's advice on method, a term that does not appear in his article. What Ross offers as a general tip, Schleiden makes into an elaborate method, laying out a system to formalize this variety of approach. Opaque objects can be simply "taken up between the small forceps," allowing observers to rotate the object physically and "view it on all sides." Transparent objects present more of a challenge; Schleiden recommends "wetting" an object in various liquids and sectioning it by hand, which requires considerable skill (Schleiden 1849, 589). He includes multiple sketches of a diatom, *Navicula viridis*, that was too small to section, but for which he could obtain "anterior" and "lateral" views (593–4) (Fig. 6.1). Overall,

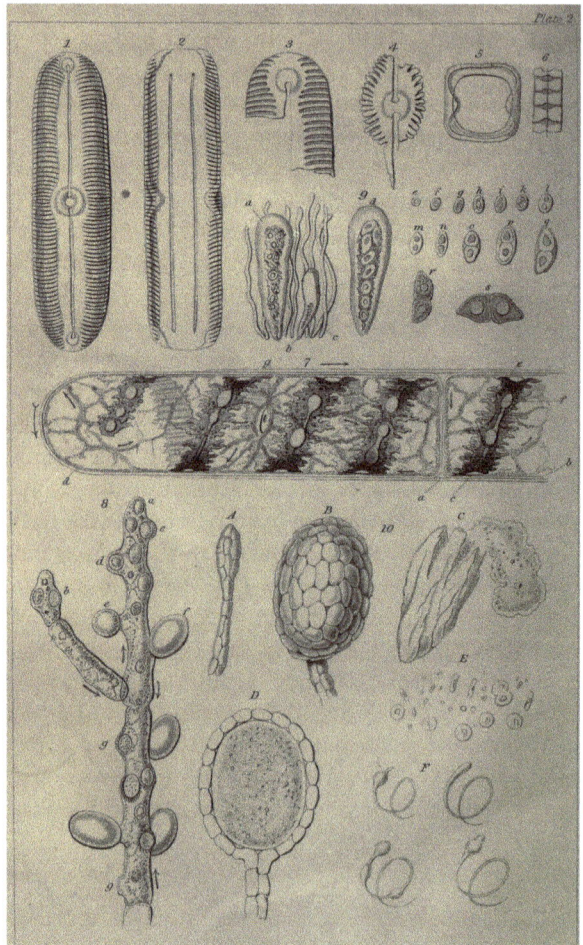

**Fig. 6.1** Matthais Jakob Schleiden. *Principles of scientific botany; or, Botany as an inductive science*. Trans. Edwin Lankester. London: Longman, Brown, Green, and Longmans, 1849. Plate II: Anterior and lateral views of the siliceous shield of *Naviculat viridis* (see pp. 593-4). Illustrations for this translation are taken from the second German edition (1846). Source: Hathitrust/Internet Archive: University of California

observers should "obtain from [an] object, properly prepared, as many views as can possibly be obtained from it, in order to construct a clear image from the combination of the individual conceptions" (589). This approach requires skill in microscopical manipulation. Schleiden thus provides not only a systematic list of interventions but a powerful reason why the microscopist should master them.

Schleiden concludes this section with a formal plan of action for microscopists: for any object, practice changing the image experimentally by varying four types of intervention, which he terms "microscopical reagents" (*mikroskopische Reagentien*) (Schleiden 1849, 590; 1842, 154). The term "reagents" in microscopy usually refers to physical substances that change the image of an object under the lens. Schleiden's use of the term includes changes to the light and apparatus as well, and emphasize that any of these manipulative interventions can yield increased visibility under the lens. His "optical, mechanical, chemical, and physical auxiliaries" change the appearance of the object by adjusting the use of light and magnification, pressure, chemical reagents, and heat or electricity, as with these images of carbonate of lime crystals presented by the British physician Lionel Beale (Fig. 6.2). By this approach "the object under investigation may be placed in as many circumstances as possible, in order thereby to increase the number of points of view." Schleiden's microscopical reagents promote many-sided comprehension by allowing the observer to access extra perspectives on an object. Schleiden closes this portion of his methodological introduction by enunciating his "fundamental rule" of many-sided (multiple-view) comprehension once again: the microscopist must "observe frequently and with the most profound attention… for *seeing* is a very difficult art" (590–91).

Overall, in his treatise on inductive method in microscopy, Schleiden emphasizes the following three main principles:

1. *Methodical practice* is the most important element contributing to the value of scientific work, so the observer must work with great patience and perseverance.
2. Seeing is a difficult art, so we must *train the eye*. The microscope is innocent; it is the human observer who errs in misinterpreting the image. Difficulties arise from using only one eye and seeing only one plane of the object at a time. Deceptive images also arise from the optics of lenses and from dust, air-bubbles, oil-bubbles, and other materials that do not properly belong to the object.
3. Observers must pursue many-sided comprehension by *multiplying views* of the object. Observers should train and experiment with different interventions of light, mounts, chemicals, sectioning, and the like, to gain facility with these actions and become familiar with their effect on the object. The observer works inductively, building up a full picture of an object by combining many different interactions with it.

These precepts offer a vision of what scientific practice should be (methodical, patient, and persevering); provide warnings (both general and specific) that microscopical vision requires training; and suggest a specific approach to inductive work (combining multiple views). Overall, Schleiden identified significant weaknesses in contemporary microscopical practice and offered an effective counter-program to

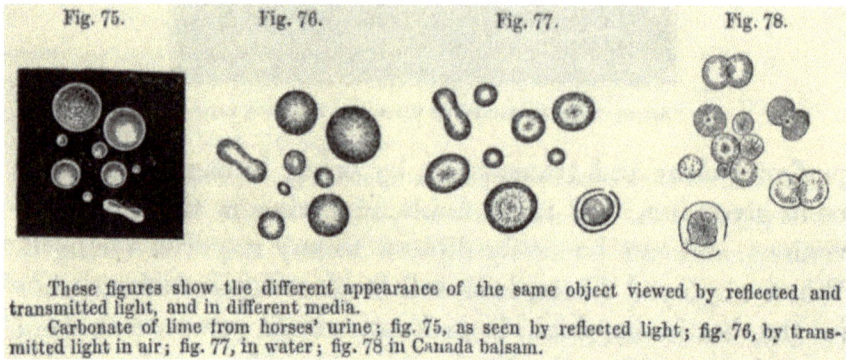

**Fig. 6.2** Lionel S. Beale. *The microscope and its application to practical medicine*, 2d edn. London: Samuel Highley, 1858. Crystals examined in different media, p. 29. Source: Hathitrust/Internet Archive: University of California

correct these weaknesses. His approach addressed problems that affected all microscopists, not just botanists. His program combines a philosophical commitment to inductive methodical practice (rather than speculation) and a checklist of actions that should help correct faulty practice.

## Practical Method: Schacht

The Danish pathologist Adolphe Hannover wrote a book *Om Mikroskopets Bygning og dets Brug* (1847) that was translated by John Goodsir for English-speaking readers under the title *On the Construction and Use of the Microscope* in 1853. Although Schleiden's *Grundzüge* had been available for 5 years, Hannover's brief (hundred-page) book says little of method; it mostly describes individual instruments and apparatus. One chapter does provide practical instructions on using the instrument, but he speaks very little about method. At the outset of the chapter giving "Directions for the use of the dioptric compound microscope," Hannover warns against book-learning, saying, "The practical management of the microscope is much more easily acquired" by working with a teacher in person and practicing at home "than by attention to a number of rules." Ironically, he then launches into "the rules to be observed for the preservation and arrangement of the microscope, for the illumination and the choice of magnifying powers, and for the selection, preparation, observation, and explanation of the object" (Hannover 1853, 45). But he chooses not to discourse on method or lay out a structure for systematic practice (other than the basic advice to "commenc[e] with the examination of easy familiar objects, and afterwards proceeding to newer and more difficult ones" (64). What little commentary on method is offered does align with Schleiden's advice in most ways. Hannover suggests the importance of both a trained eye and patience when he says, "[I]t is most essential to possess an intimate acquaintance with the instrument, great industry and perseverance, and that keenness of sight which can only be acquired by long-continued practice" (45). And Hannover offers a variation on Schleiden's principle of multiple views as a general proceeding. Instead of encouraging observers to execute many different interventions and combine the results inductively, Hannover suggests that many trials with different techniques might be needed to identify the *one best technique* for a particular object. Discussing modes of observation, he comments, "it will often happen that much time and many different methods of preparation will be employed before the right one is discovered by which we may be enabled to investigate the structure of different parts with perfect distinctness." Hannover thus offers a method by which to find the "right" approach. He acknowledges that "two observers may also arrive at the same conclusion by different methods" but does not consider comparing results to reach a fuller understanding of the object (53). So while Hannover endorses inductive practice generally and multiple views in exploratory work, he encourages observers to find the one best approach for each object—that which achieves the best image—rather than combining many views to maximize access to the object and minimize error. His book points to Schleiden's influence upon other European workers, but its translation in English does little to promote the specifics of Schleiden's program.

A very different image appears in the work of the German botanist Hermann Schacht, Schleiden's student, whose 1851 book *Das mikroskop und seine Anwendung inabesondere für Pflanzen-anatomie und Physiologie* does much more to promote method amongst microscopical workers. Schacht's book became available to English-speaking readers when Frederick Currey, a British lawyer and mycologist, translated the book into English and published it under the title *The Microscope in its special application to Vegetable Anatomy* in 1853 as part of Samuel Highley's series of textbooks, "Highley's Library of Science and Art." The series was designed to increase access to such books at affordable prices for a wider readership; the book cost only five shillings, even with woodcuts. Lankester's translation of Schleiden, at twenty-one shillings, was much more expensive. Currey abridged Schacht's book, much as Lankester had with Schleiden, but Currey removed only Schacht's discussion of specific makes of microscope—all Continental (v). The book sold well enough that an enlarged second edition appeared 2 years later.

As Schacht's title suggests, the book is very much about how to use the microscope in botanical practice specifically. He provides some general rules for investigation—in his introductory remarks, in Chapter Three ("General rules for the use of the microscope, and for the arrangement of the objects"), and in Chapter Five ("Concerning the method of investigation") (Schacht 1853, 11, 44). His instructions on microscopical method include instructions on reagents and other techniques of preparation, techniques of preservation, and on drawing as a form of note-taking (Fig. 6.3). Much of the book, however, consists of instructions on how to observe specific objects: "the dicotyledonous stem," "the bract and the calyx," "course of the pollen-tube," and the like (ix). These passages model methodical practice on discrete tasks but would not be as useful for investigators working on human, animal, or nonorganic specimens.

**Fig. 6.3** Hermann
Schacht. *The microscope in
its special application to
vegetable anatomy*, 2d edn.
Trans. Frederick Currey.
London: Samuel Highley,
1855. Box of chemical
reagents, p. 45. Many
illustrations were added to
the second edition of the
translation, this among
them. Source: Wellcome
Collection

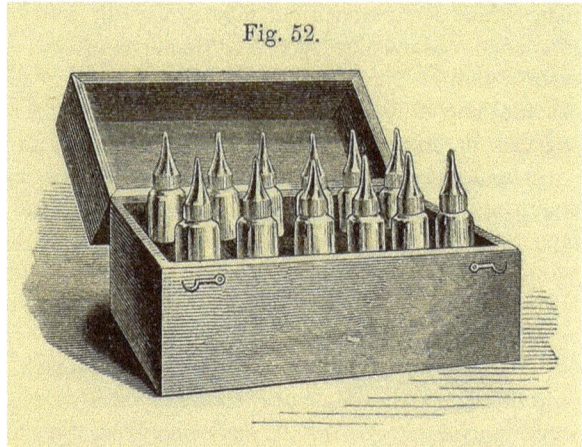

Schacht's interest in method and practice reflects his training under Schleiden from 1847–50 at the University of Jena; he reiterates Schleiden's positions in many ways. Schacht shows deference in citing his teacher, but he also supplements his work. For example, he says, "Schleiden, in his 'Grundzüge,' has treated fully of the method of preserving objects in fluid between two pieces of glass. I need, therefore, say very little about it" (Schacht 1853, 124). Despite this nod to his mentor, Schacht continues on for four pages of instruction. More significantly, he notes in his Introduction, "For the mode of proceeding in general, Schleiden has given several excellent hints, but I know of no guide for research in the field of scientific botany, notwithstanding the numerous writings which exist on the subject of the microscope" (4). Such a statement recalls Schleiden's exclamation that "no one has as yet occupied himself with the theory of microscopical observation"—although Schacht shifts the complaint from a lack of microscopical theory to a dearth of practical method (Schleiden 1849, 586). Here again, Schacht echoes—even cites and compliments—Schleiden's work but moves to supplement it, pressing additional practical advice on the reader. The older man's book, it is true, focuses on debating true structure in plants rather than providing step-by-step instruction how to examine the stem, the pollen-tube, and the like; however, Schacht uses the somewhat dismissive term "hints" (*Winke*) to describe his mentor's extensive Methodological Introduction and system of approach (Schacht 1853, 4; 1851, 5). Schacht offers himself and his own little book as a comprehensive practical "guide to all those who desire earnestly and zealously to cultivate the science of botany" (1853, 4). That said, in the phrase "science of botany" Schacht implicitly endorses Schleiden's view of botany as an inductive microscopical science of observation that attends closely to minute plant structures and their role in plant development. His vision of "the science of botany" thus aligns in many ways with that of Schleiden, even if he does frame his book as a more complete resource on practical method.

Schacht's rules of microscopical practice also in many ways recapitulate those of Schleiden: the focus on *patient, persevering method* above all, the need to *train the eye* to see through the difficulties of the microscopical image, and the pursuit of an inductive practice, *combining multiple views from different approaches*. For example, Schacht, like Schleiden, urges readers to adopt a more rigorously *methodical practice* in order to improve the science of botany. He peppers his book with aphorisms expressing this, such as "information strung together without method leads to no real enlightenment" or "if the method is accurate, the result will be valuable; if... the method be erroneous, the result will prove nothing" (Schacht 1853, 4, 44). Adopting these techniques requires patience and perseverance, something that both authors stress. Schacht warns, "*Work* without *method* will seldom lead to any result" (3). He repeats this philosophy a page later, more optimistically: "Without perseverance nothing can be effected with the microscope; whilst, on the other hand, by perseverance and determination certain results are to be attained" (4). This emphasis on method suffuses all the advice of both Schleiden and Schacht.

Second, Schacht emphasizes the difficulty of microscopical vision and the importance of *training the eye*. In fact, he quotes Schleiden at the outset of his book, saying, "*Seeing*, as Schleiden very justly observes, is a difficult art; seeing with the microscope is yet more difficult." (2). He notes many of the problems raised by Schleiden, especially the restriction to one plane surface only and the deceptive appearances produced by contaminants or our own interventions with the object. He says, "In microscopic observation two things must be remembered,--1st. That in the microscope, especially with high powers, we see *surfaces*, not *bodies*. ... 2nd. That we seldom see the objects under the microscope in their natural condition; that we consequently must take into consideration the changes which we ourselves partly produce." Thus, the observer must learn to attend to the difference between "the natural appearance of the object" and its appearance under the lens; deceptive appearances "originating in the eye itself," such as "Mouches volantes," or from "small particles of dust," "air-bubbles," Brownian motion, or currents of fluid; and other changes due to "the medium in which the object is placed, or ... the use of the knife or other influences." In recognizing "the wide difference between common vision and vision through a microscope," Schacht gestures to concerns he shares with Schleiden. He argues that microscopical vision thus requires "[l]ong and thorough practice," implicitly endorsing Schleiden's program of methodical training with four interventions ("reagents") as well (Schacht 1853, 2–3).

Finally, Schacht strongly encourages readers to *combine multiple views from different approaches* to an object, an approach to inductive practice that, as noted above, Schleiden calls "many-sided comprehension." Schacht discusses this sometimes in very literal terms. For instance, he says, "The microscope only affords a view of one surface of an object," and in studying transparent objects, "we are unable to make them out to be bodies until we have changed their position, and ascertained their dimensions in three different directions; this, in many cases, from the nature of the object itself, is a matter of great difficulty" (Schacht 1853, 23, 2). Interpreting multiple views more figuratively, Schacht reminds readers that while "the method of induction [usually] leads from individual to general results," in microscopical questions one should first glean "a superficial general knowledge of the object," examining "the entire perfect plant" using "the naked eye, or ... a magnifying-glass." The investigator will then move from "generalities to details," studying the plant "within and without" using "different magnifying powers" of the microscope. Finally, the investigator will build these details back up inductively to construct "an accurate acquaintance with the whole" (45). The initial general survey of an object—like the common advice that microscopists should use low powers for the first observation of an object—provides a working overview of that object and helps prevent mistakes at higher powers. Similarly, botanists should begin by studying the simplest plants, those in "the lower orders," then progress to more complex, "highly organized plants" (49).

Schacht promotes his own take on the need for multiple views. He explains that different kinds of research questions require different approaches to the work. Investigations of morphological questions (having to do with "outward form") and of anatomical ones (regarding internal structure) "require a different mode of proceeding." That is, studies of the major parts of plants "must be conducted differently from those connected with the development of the cells." Indeed, "it is hardly possible to point out an accurate method of proceeding for all possible cases." Furthermore, while accumulating and collating all these different observations, it may happen "that the inquirer is led aside to a collateral question" or "the principal question itself becomes essentially changed." The principal question must "never be lost sight of" while the investigator strives to collect information from any tangential queries. Schacht explains that "particular care must be taken to endeavour to throw light upon the principal question from all possible sources"—another version of "multiple views." (Schacht 1853, 46–47). The principle of multiple views should thus shape even the overarching plan of research. Overall, Schacht enthusiastically reiterates Schleiden's advice to combine multiple views from different approaches. However, while Schleiden provides a system of microscopical interventions, listing specific actions to try with each object (adjustments to the optical, chemical, mechanical, and physical constraints on the object), Schacht instead models how to plan a multiple-views approach to induction, where a variety of research questions shape everything from the most general research design to the day-to-day accumulation of observations.

## Practical Technique: British

The 1830s and 1840s were developmental decades for British microscopy, following up on technological improvements to the compound (multiple-lens) microscope that made the instrument more reliable. The simple microscope was still in use, especially for botanical work, field work, or dissections (Fig. 6.4), but much of the writing on method centers on practice with the compound microscope (Fig. 6.5), which was notoriously tricky to master. British writers on microscopy increasingly offer instruction on specific matters of *practical technique*, but a review of the major authors in this period shows that they do not offer a *systematic method* of microscopical practice until the 1850s. An early gesture toward method appears in David

**Fig. 6.4** Hermann Schacht. *The microscope in its special application to vegetable anatomy*, 2d edn. Trans. Frederick Currey. London: Samuel Highley, 1855. Dissecting microscope, p. 14. This illustration was added to the second edition of the translation. Source: Wellcome Collection

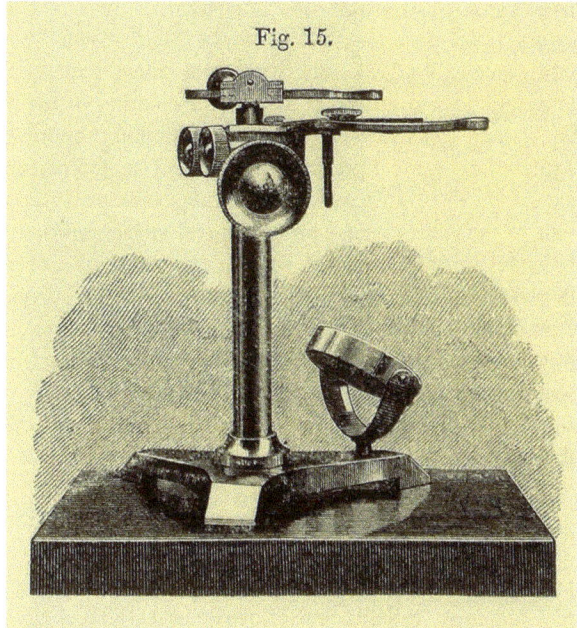

Fig. 15.

**Fig. 6.5** John Quekett. *Practical treatise on the microscope, including the different methods of preparing and examining animal, vegetable, and mineral structures.* London: H. Baillière, 1848. Frontispiece: compound microscope. Source: Wellcome Collection

Brewster's 1830 article on "Microscope" in *The Edinburgh Encyclopaedia*. He offers a set of eight "rules" to be "laid down" for microscopical observation (Brewster 1830, 228). The list format shows a methodical treatment of the subject and doubtless contributed to the recirculation of his rules in later texts. However, Brewster's rules do not provide a systematic approach to practice. Instead, they explain how optical principles (a subject he had researched) affect the microscopical image in certain ways. Some rules discuss illumination, others address focus; half of the rules discuss physiological vulnerabilities of the eye that vary with the physical angle of the head, such as the waves of corneal fluid that distort the visual field when the head is tilted to look through the eyepiece. Although Brewster presents them as "rules," their miscellaneity and restricted scope frame them more as tips than a comprehensive method of practice. They were circulated widely, perhaps aided by the compact, accessible list format.

Other early Victorian microscopists spur the mid-century interest in practice, but they do not call for a systematic method of observation. The 1832 *Microscopic Cabinet*, published by the optician Andrew Pritchard and physician C. R. Goring, helped to establish a science of microscopy with extensive discussions of error. In their most significant discussion, they present a series of numbered "aphorisms" and propose a system of test objects (difficult-to-resolve objects) that could help identify the most reliable lenses (Pritchard 1832, 175, 135). A final chapter titled "Miscellaneous Fragments" includes practical instruction on the preparation and exhibition of objects but (as the title suggests) without any systematic approach to microscopical practice or the prevention of error. (227)

John Quekett's *Practical Treatise on the Microscope* (1848) was one of the most influential texts in mid-century microscopy, centered outright as it is on practical instruction. He notes that "since the works of Sir D. Brewster, Dr. Goring, and Mr. Pritchard, no treatises of a practical nature have been published in this country" (Quekett 1848, vii). His book offers a wealth of instruction on matters of slide preparation and manipulation of the apparatus, with accompanying notes on potential difficulties of practice. Quekett's manual is itself carefully organized by types of apparatus (Culpeper's compound microscope, Wollaston's doublet, Lieberkühn reflector, stage micrometer, Valentin knife), specific microscopical objects, and test objects, with italicized subject headings for some paragraphs ("Animalcule Cages," "Stage Micrometer") (114, 191). However, the subjects of method and error arise incidentally within these discussions rather than being addressed as organizing principles. As with Pritchard and Goring, most of Quekett's attention to error has to do with optical aberrations rather than practice. He quotes Brewster's rules on illumination and includes a final chapter of "Miscellaneous hints on the management of the microscope and microscopic preparations," (187–88, 447). This chapter largely addresses how to store, care for, and organize the instrument and its accompanying equipment. One mention of error appears here in passing: the problem of "nap or down," left on a slide from cleaning, being "mistaken for highly organized structures" (448). Overall, Quekett provides a valuable and influential compendium of practical instruction but the book does not emphasize the need for a methodical approach to practice in order to prevent error.

**Fig. 6.6** Jabez Hogg.*The microscope: Its history, construction, and applications. Being a familiar introduction to the use of the instrument and the study of microscopical science.* London: Illustrated London Library and W. S. Orr, 1854. Frontispiece with alternate title page. Source: History of Science Collections, University of Oklahoma Libraries

In *The Microscope* (1854), Jabez Hogg relies heavily on other writers' work, attributed and unattributed, including that of Quekett (Dolan 2019). He explicitly frames the book as a "*popular* account of the Microscope," with elaborate illustrations to attract lay readers (Fig. 6.6), but the volume was often cited as a good resource for students (v, his emphasis). He includes a practical chapter titled "Directions for using the instrument" that collates other microscopists' advice on practice, reproducing (among other things) Brewster's "excellent directions for viewing objects" (Hogg 1854, xiv, 51). His other instructions are organized by apparatus and aim, without any overarching concern with method or error.

Dionysius Lardner, a controversial popular lecturer and author, offers some practical instruction in the chapters on vision and microscopy that he published in *The Museum of Science and Art* (1854–56), a 12-volume reference work for general readers. In 1856, apparently hoping to capitalize on a new, broader interest in the instrument, he reprinted material from his articles as a pamphlet on "the Microscope," which was bound in an eye-catching cover featuring a gilt microscope. Lardner addresses some errors of practice, such as the "foreign matter" that would "injure the observation," as well as difficulties specific to the microscope, such as the "curious, complicated, and important" role of illumination, the varying effects of which can cause "the most fatal errors and illusions" (Lardner 1856, ¶40, 43). But he treats these issues as they arise, not in any systematic approach to methodical practice.

## Practical Method: British

Although Goring, Pritchard, Quekett, Hogg, and Lardner instruct readers on practical technique, their books do not emphasize the need for methodical work nor do they present any overall method of practice. British writers' relative inattention to method changed over the course of the 1850s, although this change, when it occurred, was uneven. The London *Literary Gazette* had actually complained in an 1849 review of Schleiden's *Principles* that "this highly interesting and valuable part of the work" had been greatly abridged in the translation (Summary 1849, 480). But other British microscopists were still not much interested in method; 2 years later, a *Literary Gazette* review of the third edition of Schleiden's *Grundzüge* remarks upon the Methodological Introduction only to say that it "will perhaps seem somewhat out of place, or at least supererogatory to an English reader," who may consider that it has "no immediate reference to the study of botany" (Reviews 1851, 478). British interest in techniques of preparation and manipulation eventually surged as investigators, impressed by practices on the Continent and hoping to distinguish British microscopy as well, strove to develop a science of microscopic practice.

The influence of Continental microscopists can be seen in the work of John Hughes Bennett, a physician who became an influential Victorian advocate for medical microscopy and who was an early British evangelist for methodical work. Bennett's life as a young physician was strongly shaped by Continental approaches to medicine and microscopy. He trained in Paris and Germany for 4 years after his 1837 medical degree, in Paris focusing on medical microscopy and clinical instruction. He studied with Donné and Gruby and was friendly with Mandl (Bennett 1841, 6–7). His obituary in the *BMJ* notes that the "clear, precise, methodical style pursued by the French physicians" directed Bennett's own lifelong habits of practice and instruction (Obituary 1875, 473). Bennett became an influential figure in British medicine; he was friendly with major figures in human anatomy and physiology, such as William Sharpey, Martin Barry, and the Goodsir brothers, as well as researchers in botany such as Balfour and Edward Forbes. His work is amongst the earliest texts describing Continental clinical and microscopical practice for English-speaking readers. Jacyna has discussed how, partly due to Bennett's efforts, Edinburgh became an early center of British medical microscopy (Jacyna 2001).

In 1841 Bennett published *On the Employment of the Microscope in Medical Studies*. This brief (22-page) essay was the first lecture for a course in histology that he offered for medical students at the University of Edinburgh; a version of this lecture was also published in the *Lancet* in 1845. The *British Medical Journal* comments that although others were teaching histology at the time, Bennett's course was "the first to teach the use of the microscope systematically," and in all his courses, "the leading idea which prevailed in his mind was to teach … method" (Obituary 1875, 473, 475). Bennett claims as inspiration the fact that "in Germany… numerous private courses of lectures are given" on microscopical technique (Bennett 1841, 6). Indeed, *On the Employment of the Microscope*, drawing from Bennett's experiences in Paris and Germany, calls for microscopists to pursue method above all else, a year before Schleiden would do so in his *Grundzüge*. This essay thus represents an outlier in British writing of the 1840s.

Like Schleiden, Bennett argues that the microscope is less at fault than the observer, calling for microscopists to develop a *trained eye*. He says, "the contradictory statements of former microscopic observers arose more from their want of knowledge, than from any fault in the instrument" and that "the sense of sight as applied to the microscope must undergo a new education" (1841, 11, 13). He likewise argues that microscopical work requires *patience and perseverance*, warning, "In no case can this be accomplished without practice and industry" (13).

However, Bennett's approach differs from Schleiden's in important ways. Rather than urging observers to *collate multiple individual views* on an object, he emphasizes integrating one's own efforts with that of earlier authorities. Although he remarks on "how necessary it is for individuals to examine for themselves," his approach is not entirely empirical or inductive (11). He advises the observer to examine "previous opinions and doctrines" in the field, to "distinguish and separate what is false" from what is true, then "unite them with the result of his own labours, and thus from the whole build up a system of pathology" (27). Most important, however, is that despite Bennett's rhetorical emphasis on the importance of method, this introductory lecture—as printed—does not provide any system to ensure it. He says, for example, that dust and debris cause error but mentions this only in passing and does not integrate this concern into any kind of general rule (6, 12). Bennett did apparently draw up a list of rules for practice; he says, "At the close of this lecture, I shall indicate those rules which it is necessary should be observed, in order that every individual present may successfully view the object placed before him" (1841, 6). However, he did not include these rules in the published version of the lecture, which undercuts its proclaimed dedication to method.

The 1850 serialization of Bennett's *Lectures on Clinical Medicine*, which includes a version of this lecture on the microscope, reached a much wider audience than the little pamphlet of 1841. This later text suggests he continued to be ambivalent at best concerning detailed instructional guidebooks of the sort that others had published, but his advice does align with Schleiden's in most ways. He recognizes the need for instruction on practical matters but argues that "[t]he art of demonstrating under the microscope is only to be acquired by long practice, and, like everything requiring practical skill, cannot be learnt from books or systematic lectures" (Bennet 1850, 196). De Chadarevian has shown how Schleiden's book also paradoxically recommends practice over book reading (536). Despite Bennett's caution about book-learning, his own book became one of the most well-known Victorian texts on medical microscopy, reaching multiple editions in Britain and America.

In line with this concern, Bennett offers only "very general directions" for practice: one page on preparation, about three pages on "how to observe with a microscope," with the caveat (echoing Schleiden) that "the art of observation is at all times difficult, but is especially so with a microscope." Here he does provide a list of "the physical characteristics which distinguish microscopic objects" (shape, color, edge, size, transparency, surface, contents, and effects of reagents), laying out a basic structure for methodical observation. He endorses the idea of *method* ("rigid and exact observation... methodically cultivated from the first") and introduces the book by emphasizing repeatedly the need for a *trained eye* in any kind of observation, since "the senses require to be educated before they can receive proper impressions" (Bennet 1850, 196-97, 6). However, in this edition he again does not specifically introduce the notion of *multiple views* or perspectives on an object.

By the mid-1850s, British microscopists were publishing more work arguing for methodical practice. Lionel Beale became one of the most cited of these authors, who both influentially promoted method and wrestled at first with how to enforce it, given that microscopical work embraced such a vast and diverse array of objects. In his early book, *The Microscope and its Application to Practical Medicine* (1854), he offers a technical treatise with instruction in practical matters, organized by task and type of specimen. He cautions readers multiple times to practice cleanliness, warning them not to mistake environmental fibers (dust) for "tube casts" in the urine, or misidentify treated tissues as "morbid growths." He warns the reader that these will "run the risk of bringing discredit not only upon himself as an observer, but also upon microscopical investigation generally" (Beale 1854, 210, 124; see also 41 [Fig. 6.7]).

However, these comments are pegged to specific tasks, not in a set of systematic cautions. In fact, Beale cites Hannover as a recommended source on microscopy, which may account for why he echoes Hannover's approach in some ways in this early text (Beale 1854, 24). (He cites Schacht as a general source on microscopy but does not include him in the list of recommended sources [94]). Beale does remark that "[m]any substances should be examined in two or three different media" to compare the appearance in each, here encouraging multiple approaches to the object, as Schleiden and Schacht do (74). However, Beale (like Hannover) argues, "It is extremely difficult to lay down rules which will enable the observer to tell which method is best adapted to display the microscopical characters of any particular substance to the greatest advantage," and "[i]t is very difficult to lay down rules which will enable the observer to choose a preservative fluid for any particular specimen" (75, 81). The phrase "which method is best adapted" suggests that the observer should try to identify a single best approach to viewing any substance—although it is often difficult to do so. Again, he offers "general rules to be observed in Injecting" even as he backs away from them, saying that "no absolute rules can be given" about the timing of such interventions (109). Overall, this early book reveals Beale struggling with an ambivalence about method. He recognizes that "general rules" might improve the quality of research but is reluctant to enforce any absolute system due to the complex nature of microscopical work and the dizzying variety of substances to be analyzed.

**Fig. 6.7** Lionel S. Beale. *The microscope and its application to practical medicine*, 2d edn. London: Samuel Highley, 1858. Extraneous substances, p. 299. Source: Hathitrust/ Internet Archive: University of California

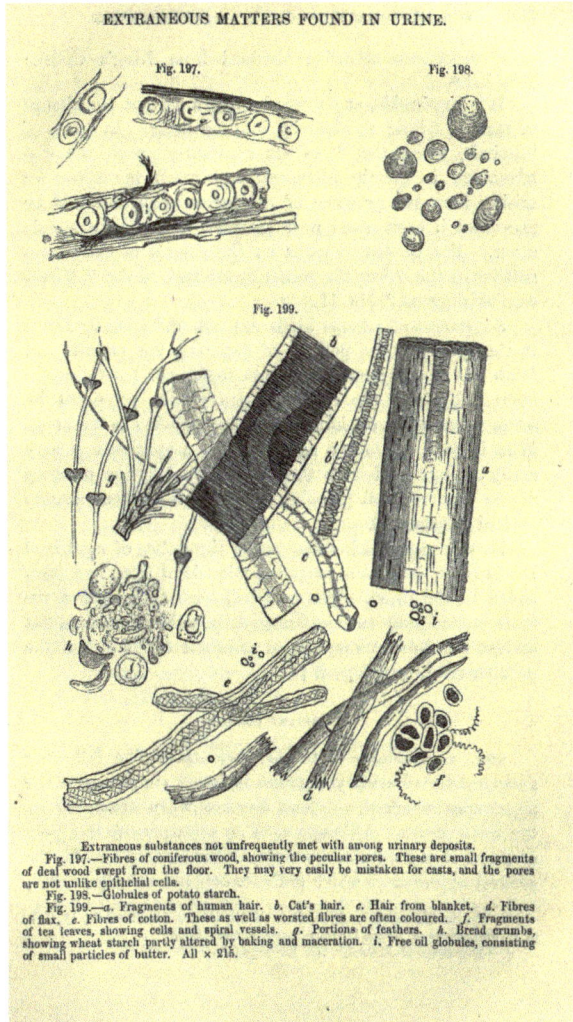

EXTRANEOUS MATTERS FOUND IN URINE.

Fig. 197.

Fig. 198.

Fig. 199.

Extraneous substances not unfrequently met with among urinary deposits.
Fig. 197.—Fibres of coniferous wood, showing the peculiar pores. These are small fragments of deal wood swept from the floor. They may very easily be mistaken for casts, and the pores are not unlike epithelial cells.
Fig. 198.—Globules of potato starch.
Fig. 199.—a. Fragments of human hair. b. Cat's hair. c. Hair from blanket. d. Fibres of flax. e. Fibres of cotton. These as well as worsted fibres are often coloured. f. Fragments of tea leaves, showing cells and spiral vessels. g. Portions of feathers. h. Bread crumbs, showing wheat starch partly altered by baking and maceration. i. Free oil globules, consisting of small particles of butter. All × 215.

Three years later, Beale published a general treatise *How to Work with the Microscope* (1857) that recapitulates some of the ambivalent remarks from his earlier book. Here, however, he lists both Hannover and Schacht amongst his list of recommended texts, and he demonstrates significant alignment with all of Schleiden's (and Schacht's) main tenets. Like Schleiden and Schacht, he puts particular emphasis on method and error. Beale structures his practical instruction as an elaborated system. His paragraphs are labeled with italicized headings and in some cases (as in his list of types of illumination) numbered (Beale 1857, 14–15). He even provides an appendix with a numbered list of microscopical apparatus, each item also pegged to specific pages of description in the book. He explains that "by transmitting" this number "to any instrument-maker, the observer will be furnished with the apparatus required" (viii). Beale dedicates a section of his final chapter to collating "Fallacies to be guarded against in microscopical observation" (98). Here he warns (for example) that it is difficult to make out the true structure of transparent or very delicate structures, or that "Extraneous Substances" may mislead the observer, providing a list of such substances for practice (101, 102). Concerned about such "errors of observation," Beale argues, "The eye of the observer requires much careful education before he is able to appreciate fully the character of the structure" before him (99, 97). He here proposes a *trained eye* as a requirement for microscopical work.

In this book, Beale repeatedly encourages observers to test specimens by means of many different interventions in order to acquire *multiple views*. Beale explains that "muscle is composed of several elementary structures, each … differing from the others in physical characters." Thus any particular preparation will show up different structures in muscle tissue (Beale 1857, 48). He later dedicates part of a chapter to elaborating this concept, titling it "Of the Importance of Examining the Same Objects in Various Ways"—specifically, by reflected, transmitted, and polarized light; mounted in air, water, glycerine, oil, turpentine, or Canada balsam; with various chemical reagents; and with high and low powers of magnification. To model this process, Beale works through the appearance of an object ("spherical crystals of carbonate of lime") under these different conditions (57–58). He later adds, "by a consideration and comparison of the different facts observed" through these diverse interventions, "one is enabled at length to embody the results arrived at in several different inquiries, and form an idea of the real structure of the part" (100). Finally, Beale emphasizes *patience and perseverance* in microscopical practice, noting that this process of "examining a specimen under very different circumstances" requires "a very long and patient investigation" (97). Like Schleiden, Beale sets up a series of exercises for the novice to practice various interventions (57–58 and Table II, 105). Beale and Schacht also emphasize the importance of note-taking and sketching during observation and offer extended instruction on these matters.

In another classic text, John William Griffith and Arthur Henfrey included a 32-page section on how to use the microscope in their *Micrographic Dictionary* (1856). Here they endorse the idea of *multiple views*, pointing to the errors that arise when the microscopist is satisfied with "simple inspection" (Griffith and Henfrey 1856, ix). Like Beale, they demur that "[a] general method for determining the structure of objects can hardly be laid down; it must vary so greatly" from one object to another. However, they then go on to lay out an elaborate, nested system of methodical practice, stating that "[t]he examination of a microscopic object must comprise,—*a*, the *microscopic analysis*, including,—1, the form; 2, the color; 3, the structure of the surface; and 4, the internal structure: *b*, the *histological analysis*… *c*, the *qualitative chemical composition*; and *d*, the *measurement*" (xxviii, their emphasis). Many of these elements are subdivided again in the exposition that follows; for example, "2. *The Color*" is particularized into four subheadings, one of which is further divided into a and b (xl). The introduction concludes with a précis of these elements in outline form that "may serve to recal [sic] to the observer the most important points to be looked for" (xl). This serves as a template for what

Schleiden had called many-sided observation. Griffith and Henfrey touch on other elements mentioned by Schleiden and others, such as the notion that the microscope is innocent or the problems of monocular vision (x, xi). "Above all," they urge, "it must never be forgotten, that microscopic investigations require more time and patience than perhaps any others" (xi). Overall, this Introduction "enter[s] minutely into … the special education of the eye" needed for microscopical vision (iv). In their emphasis on method and their template for multiple views especially, Griffith and Henfrey demonstrate how British texts shift toward a Continental approach to microscopical practice—one essentially similar to Schleiden's—in the 1850s.

William Benjamin Carpenter, who had trained at Edinburgh, published probably the most influential Victorian text on microscopical practice with *The Microscope and its Revelations* (1856), but his approach to method there is less systematic than that of Schleiden or Schacht, Beale, or Griffith and Henfrey. Although the book itself is well-organized, even to the numbering of paragraphs, Carpenter does not overtly present method as a framing category, as the others do. Indeed, he uses the word only to describe specific practices. However, he does implicitly call for better method in order to make British microscopy more productive, and his advice accords with Schleiden's overall. In his Preface, he argues that "there is a large quantity of valuable *Microscope-power* at present running to waste in this coun-try,—being applied in such desultory observations as are of no service whatever to science, and of very little to the mind of the observer." Accordingly, he hopes that "this Manual" will "tend to direct this power to more systematic labours" (Carpenter 1856, viii, his emphasis). Carpenter's vision—an efficient cadre of microscopists engaged in "systematic labours" with their instruments—implies a devotion to greater method, although he does so only indirectly, by proposing this factory-like ideal.

Similarly, Carpenter does not explicitly call for a trained eye but rather leaves this aim implicit in many passages, where his concerns often overlap with those expressed by Schleiden and others. For example, he warns about "misinterpreta-tions … peculiar to" microscopical vision due to diffraction and other optical effects, air-bubbles, monocular vision, Brownian motion, a restricted field of vision or narrow plane of focus, and the like (Carpenter 1856, 185). While authors like Schleiden, Schacht, Griffith and Henfrey, and Beale had provided tables or lists of exercises for the student to work through, Carpenter simply informs readers that "[n]o rules can be given for the avoidance of" some of these errors, "since they can only be escaped by the discriminative power which education and habit confer." However, he says later, "No experienced Microscopist could now be led astray by such obvious fallacies as those alluded-to; but it is necessary to dwell upon them, as warnings to those who have still to go through the same education" (Carpenter 1856, 186, 189). These comments mention "education and habit" almost in passing, as aims that both author and reader take for granted, without a systemic outline of how to achieve them.

Finally, Carpenter does not emphasize the value of multiple views as a general approach to the difficulties of microscopical vision; yet again he offers a partial or tacit recommendation rather than a full-throated one. He recommends a version of

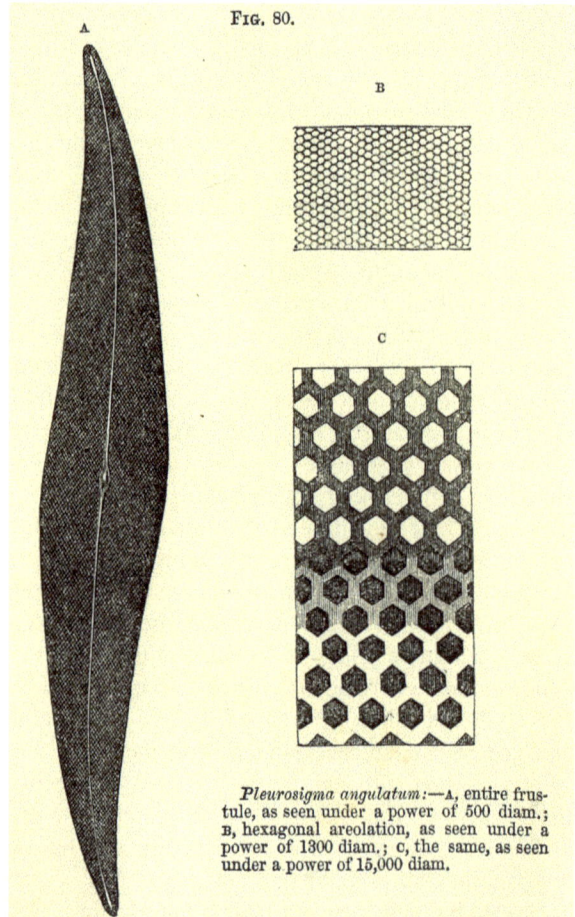

FIG. 80.

*Pleurosigma angulatum:*—A, entire frus-
tule, as seen under a power of 500 diam.;
B, hexagonal areolation, as seen under a
power of 1300 diam.; c, the same, as seen
under a power of 15,000 diam.

multiple views "in all doubtful cases" such as delicately patterned test objects. With such objects, the observer should "have recourse to every variety of oblique illumination that shall present the object under a different aspect" and "examine the object in several different positions, so that the appearance it presents in each may be compared" (Carpenter 1856, 129–30, 177). His illustration of the diatom *Pleurosigma angulatum*, a common test object then thought to be a "vegetable," demonstrates the different appearance of this object under changes of magnification and focus (244) (Fig. 6.8). Similarly, when using a polarizing apparatus, observers should seek to obtain "the greatest variety of coloration" (137). Carpenter explains that polarized light can reveal unsuspected structures and advises the observer to use it at certain times—"whenever he may have the least suspicion that its use can give him an additional power of discrimination" (178). However, Carpenter does not propose a multiple-view philosophy of observation for general purposes as Schleiden and Beale do. If anything, he directs observers rather to use a specific magnification: "the *lowest* power with which the details of structure can be clearly made-out" (162, his emphasis). This, however, would require multiple views to determine. Overall, Carpenter was as concerned about error as any of the earlier authors—he dedicates many pages to the exposition of specific errors—but he offers his advice more diplomatically and with great attention to counterarguments and to distinctions between specific objects or contexts. In discussing the difficulty of microscopical vision or the fatigue of the eye, for instance, he is careful to consider opposing views before presenting his own nuanced conclusions (9, 158).

Carpenter thus endorses many of the elements of method proposed by Schleiden and Schacht, but he often does so obliquely. One exception is the axiom, which he emphasizes, that microscopical observation requires patient perseverance. He promises that microscopical science would prosper if only "those Microscopists who spend their time in desultory observation… would but concentrate their attention upon some particular species or group, and work-out its entire history with patience and determination" (Carpenter 1856, 25). Indeed, he argues, "[T]he value of the results of Microscopic enquiry will depend far more upon the sagacity, perseverance, and accuracy of the Observer, than upon the elaborateness of his instrument" (35). The length of Carpenter's manual—about 800 pages—suggests just how much the microscopist must labor to master the practice.

## British Responses to Schleiden and Schacht

Schleiden was generally respected for his early work on cell theory. He was even mythologized for this work: a reader of Charles Dickens's periodical *Household Words*, for example, would read that Schleiden "travelled through a tangled forest of prickly and entwined facts, till at last he saw the light, and could proclaim it" ([Hart] 1857, 511). His essay on method should have been valuable as a window into the habits of practice that supported such work. But British microscopists writing after the publication of the *Grundzüge* largely ignore Schleiden's work on method even as they cite him on cell theory; on method, they tend to cite his student Schacht instead.

Some 1850s writers on microscopical method (Bennett, Hannover) cite neither Schleiden nor Schacht on research or method. However, the others I've examined here cite Schacht but not Schleiden, or they cite Schleiden on plant structures but not on method. Hogg cites Schleiden on specifics of plant structure; but on method (the outline of chemical reagents), he cites Schacht at length and does not mention Schleiden (Hogg 1854, 414, 421; 83-85). Beale does not cite Schleiden in either book, but he cites Schacht (the English translation) as a general reference in both books (Beale 1854, 94, fn; 1857, 4, fn). Carpenter cites the research of both Schleiden and Schacht but he does not cite either on method, referring readers instead to Quekett, Beale, and Griffith and Henfrey (Carpenter 1856, 12, 457, 459, viii). In an especially telling case, Henfrey, an active translator and editor, was an expert on plant anatomy who would have been familiar with Schleiden's work—he had translated a book of essays by him, *Die Pflanze und ihr Leben* (1848), that

**Fig. 6.9** Matthias Jakob Schleiden. *The Plant; a biography. In a Series of Popular Lectures*. Trans. Arthur Henfrey. London: Hippolyte Bailliere, 1848. Color frontispiece/alternate title page. Source: Hathitrust/Internet Archive: University of California

addressed some aspects of microscopical method. Henley's translation of that book (*The Plant, a Biography*) was published with the publisher Hippolyte Baillière also in 1848. *Die Pflanze* advertised its popular aims with attractive illustrations, a few color plates (opening with a lavish, colorful frontispiece), and flowery prose (Fig. 6.9)—but the first two chapters focus on questions of optics and microscopical manipulation (Fig. 6.10), so Henfrey would certainly have known of Schleiden's interest in method. Henfrey also helped Lankester in translating some portions of Schleiden's *Grundzüge* (Lankester 1849, iv). Griffith and Henfrey cite Schleiden's research, referencing both the *Grundzüge* and its translation. However, they do not refer to Schleiden's work on method, despite Henfrey's certain knowledge of it. Instead, their short reference list of general works on "Microscopes, Apparatus, and Observation" cites (among others) Quekett, Brewster, Ross, Dujardin, Mandl, Chevalier, von Mohl, Carpenter, and Schacht (Griffith and Henfrey 1856, xl).

The same pattern appears in books that, while they do not substantively address method themselves, provide a list of sources on microscopical practice. For

**Fig. 6.10**  Matthais Jakob Schleiden. *The Plant; a biography. In a Series of Popular Lectures*. Trans. Arthur Henfrey. London: Hippolyte Bailliere, 1848. Chapter heading for Second Lecture, p. 38: microscope apparatus. Source: Hathitrust/Internet Archive: University of California

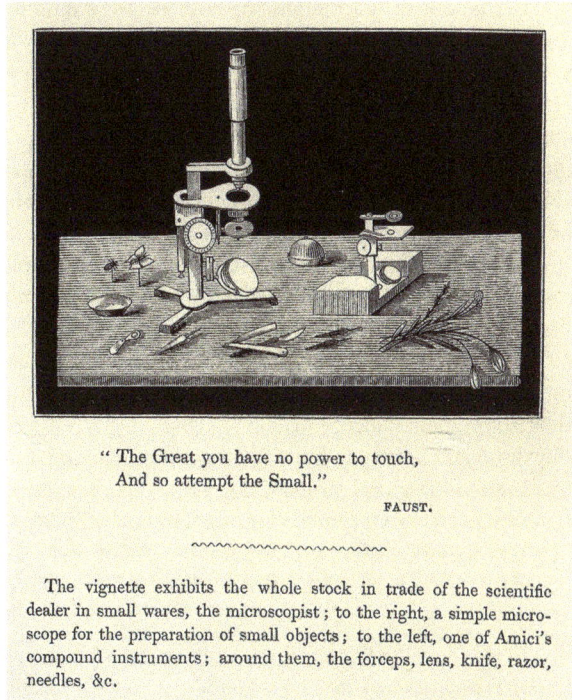

" The Great you have no power to touch,
And so attempt the Small."

FAUST.

The vignette exhibits the whole stock in trade of the scientific dealer in small wares, the microscopist ; to the right, a simple micro-scope for the preparation of small objects ; to the left, one of Amici's compound instruments ; around them, the forceps, lens, knife, razor, needles, &c.

example, the *Outlines of Physiology* published by Bennett's colleague Allen Thomson in 1848 includes eight pages on practice, mostly discussing apparatus. On method and practice he recommends texts in English, French, and German by Brewster, Carpenter, Purkinje, Mandl, Dujardin, Henfrey, and (not yet published) Quekett, but not Schleiden's *Grundzüge* (90). Similarly, the botanist and physician John Hutton Balfour published a little book in 1860 titled *The Botanist's Companion: Directions for the Use of the Microscope, and for the Collection and Preservation of Plants* (although it largely discusses the construction of microscope lenses and specific makes of microscope rather than practice). Balfour lists a number of "works" to be "consulted by the student" seeking practical instruction. He includes (among others) Quekett, Carpenter, Hannover, Beale, Hogg, Ross, Bennett, Griffith and Henfrey, Pritchard—and Schacht but not Schleiden (Balfour 1860, 40). More specifically, introducing his short section on practice, Balfour notes, "The following details are partly condensed from Schacht's treatise on the microscope, and from the works of Hannover, Quekett, Jabez Hogg, and Beale" (18). Although Balfour credits Schleiden for a "method of preserving minute structures," he does not mention Schleiden's instructions on method (28).

Why do British microscopists of the 1850s seem to ignore Schleiden's work on method, citing Schacht's version of that work instead? The Methodological Introduction in Schleiden's *Principles* should have been a very useful resource for these writers, who hoped to stimulate a more meticulous and thorough empirical practice amongst Victorian microscopists. They should certainly have been aware of the Methodological Introduction, as the *Principles* was widely reviewed. Henfrey and Balfour were botanists with presumably the most interest in the *Principles*, but Hogg, Carpenter, and Griffith and Henfrey all discuss botanical matters at length, and Beale, as a fervent vitalist, would have been closely following debates on cell theory. As these men were all prominent microscopists, they would likely have sought access to such an important text and may even have written one of these anonymous reviews.

Furthermore, the reviews usually make some note of Schleiden's interest in method. For example, an 1849 essay review of botanical books in the *British and Foreign Medico-Chirurgical Review* is critical of a number of claims in Schleiden's *Principles of Botany* but concludes, "We regard the plan of his work, and the method of investigation on which it is based, as far superior to that of any other systematic treatise on the subject" (Article II, 1849, 348). Similarly, an 1850 essay review on physiological botany in the *Edinburgh Medical and Surgical Journal* comments, "If the [methodological] precepts of Dr. Schleiden be faithfully adopted... a most favourable change for the better ought to be the result. Facts, however small in number, will then be truly facts, and not the offspring of the fancy and delusions of the observer" (Article I, 1850, 420). And the *Eclectic Review* concludes an 1850 review of Schleiden's *Principles* by noting, "for the benefit of the microscopical student" that the book "contains in an appendix some good and ample instructions for the use of the microscope" ("Brief Notices," 1850, 781). While the latter review also condemns Schleiden's "infidel" approach to popular botany in *The Plant* (781), this issue should not have greatly affected British microscopists' willingness to cite the *Principles* on inductive practice and the importance of method.

Cost is unlikely to have greatly restricted the access to Schleiden's book for the working microscopists who were writing these manuals. Schleiden's *Principles*, at 21 shillings, did cost more than four times the price of Schacht's five-shilling manual. However, although some manuals (most of those noted above, in fact) were published more cheaply, the cost for the *Principles* is in line with books such as Quekett's *Practical Treatise*, which Baillière published for 21 shillings. Griffith and Henfrey's *Micrographic Dictionary*, which these texts often cite as a resource, cost more than twice that. Although unlike Schleiden's collaborative work with Schwann the *Principles* does not appear in the library of the Quekett Microscopical Club, it is listed in the 1860 library catalogue for the Microscopical Society of London (Searson 1860, 24). Since many of the British authors I examine here were members of the MSL, or were likely to be friendly with members, they would have enjoyed access to the Society library. Furthermore, the *Principles* was also available to any subscriber to Lewis's Medical and Scientific Library, a specialty circulating library in London ("Catalogue" 1888, 174).

One possibility is that the uptake of Schleiden's essay on method may have been depressed not just by any scientific errors in his work but also by his constant embroilment in scientific disputes over cell theory. Earlier in his career he had engaged in very public disagreements with colleagues about cell development. He had by 1849 modified his views to agree with other authorities on some matters, but he continued to differ loudly from other investigators on a number of matters. Many references to Schleiden demonstrate an attempt to temper praise with an acknowledgement of his disputatious ways. A review of the *Principles* in the *Annals and Magazine of Natural History* begins with the words, "Whatever may be the opinion as to the correctness of Professor Schleiden's views upon certain questions" about cell theory and plant structures, "there can be no doubt that he ranks among the first original observers of the present day, and this work is undoubtedly the most valuable systematic exposition" of structural botany at the time. This equivocation acknowledges Schleiden's frequent clashes with other investigators over points of structure—so many that, the reviewer explains, "[w]e cannot afford space… to enter upon the discussion of the many points on which Prof. Schleiden is at issue with many other celebrated botanists" (Bibliographical Notices 1849, 442).

This ambivalence about Schleiden's research, at once appreciative and guarded, appears in other microscopists' texts as well. Hogg acknowledges Schleiden's status as an investigator, but he does so by quoting the remarks of T. H. Huxley, who challenged his work. Huxley says, "To Schwann and Schleiden are we indebted" for presenting "the cell-theory as a colligation of the facts" that "if not absolutely true, … was the truest thing that had been done in biology for half a century" (Hogg 1854, 336). Such mixed praise certainly tempers the endorsement. Carpenter also both acknowledges the import of Schleiden's research and questions that work. He says, "It was by Schleiden that the fundamental truth [was] first broadly enunciated, that … the *life-history of the individual cell* is … the very first and absolutely indispensable basis… for Comparative Physiology in general" (Carpenter 1856, 12, his emphasis). However, Carpenter (like Huxley) prefaces this praise with a more qualified statement: "whatever may be the errors with which his [Schleiden's] statements … are chargeable, there cannot be any reasonable question as to the essential service he has rendered to science, in pointing out the way to others, on whose results greater reliance may be placed" (Carpenter 1856, 12). Such comments recognize Schleiden's innovative work while declining to endorse all his conclusions. His mixed record on scientific matters, and his eagerness to differ from his peers, might discourage readers from following his method too closely; yet these mid-century British writers do insist on the importance of Schleiden's work for botanical science.

Perhaps just as significant as Schleiden's willingness to dispute others' work is that in these conflicts he adopted a combative, sarcastic tone that chastises his fellow investigators for reaching conclusions different from his own. It may seem that in adopting this challenging, confrontational tone Schleiden simply aligned himself with his countrymen's approach to what Raf de Bont calls "the etiquette of disagreement"; at the time German science was tolerant, even encouraging, of open disagreement while in the British context gentlemanly civility was paramount (de Bont 2013, 310, 312-4). But Schleiden's outrage outstrips even German norms, and it colors much of the *Principles*. He grouses, for instance, "Almost all the works that have hitherto appeared on the lower *Fungi* are wholly useless, and may, without farther consideration, be cast aside, since the work must again be commenced from the beginning" (Schleiden 1849, 153). He dismisses a book by Theodor Hartig, saying that its "complete uselessness" derives from the author's "total ignorance of the literature of the subject" and his lack of "the necessary skill in manipulation" or even "a correct method" (410). He quotes a definition used by the eminent botanist Heinrich Friedrich Link, a botanist of an earlier generation, then mocks it with a reference to mechanical technologies of mass publication, saying, "No science will be advanced in this way, but merely groundless chattering stereotyped" (262). And discussing the "milk-vessels" of plants, Schleiden fulminates that

> owing to the total neglect of a correct, scientific method, and the puerile sporting with hypotheses, without any foundation or guiding principle, the question respecting them is loaded with such a heap of nonsense, that the best way in beginning upon it is, in the first place, to throw overboard all that has hitherto been done and commence entirely *de novo*, instead of undertaking the thankless task of cleansing this true Augean stable. (114)

Schleiden launches his critique here with complaints about "neglect of a correct, scientific method," a theme that concerned him greatly. Indeed, the *Edinburgh Medical and Surgical Journal* review suggests that this vitriolic tone generally derives from Schleiden's commitment to scientific methodology. The reviewer comments, "he becomes angry at the immense number and rash unfounded character of the hypotheses in which botanists have indulged their fancy in speculating on this subject" (Article I 1850, 387). In other words, Schleiden's ire is aroused by the careless work of his colleagues, and his conviction that botany requires a more rigorous practice and mode of reasoning. The review notes a "remarkable" feature of Schleiden's treatise: its suggestion "that all botany hitherto taught has been worse than useless" in accumulating mothing more than "an immense and confused mass of badly determined facts, fictitious representations, rash inferences, and unfounded hypotheses." In Schleiden's view, the reviewer explains, "None has observed carefully; none has reasoned correctly," with but a few exceptions (Article I, 1850, 419–420).

It is evident that Schleiden's quarrelsome prose affects the scientific reception of his work. Multiple reviews condemn his practice of lashing out at other investigators. In a review of the state of "physiological botany" in 1845, Link calls out Schleiden and two others (Justus von Liebig and Charles Gaudichaud-Beaupré) for the "violence" of their phrasing and tone, which is "injurious to the cause" of a more scientific botany (Link 1849, 193). (As noted above, Link had been swept up in some of these attacks). Schleiden, Link says, "is the most violent" of these, condemning other perspectives almost automatically and "trampl[ing] upon all who differ from him." Schleiden has thus "excited all botanical writers against him." The first edition of the *Grundzüge*, indeed, "was not incorrectly called a libel," and while the second edition is better, "the same violence is manifest." Link warns Schleiden that his love of controversy does not serve him well: "[w]here there is a noise, there flock boys and idlers" (206, 208–09). Despite the import of Schleiden's work, the reputation of his *Principles of Scientific Botany* was damaged by his fiery invective.

Scientific work has long been regulated by norms of civil discourse and community (Shapin 1994). Appropriate discourse was an important element of the British "gentleman of science," a role that was in transition during these years of increasing attention to method. James Secord has noted that "[t]he status of the practitioner of knowledge was at a point of maximum flux in the 1840s" but "virtues of disinterestedness, piety, and independence were paramount." Scientific speech should, then, be both "polite and authoritative" (Secord 2001, 405–06, 421). Others who transgressed these implicit principles had run up against resistance to their work, as with the embryologist Barry's exuberant claims (see Jacyna 2003, 49).

In the *Principles* Schleiden repeatedly violated these gentlemanly norms. British reviewers called out his scathing critiques, saying that his textbook added to botanical knowledge but debased the scientific conversation. The *British and Foreign Medico-Chirurgical Review*, for instance, endorses the *Principles* overall but concludes by both "express[ing] our dissent" from his views and "condemn[ing] the spirit and temper in which he has put them forth." (Article II 1849, 348). The *Literary Gazette* offers a generally favorable review of the *Grundzüge*, but it decries Schleiden's "unseemly violence of language" in criticizing his colleagues, many of them "worthily distinguished men." The reviewer is especially perturbed that this polemical language persists in later editions of the treatise, which could have been amended. The review concludes, "We regret that the author has allowed some of these passages to appear in a scientific treatise, and especially that he has retained them in the third edition of a work that must be in the hands of every botanist who would master the existing treasures of his science" (Reviews 1851, 479). The sentence twists around at the last moment to offer a favorable assessment of the text, but its condemnation remains: Schleiden's hotheaded insults sully his botanical research and the work of science overall. Similarly, the *British and Foreign Medico-Chirurgical Review* condemns the "tone and spirit" of Schleiden's prose, which are "totally unworthy of a man who occupies a high scientific position, and who professes to make truth his sole aim" (Article II 1849, 326). Reviewing the *Principles*, the *Literary Gazette* complains that while Lankester omitted portions of the Methodological Introduction, he retained Schleiden's "ungenerous vituperations" of colleagues both alive and dead, "which so much disfigure the work." Indeed, the reviewer notes, Schleiden himself argues in his original Introduction that "censure" should be avoided "because, if undeserved, it proves the incapacity of him who applies it, while, in most cases, it discloses his dishonesty and want of truth." His "sad exhibition of temper" is extraordinary "in any modern scientific publication" and must be "utterly repugnant to the public taste" (Summary 1849, 480). This mixed reception of Schleiden's work may even have colored Darwin and Huxley's response to Ernst Haeckel's *Generelle Morphologie* (1866), which they respected but thought "unsuited for the British market" in part due to its "aggressive style" (de Bont 2013, 326).

Perhaps most galling, Schleiden's harsh criticisms extended to British microscopes. After praising specific German, Austrian, and French microscope-makers, he remarks, "The more recent English instrument seem to be so much inferior to those… that they bear no comparison." He admits that "there certainly is no lack of clever observers in England" but "no microscopical botanical researches of any importance" save those of Brown. He concludes that "what observations the English have made can frequently be easily refuted by a cursory glance through our glasses," so "this deficiency can only be attributed to the defectiveness of their instruments." The translator Lankester counters this calumny in a footnote, arguing that in fact "many improvements have taken place" in microscope construction since 1842, when Schleiden first published the *Grundzüge*; and "no country in Europe" can claim as many such technological improvements as can England. The problem, as Lankester sees it, is "the reverse": "we have plenty of microscopes, and those the best in the world; but we have had but few observers," because the fine English microscopes "have been used rather as playthings than as the instruments of profound philosophical research" (Schleiden 1849, 577–78).

Many British reviews similarly point to Schleiden's dismissal of British microscopes as unfair. The *Edinburgh Medical and Surgical Journal* condemns his remarks as "unjustly severe," and the *British and Foreign Medico-Chirurgical Review* charges that his statements "show the author's ignorance of the present state of microscopes and of microscopic knowledge in this country" (Article I 1850, 420; Article II 1849, 334). Such defensive comments show that Schleiden's casual dismissal of English instruments had been understood as an attack. The quality of English microscopes was a point of pride for many British workers, and Schleiden's remarks must have stung.

Schleiden's critique embraces English practices as well as English instruments. At this time many Continental investigators worked with relatively low powers, while the most prestigious English makers (Ross, Smith & Beck, Powell & Lealand) were producing elaborate instruments and competing to construct the highest-powered objectives. Schleiden, citing as "the most distinguished opticians" only Continental makers, argued that high-powered lenses promote speculation rather than clear vision—that anything visible in such lenses that cannot be seen at lower powers must be "mere imagination" (Schleiden 1849, 579). Griffith and Henfrey defend both English makers and high-powered object-glasses against these criticisms. They argue that "these statements may be very true in regard to German object-glasses as used by Germans; but they do not apply to the English object-glasses as used here." Indeed, "we feel convinced that if Schleiden were to try to obtain a view of the hexagonal structure of the dots on the valves of a *Gyrosigma* [a test object] with his object-glasses, he would signally fail" (Griffith and Henfrey 1856, xv–xvi). This response represents British investigators' nationalistic concerns that Continental workers were far outstripping their own efforts, and it doubles down on the fascination with viewing test objects through high-powered objectives that was thought to characterize British work. Their insistent, almost hostile tone doubtless rises to meet Schleiden's own belligerent rhetoric.

Schleiden's bombast stands in stark contrast to the approach demonstrated by writers like Beale and Carpenter, who—as I have shown above—share Schleiden's concerns about error and method. However, they frame these concerns very differently. Their adoption of a more civil discourse speaks to more than just contemporary concerns about the gentlemanly status of science. They also explicitly argue for the harm that error can do to both scientific knowledge and their fellow microscopists. Beale, for instance, warns readers to avoid hasty testimony because, by it, "errors have been propagated and strengthened to an extent almost incredible, and years of laborious investigation have been spent in overthrowing statements which had never resulted from actual observation in the first instance" (Beale 1857, 97–98). Beale's comments express frustration and sorrow about these squandered, fruitless years, presenting them as a hindrance to scientific progress but also just a sheer waste of what Carpenter calls *"Microscope-power,"* and of the human hours, days, weeks, and years that produced it. Carpenter is even more explicit on this matter, when he urges observers to delay announcing any discovery until quite certain of their results. "It is due to Science," he explains, "that it should be burdened with as few false facts and false doctrines as possible." Perhaps as important, "It is due to other truth-seekers, that they should not be misled, to the great waste of *their* time and pains, by *our* errors" (Carpenter 1856, 185, his emphasis). Such remarks forcefully remind readers of the collaborative nature of collective empiricism. In their view, investigators have a duty of care not only to the scientific project but to their fellow-workers as well.

## Conclusion

Schleiden may have thought he was presenting himself as a solitary, beleaguered researcher simply trying to practice and defend good science, but it appears many microscopists read the *Principles* as telling a very different story, one where an aggrieved bully attacks the conventions of scientific discourse and the developing international community of microscopy. With Schleiden's approach to microscopical observation tidily summarized in book form through Schacht, who was a much less controversial speaker, authors like Hogg, Beale, and Carpenter may have felt disinclined to publicly recognize Schleiden, someone so disagreeable and so offensively uninformed about the state of British microscopy, on matters where Schacht or others could be cited instead. However, these writers on method largely endorse the same tenets that Schleiden emphasizes in his Methodological Essay—patience and method, a trained eye, and the inductive practice of combining multiple views—even if they do not credit him for his emphasis on these practices or for his role as an early resource on Continental approaches to microscopy.

## References

Article I.—Outlines of structural and physiological botany (1850) Edin Med Surg J 74(185): 369–421
Article II. Principles of scientific botany (1849) Brit Foreign Med-Chirurg Rev 4(8): 324–348
Balfour [JH] (1860) The botanist's companion; or, directions for the use of the microscope, and for the collection and preservation of plants. Adam and Charles Black, Edinburgh
Beale LS (1854) The microscope and its application to practical medicine. Samuel Highley, London
Beale LS (1857) How to work with the microscope. A course of lectures on microscopical manipulation, and the practical application of the microscope to different branches of investigation. John Churchill, London

Bennett JH (1841) On the employment of the microscope in medical studies: a lecture introductory to a course of histiology [sic]. Maclachlan, Stewart, Edinburgh

Bennet JH (1850) An introduction to clinical medicine. Six lectures on the method of examining patients; percussion; auscultation; the use of the microscope; and the diagnosis of skin diseases. Sutherland and Knox, Edinburgh

Bibliographical Notices: Principles of scientific botany (1849) Ann Mag Nat Hist 2,4(24): 442–443

Brewster D (1830) Microscope. In: Brewster D (ed) The Edinburgh encyclopedia. William Blackwood, John Waugh, John Murray, Baldwin & Cradock, and J. M. Richardson, Edinburgh and London, 215–232

Brief Notices: Principles of scientific botany (1850) Eclectic Rev 2, 27(June): 780–781.

Carpenter WB (1856) The microscope: and its revelations. John Churchill, London

Catalogue of Lewis's medical & scientific library (1888) London, Lewis's Library

de Bont, Raf (2013) "Writing in Letters of Blood": Manners in Scientific Dispute in Nineteenth-Century Britain and the German Lands. Hist Sci 51(3):309–335

de Chadarevian S (1993) Instruments, Illustrations, Skills, and Laboratories in Nineteenth-century German Botany. In: Mazzolini RG (ed) Non-verbal communication in science prior to 1900. Olschki, Firenze, pp 529–550

Dolan JR (2019) Dueling charges of plagiarism in the mid-19th-century world of microscopy — who was the copycat? Linnean 35(2):11–18

Griffith JW, Henfrey A (1856) The micrographic dictionary; a guide to the examination and investigation of the structure and nature of microscopic objects. John Van Voorst, London

Hannover A (1853) On the construction and use of the microscope (trans. Goodsir, J). Sutherland and Knox, Edinburgh. Danish edition: Hannover A (1847) Om Mikroskopets Bygning og dets Brug. Reitzel, Kjøbenhvn.

Hart EA (1857) Nature's greatness in small things. Household Words 16(401):511–513

Hogg J (1854) The microscope: Its history, construction, and applications. Being a familiar introduction to the use of the instrument and the study of microscopical science. Illustrated London Library and W. S. Orr, London

Jacyna LS (2001) A host of experienced microscopists': The establishment of histology in nineteenth-Century Edinburgh. Bull Hist Med 75(2):225–253

Jacyna LS (2003) Moral fibre: The negotiation of microscopic facts in Victorian Britain. J Hist Bio 36(1):39–85

Lankester E (1849) Translator's preface. In: Schleiden MJ, Principles of scientific botany (trans: Lankester E). Longman, Brown, Green, and Longmans, London, p ii-iv.

Lardner D (1856) The microscope: From "The museum of science and art". Walton and Maberly, London

Link HF (1849) Annual report on researches in physiological botany, during the years 1844 and 1845. In: Henfry A (ed) Reports and Papers on Botany. Ray Society, London, pp 193–313

Obituary: John Hughes Bennett, M.D., F.R.S.E. (1875) BMJ 2(9 Oct): 473–478

Pritchard A, Goring CR (1832) The microscopic cabinet of select animated objects. Whittaker, Treacher, and Arnot, London

Quekett J (1848) Practical treatise on the microscope, including the different methods of preparing and examining animal, vegetable, and mineral structures. H. Baillière, London

Reviews: Principles of scientific botany (1851) Lit Gazette J Belles Lettres, Arts, Sci 1799(12 July): 477–479

Ross A (1839) Microscope. In: The Penny Cyclopaedia of the Society for the Diffusion of Useful Knowledge. Charles Knight, London, pp 177–188

Schacht H (1853) The microscope in its special application to vegetable anatomy (trans: Currey F). Samuel Highley, London. German edition: Schacht H (1851) Das Mikroskop und seine Anwendung inabesondere für Pflanzen-Anatomie und Physiologie. G. W. F. Müller. Berlin.

Schickore J (2007) The microscope and the eye: A history of reflections, 1740-1870. University of Chicago Press, Chicago

Schleiden MJ (1847) Contributions to phytogenesis. In: Schwann T, Microscopical researches into the accordance in the structure and growth of animals and plants (Mikroskopische

Untersuchungen, 1839) (trans: Smith H). Sydenham Society, London, 229–263. German edition: Schleiden MJ (1838) Beiträge zur Phytogenesis. Archiv Anat Phys wiss Med 5(2): 137–176

Schleiden MJ (1848) The Plant; a Biography. In a Series of Popular Lectures (trans: Henfrey A). Hippolyte Bailliere, London. German edition: Schleiden MJ (1848) Die Pflanze und ihr Leben: populäre Vorträge. Wilhelm Engelmann, Leipzig

Schleiden MJ (1849) Principles of scientific botany; or, Botany as an inductive science (trans. Lankester A). Longman, Brown, Green, and Longmans, London. German edition: Schleiden MJ (1842) Grundzüge der wissenschaftlichen Botanik: Nebst einer methodologischen Einleitung als Anleitung zum Studium der Pflanze. Wilhelm Engelmann, Leipzig

Searson WG (1860) Catalogue of the library of the Microscopical Society of London, 1859. Trans Micros Soc J 8(1):13–28

Secord J (2001) Victorian sensation: The extraordinary publication, reception, and secret authorship of Vestiges of the natural history of creation. University of Chicago Press, Chicago

Shapin S (1994) A social history of truth: Civility and science in seventeenth-century England. University of Chicago Press, Chicago

Summary: Principles of scientific botany (1849) Lit Gazette J Belles Lettres, Arts, Sci 1693(30 June): 480

Thomson A (1848) Outlines of physiology; for the use of students. Maclachlan, Stewart, Edinburgh

# Chapter 7
# Glimpses of the Colonial Collections at the 1862 London Exhibition: The Case of the Angolan "Objects" in the Portuguese Section

Sara Albuquerque and Ângela Salgueiro

**Abstract** The 1862 London Exhibition "was a symbol of mid-Victorian aspiration" with a clear image of Britain's ambition and its empire. These exhibitions were opportunities for other empires, such as the Portuguese, to assert and highlight the potential of their colonies. The case of the Portuguese representation during the nineteenth century in world exhibitions has been examined; however, the display of colonial products remains somehow less explored, namely, those related to the African flora. This research examines the representation of Portugal and its colonies at the 1862 London Exhibition, in particular, the case of colonial objects of natural history collected from what is known today as Angola. After the loss of Brazil, Africa was seen in a mythical way as the *Eldorado*, ready to fulfill the destiny of the nation by which it could eventually recover the status of a great power. Several aspects of the exhibition were analyzed, in particular, the objects and actors involved in the preparation of the Portuguese section. Regarding the latter, two main figures were crucial for the organization of this representation: Friedrich Welwitsch (1806–1872), who performed the *Iter Angolense* expedition (1853–1860), and organized, contributed, and suggested objects that should be collected from Angola and Júlio Máximo de Oliveira Pimentel (the second viscount of Vila Maior, 1809–1884), the royal commissioner at the London Exhibition. To understand which objects were on display, Welwitsch's publications, *The Preliminary Notes on various objects from Angola* (1861) and *Explanatory Synopsis of Samples of Timber and Medicinal Drugs* (1862), were crucial to this research. Although the Portuguese representation was severely criticized by the press, Welwitsch was awarded four gold medals for the colonial objects presented.

**Keywords** Angola · 1862 London Exhibition · Friedrich Welwitsch (1806–1872) · Portugal · Viscount of Vila Maior (1809–1884)

S. Albuquerque (✉) · Â. Salgueiro
IHC—NOVA FCSH and University of Évora and IN2PAST, Évora, Portugal
e-mail: sma@uevora.pt; asgs@uevora.pt

© The Author(s), under exclusive license to Springer Nature Switzerland AG 2024
L. M. Mendonça de Carvalho (ed.), *The Victorians: A Botanical Perspective*,
https://doi.org/10.1007/978-3-031-68759-4_7

## Introduction

In the nineteenth century, great exhibitions—worldwide or international—were places of wonder, exchanges, and encounters that amplified the circulation and dissemination of knowledge, granting nations the opportunity to promote their technical, scientific, and artistic progress. Universal exhibitions were visited by an enormous public, which allowed new relations between science and the public—the popularization of science, as well as the circulation of science and technology (Matos et al. 2010; Castro 2017). The topic of international exhibitions is not novelty and has resulted in rich and abundant literature that continues to grow (Bennett 1988; Çelik and Kinney 1990; Greenhalgh 1991; Gilbert 1994; Hoffenberg 2001; Mackenzie 2008; Qureshi 2011; Blanchard et al. 2011).

The case of Portuguese sections during nineteenth-century world exhibitions has already been examined (Vicente 2003; Cantinho 2006; Alves 2006; Matos 2010; Souto 2010, 2011; Castro 2017); however, the display of colonial products remains less explored. At the time these exhibitions occurred, Portugal was facing the period of Regeneration (1851), a new progressive cycle in which the country transitioned from a phase of conflict and uncertainty (Portuguese Civil War 1832–1834) to a period of relative political stability, social pacification, and economic and technical-scientific development. For this reason, these grand events offer the possibility to present Portuguese products as part of the process of affirming the power of the Portuguese State (Souto 2010: 98). The international exhibitions of the nineteenth century "'were used as vehicles to legitimate and disseminate the colonial policies of the European empires, and to spread propaganda about them" and reflected "the worldview of a society that saw itself on a global scale and was the fruit of progress'" (Castro 2017: 309). Considering these colonial ideas and representations, this research intends to go against the grain and provide a glimpse of the products on display, spaces, and actors involved in the execution of the representation of Portugal, namely, the Portuguese colonies at the time, mainly Angola. To do this, the case of the Portuguese representation in the London International Exhibition of 1862 was analyzed. After the independence of Brazil in 1822, which was only recognized by Portugal in 1825, the Portuguese empire tried to rebuild itself, and Africa was part of this plan (Jerónimo 2018). It is important to note that Angola was considered the most promising province of the empire and was seen as the new *Eldorado* of Portuguese imperial nationalism (Alexandre 1998: 151).

## The Victorians and the 1862 London Exhibition

The Exhibition of 1862 grows as naturally out of that of 1851 as one generation proceeds from another. It is not an imitation but a consequence of its predecessor, and one, we cannot doubt, which is destined in its turn to form a link in a chain, and a step in a series, by which the progress of the arts may be periodically reckoned amongst us – arts that renew the youth of our civilization, and help to prove that all our bygone triumphs are but the earnest and the prediction of greater triumphs yet to come.[1]

The International Exhibition of 1862 was held between 1st May and 1st November at South Kensington, London, and it received more than 6 million visitors.[2] Organized by the *Society for the Encouragement of Arts, Manufactures and Commerce*, held 28,000 exhibitors from thirty-six countries (Yallop 2011a: 7).

Not only the 1851 Great Exhibition but also the 1855 Paris Exhibition, "had accustomed the Victorians to the idea of huge and eclectic displays, sparkling showcases for the most beautiful, the most efficient and the most innovative" and the 1862 London Exhibition was no exception (Yallop 2011a: 9). The London exhibition visitors were expected "to be amazed and entertained by the practical, the pioneering and the extraordinary." These expectations were also influenced by the press; for example, *The Illustrated London News* (Fig. 7.1) filled the imagination of those who had never before visited an exhibition through its engravings and detailed

**Fig. 7.1** "The International Exhibition: View of the Nave looking East" engraving from *The Illustrated London News*, p. 249, 30.8.1862. Copyrights: Public Domain, Google-digitized

---

[1] *The Illustrated London News*, 'The Exhibition of 1862', p.79, 25.1.1862.

[2] 'The Exhibition Building of 1862', in Survey of London: Volume 38, South Kensington Museums Area, ed. F H W Sheppard (London, 1975), pp. 137–147. British History Online http://www.british-history.ac.uk/survey-london/vol38/pp137-147 [accessed 13 February 2023]; Bureau International des Expositions, Expo 1862 London, https://www.bie-paris.org/site/fr/1862-london [accessed 13 February 2023].

descriptions in a special event supplement that informed about the exhibition's progress and related news.

The 1862 London Exhibition "was a symbol of mid-Victorian aspiration" with a clear image of the world of Britain's ambition and its empire. However, it was also the possibility, for millions of visitors to admire exotic and desired novelties and beautiful and unusual things (Yallop 2011a: 9; Souto 2011). The exhibition had an emphasis on the historical and the romantic, focusing the public's attention "on the beauty of medieval and Renaissance objects for the first time, introducing unknown styles and techniques" (Yallop 2011a: 11). Nevertheless, the foundation of the exhibition was the manufacture and industrial production. The construction of the South Kensington building conceived for this event followed Francis Fowke (1823–1865) project and was marked by the architecture of cast iron, highlighting the idea of industrial production.[3] The Exhibition Palace was conceived to represent the ideal of progress and the achievements of the Industrial Revolution as a space of economic, social, and political power and "for presenting and promoting the socioeconomic development of an imperial Europe, which had mapped out the world" (Castro 2017: 309). Influenced by the imperial expansion in Africa and driven by great scientific expeditions to the interior of the African continent, the London Exhibition would function as a vehicle for propaganda and dissemination of the *exotic* to a wider audience (Souto 2011):

> [...] the exhibitions provided unique opportunities to highlight new landscapes, especially those that exhibited the local indigenous artistic culture of the distant possessions belonging to the European colonial empires (Castro 2017: 309).

In this context, Portugal benefited from the results of the *Iter Angolense* expedition (1853–1860) performed by the Austrian botanist Friedrich Martin Joseph Welwitsch (1806–1872). Welwitsch became a member of the governmental commission, formed on 10th April 1861 with the mission to prepare samples of objects of the Portuguese representation, in particular the 5th section (products of the overseas territories) at the 1862 London Exhibition (Dolezal 1974: 67–68).[4]

The *Iter Angolense* expedition had as its main objective "to obtain the most extensive knowledge [...] of natural products," promoting the "development of the wealth and well-being of its inhabitants; and relations with the Metropolis" (Albuquerque, Brummitt, and Figueiredo 2009: 641).[5] Centered in Angola, the most promising province of the Portuguese empire, the expedition compiled collections, studies, and inventories of the fauna, flora, and mineral resources of Luanda, Bengo, Cuanza Norte, Malange, Benguela, Namibe, and Huila provinces (Alexandre 1998: 151; Areias 2012: 38).

---

[3] Bureau International des Expositions, Expo 1862 Londres, https://www.bie-paris.org/site/fr/1862-london [accessed 13 February 2023].

[4] Diário de Lisboa, Número 85, Página 1027, 3ª Coluna, 17th April 1861.

[5] Decreto de 10 de Abril de 1852. *Collecção Official da Legislação Portugueza redigida por José Maximo de Castro Neto Leite e Vasconcellos*. Anno de 1852, Lisboa, Imprensa Nacional, 1853, p. 56.

## Preparing the Representation of Portugal at the 1862 Exhibition

By the royal decree of 10 April 1861, King D. Pedro V (1837–1861) created the *National Products Exhibition Steering Committee/Comissão directora da exposição dos produtos nacionais* in Lisbon. This Committee aimed to organize an exhibition in Lisbon to choose the metropolitan and colonial products to be sent to London, the publication of the official programs of the event and the definition of the necessary measures for the success of the Portuguese representation.[6] With the honorary presidency of D. Fernando II (1816–1885), the father of D. Pedro V, the Commission was organized into five sections: agricultural industry; manufacturing industry; extractive industry, buildings, and steam engines; Fine Arts; and products from overseas provinces (Souto 2011: 90–91).[7] Chaired by José Rodrigues Coelho de Amaral (1808–1873), former governor of Angola, and secretariat by Simão José da Luz Soriano (1802–1891), the 5th section had ten vowels, including F. Welwitsch, who wrote the *Preliminary Notes on various objects from Angola, proper to the London exhibition* [*Apontamentos preliminares de varios objectos de Angola, proprios á exposição de Londres*] (1861).[8]

On 27 April, the Committee was permanently installed in the *Ministry of Kingdom Affairs* [*Ministério dos Negócios do Reino*] and thereafter called the *Permanent Central Commission* [*Comissão Central Permanente*] being operational until July 1863.[9] Composed of 74 individuals, it was criticized for its excessive size, which would compromise its operability (Souto 2011: 90–91).

A decree sent by the Secretary of State for Navy and Overseas Affairs to the Governor-General of Angola, Sebastião Lopes de Calheiros and Meneses (1816–1899),[10] urged him to "excite" the living forces of the Province, to send to Lisbon the main agricultural and industrial products of the region. With a pragmatic objective and clear political, economic, and financial motivations, the following was mentioned:

> It should be borne in mind that these exhibitions are not only about the high deserving and absolute perfection of the products, but also about what each country can produce, so that often a less perfect article or object can be achieved for tiny prices and which meets many of human needs, deserves a prize and can show the existence of a profitable trade source.[11]

---

[6] Diario de Lisboa, n.° 84, 16-04-1861, p. 1017

[7] *Diario de Lisboa*, n.° 84, 16-04-1861, p. 1017.

[8] Diario de Lisboa, n.° 84, 16-04-1861, p. 1017–1018. Rectificações no Diário de Lisboa, n.° 85, 17-04-1861, p. 1027.

[9] Decreto de 14 de Julho de 1863. *Collecção Official da Legislação Portugueza redigida por José Maximo de Castro Neto Leite e Vasconcellos. Anno de 1863*, Lisboa, Imprensa Nacional, 1864, p. 319.

[10] "Commissão Portugueza para a Exposição Universal de Londres no anno de 1862". *Jornal do Commercio*, n.° 2280, 08-05-1861, p. 1.

[11] Portaria de 27 de Abril de 1861. *Collecção Official da Legislação Portugueza redigida por José Maximo de Castro Neto Leite e Vasconcellos. Anno de 1861*, Lisboa, Imprensa Nacional, 1862,

As mentioned previously, Welwitsch organized the showcase of colonial products from the Angolan possessions: botanical, ethnological, commercial, industrial, and agricultural, for the London Exhibition. In addition to contributing with collections from *Iter Angolense*, he also suggested that other objects be displayed at the exhibition. To do this, Welwitsch produced a preliminary list that provided suggestions for several objects to be sent from Angola, which included several zoological specimens with the localities where these "objects" could be found more easily.[12]

The *Preliminary Notes on various objects from Angola* were published on April 30, listing 34 colonial products indispensable to the London Exhibition.[13] Although Welwitsch refers to "objects," they are in their majority natural products that have not been manufactured. Some of these listed "objects" included elephant teeth (Cassange and Benguela), hippo's teeth (Loanda and Mossamedes), zebra fur (Mossamedes), boa skins (Loanda), and ostrich feathers (Mossamedes). The list of ethnological objects (botanical and geological) was the following: weapons and suits of the "natives" from different backlands, pottery and musical instruments made by the "natives" (Golungo Alto), bricks (Huila), raw cotton (from Luanda, Ambaca, Mossamedes, and Porto Pinda), gums, resins, fruits, and seeds from different species, pineapple and banana fibers, tobacco leaves and cigars (Ambaca), coffee (Golungo Alto and Cazengo), petroleum (Libongo), and salt (Porto Pinda, Mossamedes), just to name a few.[14]

The items suggested clearly demonstrate the necessity to show the exotic and to give a glimpse of "The other" at the international exhibition. Examples include animal teeth, feathers, fur, and skins (Arnold 1996; Pratt 1992). The ethnological objects (weapons, suits, pottery, and musical instruments) "seemed to emphazise to the Victorians the gulf of racial difference," and in the 1862 exhibition, the African objects "were most commonly kept separate from familiar works of art" (Yallop 2011b: 333). It is important to emphasize that the "display of ethnographic artifacts reinforced the Victorians' view of themselves as advanced, sophisticated, and superior" (Yallop 2011b: 337).

At the time, cotton was seen as an exotic item, such as silk, which was on display in almost every country represented in the exhibition, and Portugal was no exception.[15] At Welwitsch's *Preliminary notes*, the first botanical product mentioned is cotton.[16] The gums and resins also aroused interest, in particular, copal gum/*goma copal*, whose origin was a mystery to the Austrian collector (Welwitsch 1866).

---

p. 184–185. Translated from the Portuguese: "Que deve ter em vista que nestas exposições não se atende só ao alto merecimento e perfeição absoluta dos produtos, mas também se pretende conhecer o que cada país pode produzir, de modo que muitas vezes um artigo ou objecto menos perfeito, mas que se alcança por preços diminutos, e que satisfaz muitas das necessidades humanas, merece prémio e pode mostrar a existência de uma fonte de comércio proveitosa."

[12] Portaria de 30 de Abril de 1861. *Collecção Official da Legislação Portugueza...*, p. 185–187.

[13] Portaria de 30 de Abril de 1861. *Collecção Official da Legislação Portugueza...*, p. 185–187.

[14] Portaria de 30 de Abril de 1861. *Collecção Official da Legislação Portugueza...*, p. 185–187.

[15] Jornal do Commércio, N° 2712, 21 October 1861, page 1.

[16] Portaria de 30 de Abril de 1861. *Collecção Official da Legislação Portugueza...*, p. 185–187.

Welwitsch, while in Angola, attempted to identify the species of trees from which copal gum originated and later suggested that the gum was a fossil resin (Welwitsch 1866). Although Welwitsch was focused on botanical collections, the zoological items provided the exotic factor that was needed to display the Portuguese colonies at this exhibition, Angola in particular.

Accompanying the published list were instructions for shipping the products, regarding their presentation, conservation, classification, and dispatch. The Central Commission made recommendations for exhibitors, which were published in various official bodies and press, with precise instructions for sending the goods to the metropolis—Lisbon. It was requested to send it "in excess of the quantity indicated to cover any accidental misconduct."[17] The vegetable products should be packed in 1-liter or 1-kilogram wide-mouth glass-stoppered glass bottles. The wood samples should be "bark logs of 0.15 m long and approximately 0.08 m in diameter."[18]

Setting the date of 31 October 1861 as the time limit to receive the products at the *Arsenal da Marinha/Navy Arsenal* in Lisbon, they should be accompanied by all the necessary additional information, in particular, the identification of the exhibitor; the common name of the product; the place and date of production; the destination of the products after the end of the exhibition; and the refund, sale or assignment to the government, among others.[19] These kinds of showcases were traditionally presented at the *Arsenal da Marinha* before leaving for the world exhibitions (Cayolla 1945: XIII–XIV). It was mainly an opportunity to "test" the products and to select the best ones for the international exhibition. The exhibitions in Lisbon were an essential preparatory element for the construction of the official sections in international exhibitions. This also reflects all the propaganda around the exhibitions. Considering that most of the Portuguese audience was not able to visit the international exhibitions, these collections were displayed to the Portuguese public. In this way, the public could have a sense of belonging to the Portuguese empire through its colonial collections.

Nevertheless, this process was extremely lengthy. In the press, the delays and difficulties of the district committees were mentioned. In this sense, the traditional exhibition of national products, which took place in the *Sala da Fazenda*, a special room at the *Arsenal da Marinha*, would only be held in December, not allowing an evaluation and a rigorous and timely examination of the collections.[20] According to the testimony of a journalist from the Portuguese newspaper *A Revolução de Setembro*, "the room was surrounded by shelves and showcases completely

---

[17] "Exposição Universal de Londres. Commissão Central Portugueza", *A Revolução de Setembro*, n.° 5805, 13-09-1861, p. 3.

[18] "Exposição Universal de Londres". *O Commercio do Porto*, n.° 218, 24-09-1861, p. 1.

[19] "Exposição Universal de Londres. Commissão Central Portugueza", *A Revolução de Setembro*, n.° 5805, 13-09-1861, p. 3; "Exposição Universal de Londres". *O Commercio do Porto*, n.° 158, 16-07-1861, p. 1.

[20] "Exposição Universal de Londres em 1862". *Jornal do Commercio*, n.° 2455, 05-12-1861, p. 1; "Exposição Universal de Londres. *O Commercio do Porto*, n.° 106, 11-05-1861, p. 1.

occupied by specimen collections."[21] The same newspaper also reported on a collection of colonial products, prepared by the Overseas Council [*Conselho Ultramarino*] and presented to D. Fernando.[22]

In the early months of 1862, initiatives related to the preparation for Portuguese participation intensified. On 3 March, the government appointed Júlio Máximo de Oliveira Pimentel (1809–1884), the second Viscount of Vila Maior, as Royal Commissioner for the London Exhibition.[23] His role was to "represent the economic interests of the country at the London Universal Exhibition"; "chair the study committee"; and "inspect everything concerning the exhibition of Portuguese products."[24]

The Portuguese collections finally left Lisbon on 21 March, aboard the *Vasco da Gama* steam, arriving in London on 28 March.[25] The *Central Commission* thus had only 1 month to transport, unpack, and display the products at the South Kensington Exhibition Palace. In addition, only in July, 2 months after the inauguration of the London Exhibition, was the provision of 35,000 $ 000 *reis* for expenditure on Portuguese products authorized.[26]

## The Portuguese Section at the 1862 International Exhibition

> The South of Europe is wonderfully well represented by Spain and Portugal in the wine and food class. Out of 2200 exhibitors from those twin countries, more than one-half make a display of creature comforts. These comforts have a very wide range both in fluids and solids; and amongst the latter there will be almost everything from acor-coffee to sausages. Portugal promises to show some fine modelling in wax; and Spain one exhibitor of machinery.[27]

The products from the Portuguese section that deserved more attention from the *Illustrated London News* were wine and food and "fine modelling in wax," as the

---

[21] "Productos para a Exposição de Londres", *A Revolução de Setembro*, n.° 5896, 31-12-1861, p. 2; "Visitas reaes", *A Revolução de Setembro*, n.° 5952, 12-03-1862, p. 1.

[22] "Productos para a Exposição de Londres", *A Revolução de Setembro*, n.° 5896, 31-12-1861, p. 2.

[23] Decreto de 3 de Março de 1862. *Collecção Official da Legislação Portugueza redigida por José Maximo de Castro Neto Leite e Vasconcellos. Anno de 1862*, Lisboa, Imprensa Nacional, 1863; *Relatorio do Commissario Regio...*, p. 3.

[24] "Instrucções para regular o serviço dos commissarios portugueses à Exposição Universal de Londres", 11-03-1862, p. 1 (PT-UC-FCT-BOT-VVM-L-11_0138). Arquivo da Universidade de Coimbra, Departamento de Ciências da Vida da Faculdade de Ciências e Tecnologia—Arquivo de Botânica, Fundo Visconde de Vila Maior, Secção Comissário régio à exposição de Londres de 1862, Série Correspondência Oficial do Governo.

[25] *Relatorio do Commissario Regio...*, p. 4; "Productos portuguezes", *A Revolução de Setembro*, n.° 5950, 09-03-1862, p. 2.

[26] Lei de 2 de Julho de 1862. *Collecção Official da Legislação Portugueza redigida por José Maximo de Castro Neto Leite e Vasconcellos. Anno de 1862*, Lisboa, Imprensa Nacional, 1863, p. 172.

[27] *The Illustrated London News*, 'Progress of the International Exhibition', p. 400, 19.4.1862

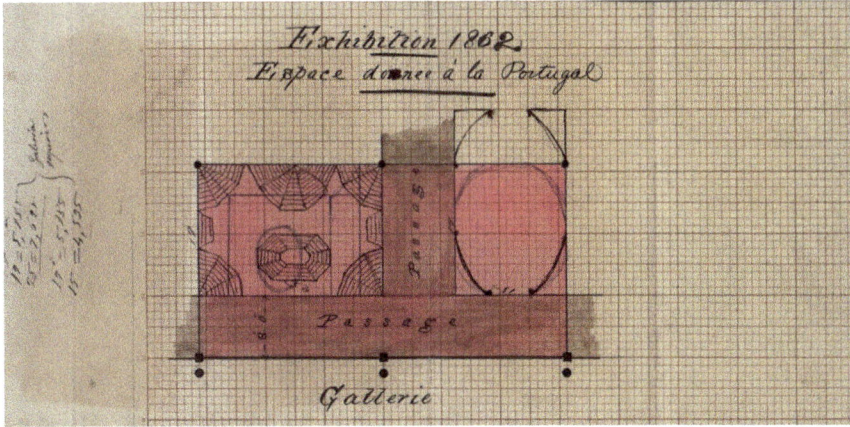

**Fig. 7.2** Sketch of the space for Portugal on the top floor of the Palace of Exhibitions in South Kensington. Source: Unidentified Author. Arquivo da Universidade de Coimbra (PT-UC-FCT-BOT/VVM/L-04)

quote above demonstrates. Although colonial products were not reported by this newspaper, they caught the attention of scientific institutions and scientists, as we will demonstrate in this section.

Arriving in London, the royal commissioner encountered some constraints. First, the small size of the space reserved for the Portuguese section, 445 m² in total (Figs. 7.2 and 7.3). Then, the products were displayed in four different spaces: two on the ground floor—flanked by the sections of Spain and Italy—and two on the upper floor, where the collections of colonial products were accommodated.[28] According to the *Report of the Commissioner* [*Relatório do Commissário Régio*], there were three sections dedicated to Angolan products, and one of them had Welwitsch's name (Fig. 7.3). The choice for a display typology based only on nationality was widely criticized, both by the various national commissions—such as João de Andrade Corvo (1824–1890), by the exhibitors themselves and by the press.

This model made it difficult to display products and evaluate them, damaging the activity of the international jury. In 1867, in reaction to this model at the Paris Universal Exhibition [*Exposition Universelle d'Arte et d'Industrie*], the system of pavilions displayed the products by country and by typology (Souto 2011: 102–103).[29] From March to April 1862, the Viscount of Vila Maior was responsible for the design, planning, and execution of the Portuguese section at the London International Exhibition, "obliged to draw up within a few hours a project that

---

[28] *Relatorio do Commissario Regio*…, p. 8 e 10–11; LONDON INTERNATIONAL EXHIBITION, 1862, p. 123.

[29] "Palacio da Exposição de Paris". *O Panorama*, n.° 30, 1867: 240.

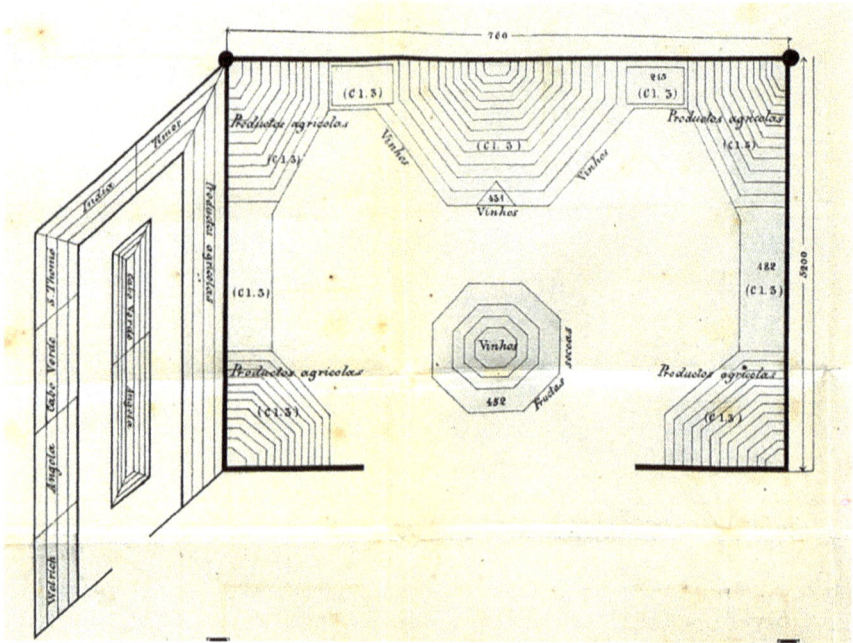

**Fig. 7.3** Definitive floor plan of the Portuguese products displayed at the 1862 Exhibition, first floor. Source: *Relatorio do Commissario Regio…*, p. 66

should be carefully considered and subjected to severe scrutiny to satisfy requirements of a methodical disposition and reasonable elegance."[30]

This planning was intricate, involving not only the hiring of specialized companies in London but also the collection, transport, unpacking, and storage of products sent from Portugal in different steamers, with transport guides addressed both for the royal commissioner and the consulate. In addition, the exhibition cabinets and tables had to be built on-site, especially for the occasion.[31] With the inauguration of the Exhibition on the first of May, the activity of the royal commissioner and the members of the study commission focused mainly on the support and clarification of the international jury, due to the poor organization of the Portuguese catalogue and the lack of additional information sent from national exhibitors.[32] In addition to

---

[30] *Relatorio do Commissario Regio…*, p. 9.

[31] Ofício do Visconde de Vila Maior para o Ministro das Obras Públicas, 28-04-1862 (PT-UC-FCT-BOT-VVM-L-11_0175). Arquivo da Universidade de Coimbra, Departamento de Ciências da Vida da Faculdade de Ciências e Tecnologia—Arquivo de Botânica, Fundo Visconde de Vila Maior, Secção Comissário régio à exposição de Londres de 1862, Série Minutas de correspondência expedida.

[32] Ofício do Visconde de Vila Maior para o Ministro das Obras Públicas, 28-04-1862 (PT-UC-FCT-BOT-VVM-L-11_0175). Arquivo da Universidade de Coimbra, Departamento de Ciências da Vida da Faculdade de Ciências e Tecnologia—Arquivo de Botânica, Fundo Visconde de Vila Maior, Secção Comissário régio à exposição de Londres de 1862, Série Minutas de correspondência expedida.

the Viscount of Vila Maior, this task was attended by João Palha de Faria Lacerda, Francisco António de Vasconcelos, and Francisco Augusto Florido da Moita e Vasconcelos.[33]

Together with the exhibition, there were some parallel initiatives involving the commission, such as a meeting organized in August in Manchester by the *Cotton Supply Association* in the context of the international cotton crisis caused by the American Civil War, resulting in a shortage of cotton supply in the world market (Alves 2006).[34] The viscount of Vila Maior was present, representing the government, and investors were sought to promote colonial cotton culture, mainly in Angola.[35] The cotton crisis was seen by the Portuguese government as an opportunity to promote the culture of cotton; a clear example of this was the work developed by Welwitsch on the cultivation of cotton in Angola (Welwitsch 1859, 1861). Concerning the Portuguese products exhibited, the press and catalogues focused mainly on agricultural products and on animal and vegetable substances used in manufacturing, similar to other countries such as Spain, Russia, and Brazil.[36]

The journal *A Revolução de Setembro* focused on the predominance of colonial products, considering them "peculiar," but criticizing the way they were on display.[37] These "peculiar" products caught the attention of scientific institutions, companies, and scientists, who during the Exhibition, contacted the royal commissioner to acquire some of these products. For example, on 30 June 1862, William Jackson Hooker (1785–1865), director of the Royal Botanic Gardens, Kew, officiated with the Viscount of Vila Maior to obtain the sale or purchase of botanical products on display in the Portuguese section to enrich the Museum of Economic Botany. Eventually, it was decided to hand over "an interesting collection of the Agricultural Products" (Fig. 7.4).[38]

On 29 September, Thomas Archer (1817–1885), director of the *Industrial Museum of Scotland*, requested that the Portuguese Commissioner send raw materials:

> It will occur that very many of the specimens especially the raw produce will not be required again in Portugal; but will if deposited in one of the National Museums of Great Britain

---

[33] "Exposição Universal de Londres". *O Commercio do Porto*, n.° 185, 13-08-1862, p. 1.

[34] *Relatorio do Commissario Regio...*, p. 40–41.

[35] *Relatorio do Commissario Regio...*, p. 40–41.

[36] "Exposição Universal de Londres". *O Commercio do Porto*, n.° 106, 09-05-1862, p. 2; "Exposição Universal de Londres". *O Commercio do Porto*, n.° 112, 16-05-1862, p. 1; London International Exhibition, 1862, p. 118.; THE INTERNATIONAL EXHIBITION, 1863, p. 313.

[37] A.J.S., "Impressões da Inglaterra e da Exposição", *A Revolução de Setembro*, n.° 6100, 11-09-1862, p. 3.

[38] Ofício de W.J. Hooker, director of Royal Kew Gardens, 30-06-1862, p. 1 (PT-UC-FCT-BOT-VVM-L-02). Arquivo da Universidade de Coimbra, Departamento de Ciências da Vida da Faculdade de Ciências e Tecnologia—Arquivo de Botânica, Fundo Visconde de Vila Maior, Secção Comissário régio à exposição de Londres de 1862, Série Correspondência.

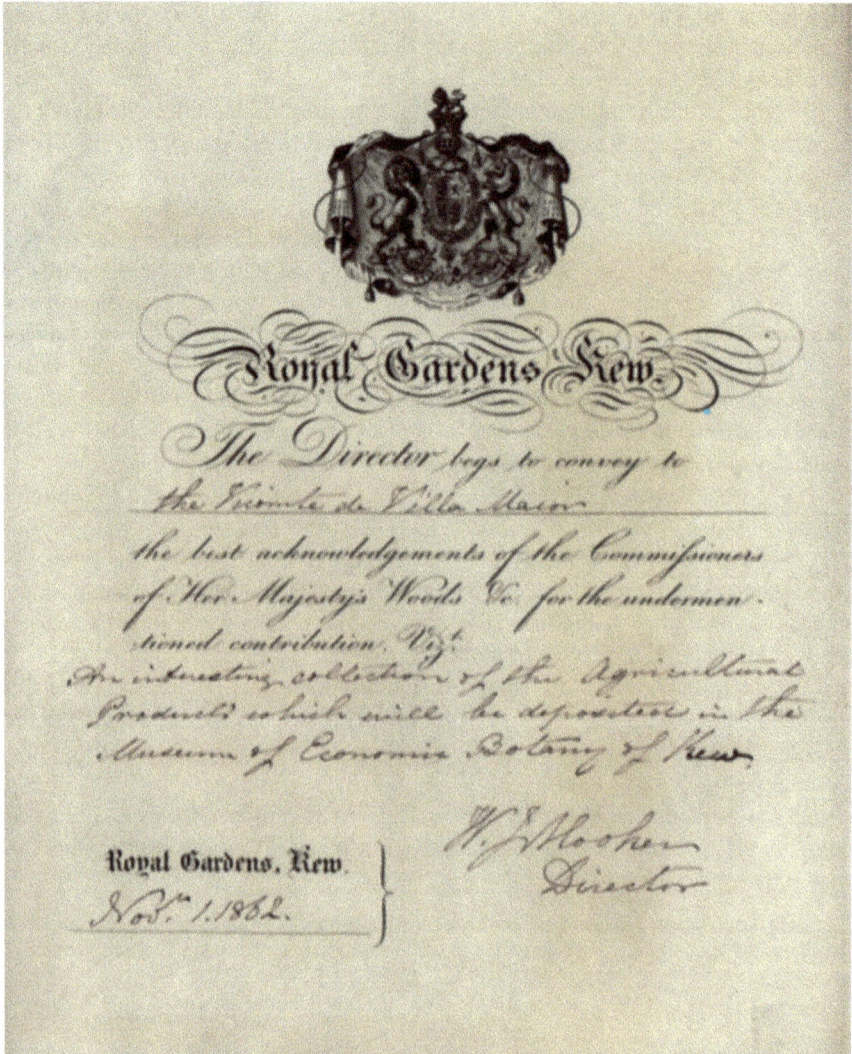

**Fig. 7.4** Official document from W.J. Hooker acknowledging the Viscount of Vila Maior (November 1, 1862). Source: Arquivo da Universidade de Coimbra (PT-UC-FCT-BOT/VVM/L-08) (Ofício de W.J. Hooker, director of Royal Kew Gardens, 01-11-1862 (PT-UC-FCT-BOT-VVM-L-08). Arquivo da Universidade de Coimbra, Departamento de Ciências da Vida da Faculdade de Ciências e Tecnologia—Arquivo de Botânica, Fundo Visconde de Vila Maior, Secção Comissário régio à exposição de Londres de 1862, Série Correspondência Expo 1862)

continue to further the chief object of the International Exhibition by keeping before the public the quality and peace of production of those articles.[39]

---

[39] Ofício de Thomas Archer, director of the Industrial Museum of Scotland, 29-09-1862, p. 1 (PT-UC-FCT-BOT-VVM-L-02). Arquivo da Universidade de Coimbra, Departamento de Ciências da Vida da Faculdade de Ciências e Tecnologia—Arquivo de Botânica, Fundo Visconde de Vila Maior, Secção Comissário régio à exposição de Londres de 1862, Série Correspondência.

The official document included a list of articles of interest to the *Industrial Museum of Scotland*, namely, "Pharmaceutical products and preparations" (class 2) and "acorn coffee" (class 3). However, Júlio Máximo de Oliveira Pimentel had great difficulty dealing with the requests received, as many exhibitors did not authorize the handover or sale of the collections.[40]

Following Portuguese participation in the London Exhibition, the exhibitors received 165 bronze medals and 240 honorable mentions from the international jury, privileging unprocessed raw materials and food products (Souto 2011: 94).[41] The rewards were given to the national representatives at a public ceremony held on July 11, 1862, at the Exhibition Palace.

Austrian botanist F. Welwitsch received four medals for his colonial collections: in class 2 (substances, chemicals, and pharmaceutical processes), sections A and B, chaired by Ballard, were awarded two "for interesting medicinal substances collected in Angola."[42] Class 3 (food substances, including wines), directed by Boussingault, awarded the "Agricultural Product Collection – for excellence in quality"[43] (section A—agricultural production), and class 4 (animal and vegetable substances employed in manufacturing), chaired by Chevalier de Schwarz, recognized the quality of the "woods and gums of Angola" collected by the same naturalist (section C—vegetable substances).[44]

It is important to consider that the jury evaluation took place between 7 May and 17 June, at a time when many exhibitors were still organizing their exhibitions. The press was quite incisive in this regard, criticizing the delay in dispatching the objects to London, the lack of national representation and the poor layout of the collections.[45]

Another constraint was the organization of national catalogues, indispensable tools for judges' work and the dissemination of products to visitors, traders, and industry. These were poorly structured and incomplete—partly due to the submission of incomplete product sheets, undermining the assessment of the Portuguese section, as it had been established "that medals should be awarded without reference to nationalities, i.e., the rewards should fall solely on the absolute merit of the products."[46]

---

[40] *Relatorio do Commissario Regio...*, p. 44.

[41] Ofício de 16 de Julho de 1862. *Diario de Lisboa*, n.° 167, 28-07-1862, p. 1983–1985; "Exposição Universal de Londres". *O Commercio do Porto*, n.° 173, 30-07-1862, p. 2.

[42] *Relatorio do Commissario Regio...*, p. 108.

[43] *Relatorio do Commissario Regio...*, p. 109.

[44] *Relatorio do Commissario Regio...*, p. 112; SAMPAIO, A.R., "Política estrangeira—Londres, 2 de Agosto de 1862", *A Revolução de Setembro*, n.° 6073, 12-08-1862, p. 1; SAMPAIO, A.R., "Política estrangeira—Londres, 3 de Agosto de 1862", *A Revolução de Setembro*, n.° 6077, 14-08-1862, p. 1.

[45] Machado, J.C., "Três cartas III". *Archivo Pittoresco*, 5.° ano, n.° 26, 1862, pp. 202–203; SAMPAIO, A.R., "Política estrangeira—Londres, 2 de Agosto de 1862", *A Revolução de Setembro*, n.° 6073, 12-08-1862, p. 1; A.J.S., "Impressões da Inglaterra e da Exposição", *A Revolução de Setembro*, n.° 6100, 11-09-1862, p. 3.

[46] "Relatorio do Sr. Visconde de Villar Maior, Comisario Regio na Exposição de Londres", *A Revolução de Setembro*, n.° 6084, 23-08-1862, p. 3; "Exposição Universal de Londres". *O Commercio do Porto*, n.° 185, 13-08-1862, p. 1.

Delivered to Francisco de Almeida Portugal, 2nd Earl of Lavradio (1797–1870), the medals and honorable mentions would later be given to the Portuguese exhibitors. Regarding the destination of the rewards obtained by the producers in Angola, a decree, dated 5 May 1866, indicated that "the Medals and Diplomas... awarded to the Exhibitors of the Province who were most distinguished at the London Universal Exhibition" would be forwarded to the General Governor. These would be delivered at a formal session, which would be attended by members of the Government Council, the City Council, and the main authorities of the province.[47]

## Synopsis of Samples of Timber and Medicinal Drugs

At the London exhibition, Welwitsch himself contributed with 122 objects from Angola, of which 52 were timber samples (Dolezal 1974: 67–68). These specific items were collected by the Austrian botanist during the expedition of *Iter Angolense*, between 1853 and 1860. Welwitsch, as the delegate of the Portuguese Government, accompanied the showcase to London and its return to Lisbon; however, after that, the itinerary of the items is not clear (Cayolla 1945: XIV). As mentioned previously, as a result of the products on display, the naturalist was awarded four gold medals for the objects presented.[48] According to the jury report, the samples of Angolan wood "consisted of specimens full of interest and novelty, all unknown here" (Dolezal 1974: 67–68).

A detailed description of the objects on display was published at *Synopse Explicativa das Amostras de Madeiras e Drogas Medicinaes* [*Explanatory Synopsis of Samples of Timber and Medicinal Drugs*] (Welwitsch 1862).[49] This publication also includes a collection of African medicinal drugs offered to the *Gabinete Pharmacologico da Escola Médico-Cirurgica de Lisboa* [*Pharmacological Office of the Medical-Surgical School of Lisbon*] (Welwitsch 1862).

A copy of the *Synopse Explicativa,* which belonged to Welwitsch, is housed at the *Museu Nacional de História Natural e da Ciência* [*National Museum of Natural History and Science*], Lisbon (MUHNAC-UL). This publication has innumerable hand notes made by the Austrian botanist, adding extra information related to the species, habitat, uses and, in some cases, also corrects the published information. According to Conde de Ficalho (Francisco Manuel de Melo Breyner, 1837–1903), this particular *Synopse Explicativa* was offered to him by William Philip Hiern

---

[47] Portaria de 5 de Maio de 1866. *Annaes do Conselho Ultramarino. Parte official*, Série VII, 1866, p. 20.

[48] The medals distributed to the Portuguese exhibitors were delivered personally by the king in Lisbon at 11th May 1863 (Dolezal 1974: 68).

[49] The manuscript drafts for this publication are housed at Academia de Ciências de Lisboa/ Academy of Sciences, Lisbon, Portugal (Manuscritos de Frederico Welwitsch, Série Azul, Cota 907, "F. A. W. Enumeração das amostras de Madeira que destina Frederico Welwitsch à Exposição Internacional de Londres em 1862").

(1839–1925), a British mathematician and botanist (Ficalho 1947: 161). This is not surprising, considering that Hiern (1839–1925), after Welwitsch's death, was employed to divide Welwitsch's African Collections and had access not only to the herbarium specimens but also to notes, diaries, and books that belonged to the Austrian botanist (Hiern 1896; Albuquerque et al. 2009: 642). In this way, when Ficalho published *Plantas Úteis Da África Portuguesa* [*Useful Plants From Portuguese Africa*], he considered not only Welwitsch's publication but also his handwritten notes at the *Synopse* (Ficalho 1947).

In *Synopse Explicativa*, 52 wood samples and 96 medicinal drugs are listed (Table 7.1) (Welwitsch 1862). In some cases, the same species appeared more than once, and there are examples of introduced species, such as *Melia azedarach* L. (no. 24), *Spondias mombin* L. (no. 28) and *Ceiba pentandra* (L.) Gaertn. (no. 52) (POWO 2023; Ficalho 1947). Of the wood samples on display, it is worth mentioning: *Maerua angolensis* DC. (no. 39) and *Gardenia ternifolia* subsp. *jovis-tonantis* (Welw.) Verdc. (no. 20 & no. 44), both of which are native species (POWO 2023).

The species *Maerua angolensis* was the first Angolan plant described in the scientific literature (1824). It was collected near Benguela by Joaquim José da Silva, a native of Rio de Janeiro who studied in Coimbra. Silva was sent to Angola, by a royal decree, from 1783 to 1787 in a double role as Secretary of Angola and as a scientific explorer (Huntley 2008; Kananoja 2015).

The other species, *Gardenia ternifolia* subsp. *jovis-tonantis* (initially named by Welwitsch as *Decameria jovis-tonantis* Welw.) reflects the cross-cultural encounters between the collector and the local people. The specific epithet "*jovis-tonantis*" is related to the plant properties, as the species branches were placed on the top of the straw-roofed homes (*cubatas*) as a lightning rod because it was believed that they could protect from electrical discharges. Considering this particular virtue, Welwitsch derived the species name "*jovis-tonantis*," which is dedicated to the God of Thunder (Ficalho 1947: 197).

The case of *tacula—Pterocarpus tinctorius* Welw. (no. 5), from the red wood, it was possible to obtain a powder by friction over a stone (Ficalho 1947: 143–144). This *tacula* powder could be mixed with oil or water to prepare red pigments for use on hair or skin. Welwitsch mentioned that on special occasions, the feet of the local people were painted red to imitate shoes (Welwitsch 1862: 33). According to Kananoja, *tacula* was expensive, and in addition to being used as a dyestuff, it was also used in medicine and during female coming-of-age ceremonies (Kananoja 2015: 51–52). Kananoja also added that "the widespread use of *tacula* in West Central Africa reflected the symbolic association between color, the ancestral world, and liminal states of initiation and rites of passage" (Kananoja 2015: 54). Another species of *Pterocarpus*, which has the common name *mirahonde—Pterocarpus angolensis* DC. (no. 48) was used to produce objects and weapons. A blood-colored resin was extracted from the wood trunk, and according to Welwitsch, the natives, in addition to using it to treat their wounds, would sell it to the pharmacists at the Angolan coast as *Dragon blood*, although this was not the real *Dragon blood* because it was not from the *Dracaena* (Figueiredo and Smith 2017: 84; Ficalho 1947: 144–145; Welwitsch 1862: 37).

**Table 7.1** The table compiles the common names and numbers given by Welwitsch to the objects on display in 1862 and the currently accepted names. It is important to state that this is merely a preliminary list based on Welwitsch's *Synopsis*. Although several references were consulted to confirm the currently accepted names, it is still necessary to compare the information with the herbarium specimens. In addition, many of the common names have changed over the decades; they also differ among different ethnic groups, and there are cases in which fruits, flowers, roots, and leaves have different names for the same species (Figueiredo and Smith 2017; WCSP 2023; IPNI 2023; GBIF.org 2023; ACTD 2023; NHM 2023; Ficalho 1947; Welwitsch 1862)

| No. Synopsis | Common name | Accepted name |
|---|---|---|
| **Wood samples from Angola (no. 1 to 52)** | | |
| 1 | *Calôlo* | *Phoenix reclinata* Jacq. Arecaceae |
| 2 | *Mangue do monte* *Mangue branco* | *Corynanthe paniculata* Welw. Rubiaceae |
| 3 | *Mafura* (Mozambique) *Guimbi* (interior of Angola) | *Trichilia emetica* Vahl Meliaceae |
| 4 | *Cosanza* | *Memecylon* sp. Melastomataceae |
| 5 | *Tacula* *Hûla de Golungo Alto* | *Pterocarpus tinctorius* Welw. Fabaceae |
| 6 | *Mucamba-camba* *Moreira* (Portuguese settlers) | *Milicia excelsa* (Wel.) C.C.Berg Moraceae |
| 7 | *Tacula do Zenza* | *Pterocarpus* sp. Fabaceae |
| 8 | *Musalengue* | *Premna angolensis* Gurke Lamiaceae |
| 9 | *Quiseco* *Quisécua* *Caseco* | *Millettia* sp. Fabaceae |
| 10 | *Mufufutu* | *Albizia ferruginea* (Guill. & Perr.) Benth. Fabaceae |
| 11 | *Mufufutu* | *Albizia angolensis* Welw. Fabaceae |
| 12 | *Mungundo* | *Symphonia globulifera* L.f. Clusiaceae |
| 13 | *Muriambambe* | *Coffea canephora* Pierre ex Â.Froehner Rubiaceae |
| 14 | *Moreira* *Mucamba-camba* | *Milicia excelsa* (Welw.) C.C.Berg Moraceae |
| 15 | Root of *Tacula* | *Pterocarpus* sp. Fabaceae |
| 16 | *Caseque* | *Milletia* sp. Fabaceae |
| 17 | *Quipuculo cafele* | *Vernonia doniana* DC. Asteraceae |

(continued)

**Table 7.1** (continued)

| No. Synopsis | Common name | Accepted name |
|---|---|---|
| 18 | *Dendo* | *Diospyros dendo* Welw. ex Hiern Ebenaceae |
| 19 | *Calusange* | *Steganotaenia araliacea* Hochst. Apiaceae |
| 20 | *Unday* *N-Day* | *Gardenia ternifolia* subsp. *jovis-tonantis* (Welw.) Verdc. Rubiaceae |
| 21 | *Quibaba* | *Entandrophragma angolense* (Welw.) Panshin Meliaceae |
| 22 | *Mucaça-Ncumbi* | *Carapa procera* DC. Meliaceae |
| 23 | *Calalanza* *Tacula falsa* (Portuguese settlers) | *Annea laxiflora* (Benth.) Mackinder&Wieringa Fabaceae |
| 24 | *Bombôlo* | *Melia azedarach* L. Meliaceae |
| 25 | *Quibaba roxa* | *Celtis africana* Burm.f. Cannabaceae |
| 26 | *Quibaba do Mussengue* *Quibaba do Hungo* | *Khaya anthotheca* (Welw.) C.DC. Meliaceae |
| 27 | *Mutune* | *Harungana madagascariensis* Lam. ex Poir. Hypericaceae |
| 28 | *Munguengue* | *Spondias mombin* L. Anacardiaceae |
| 29 | *Mutála-menha* | *Millettia nudiflora* Welw. ex Baker Fabaceae |
| 30 | *N-caça n-cumbi* | *Carapa procera* DC. Meliaceae |
| 31 | *Mulumba* | *Pterocarpus rotundifolius* (Sond.) Druce Fabaceae |
| 32 | *Mussondo* *Muçondo* | *Pseudospondias microcarpa* (A.Rich.) Engl. Anacardiaceae |
| 33 | *Cafequesu de monte* *Quisunhunga* | *Mimusops* sp. Sapotaceae |
| 34 | *Muance* | *Albizia welwitschii* Oliv. Fabaceae |
| 35 | *Quibosa iã mugito* | *Cordia* sp. Boraginaceae |
| 36 | *Mugongue* | *Premna angolensis* Gurke Lamiaceae |
| 37 | *Muzumba* | *Millettia versicolor* Welw. ex Baker Fabaceae |

(continued)

**Table 7.1** (continued)

| No. Synopsis | Common name | Accepted name |
|---|---|---|
| 38 | *Cafequesu* | *Mimusops* sp. Sapotaceae |
| 39 | *Muriangombe* | *Maerua angolensis* DC. Capparaceae |
| 40 | *Quitundo* | *Ozoroa insignis* Delile Anacardiaceae |
| 41 | *Pau Quicongo de Huilla* | *Tarchonanthus camphoratus* L. Asteraceae |
| 42 | *Maboca* | *Strychnos* sp. Loganiaceae |
| 43 | *Noxa/Nocha* | *Parinari curatellifolia* Planch. ex Benth. Chrysobalanaceae |
| 44 | *Unday de Huilla* *Mulábi* | *Gardenia ternifolia* subsp. *jovis-tonantis* (Welw.) Verdc. Rubiaceae |
| 45 | *Mueia* | *Terminalia sericea* Burch. ex DC. Combretaceae |
| 46 | | *Faurea rochetiana* (A.Rich.) Chiov Proteaceae |
| 47 | *N-panda* *Umpanda* *Mupanda* (Humpata) | *Brachystegia spiciformis* Benth. Fabaceae |
| 48 | *Mirahonde* | *Pterocarpus angolensis* DC. Fabaceae |
| 49 | *Figueira brava dos colonos de Huilla* | *Apodytes dimidiata* E.Mey. ex Arn. Icacinaceae |
| 50 | *Bimba* | *Aeschynomene elaphroxylon* (Guill. & Perr.) Taub. Fabaceae |
| 51 | Coffee tree trunk cross-section | *Coffea canephora* Pierre ex A.Froehner Rubiaceae |
| 52 | Plates made of *Mufumeira* wood (*gamelas pequenas*) *Mufumeiras* (portuguese adaptation for Mufuma) | *Ceiba pentandra* (L.) Gaertn. Malvaceae |
| **Medicinal drug samples (no. 53 to 149)** | | |
| 53 | Micaceous iron ore | Mineral origin |
| 54 | Micaceous iron powder | Mineral origin |
| 55 | Iron pyrites | Mineral origin |
| 56 | *Pemba* Stone | Mineral origin (variety of clay) |
| 57 | *Losna de Humpata* | *Artemisia afra* Jacq. Asteraceae |
| 58 | *Cachinde-Candange* *Alecrim das paredes* (denomination by the Portuguese settlers) | *Myrothamnus flabellifolius* Welw. Myrothamnaceae |

(continued)

**Table 7.1**  (continued)

| No. Synopsis | Common name | Accepted name |
|---|---|---|
| 59 | *Fel da terra de flor branca* | *Swertia welwitschii* Eng. Gentianaceae |
| 60 | *Avenca* | *Adiantum* sp. Adiantaceae |
| 61 | *Encotahóte (N-cotahóte)* | *? Andropogon stypticus* Welw. |
| 62 | *Catete Bulla* | *Tinnea antiscorbutica* Welw. Lamiaceae |
| 63 | *Barbas de Mulemba* | *Ficus thonningii* Blume Moraceae |
| 64 | Fruits and seeds of *Sacalaséne* | *Aframomum angustifolium* K.Schum. Zingiberaceae |
| 65 | *Sabongo* (fruits) | *? Xylopia aethiopica* A.Rich. Annonaceae |
| 66 | Fruits of *Butua* (*Butua* seeds) | *Tiliacora chrysobotrya* Welw. ex Ficalho Menispermaceae |
| 67 | Roots and stems of *Butua* | *Tiliacora chrysobotrya* Welw. ex Ficalho Menispermaceae |
| 68 | Trunk of *Butua* | *Tiliacora chrysobotrya* Welw. ex Ficalho Menispermaceae |
| 69 | Cross section of a *Butua* trunk | *Tiliacora chrysobotrya* Welw. ex Ficalho Menispermaceae |
| 70 | *Solanum tinctorium* | *Solanum scabrum* Mill. Solanaceae |
| 71 | *Dongos de Congo* | *Aframomum melegueta* (Roscoe) K. Schum. Zingiberaceae |
| 72 | Bark of *Mucumbi* | *Lannea* sp. Anacardiaceae |
| 73 | Trunk and bark of *Molungo* | *Erythrina abyssinica* Lam. Leguminosae |
| 74 | *Pepe* fruit *Gisepe* fruit | *Monodora myristica* Dunal Annonaceae |
| 75 | Bark of *Quibaba* (*Quibaba de Queta*) | *Entandrophragma angolense* C.DC. Meliaceae |
| 76 | *Gipepe de Songo* *Jipepe de Songo* *Xipepe de Songo* | *Monodora angolensis* Welw. Annonaceae |
| 77 | Bark and fruit of *Mulôlo* | *? Piliostigma* thonningii (Schumach.) Milne-Redh. Leguminosae |
| 78 | Root bark of *Mubango* | *Croton mubango* Müll.Arg. Euphorbiaceae |
| 79 | Root of *Mundondo* | *Mondia whitei* Skeels Asclepiadaceae |

(continued)

**Table 7.1** (continued)

| No. Synopsis | Common name | Accepted name |
|---|---|---|
| 80 | *Quibaba de Mussengue* | *Khaya anthotheca* C.DC. Meliaceae |
| 81 | Bark of *Musuemba* | *Albizia coriaria* Welw. Leguminosae |
| 82 | Bark of *Musoso* | *Entada abyssinica* Steud. Leguminosae |
| 83 | *Pau Quicongo de Huíla* | *Tarchonanthus camphoratus* L. Asteraceae |
| 84 | Powder of *Pau Quicongo de Huíla* | *Tarchonanthus camphoratus* L. Asteraceae |
| 85 | Root of *Tacula* | *Pterocarpus* sp. Fabaceae |
| 86 | Powder of *Tacula* | *Pterocarpus* sp. Fabaceae |
| 87 | *Umpeque* fruits | *Ximenia americana* L. Olacaceae |
| 88 | *Maboca* fruits | *Strychnos cocculoides* Baker Loganiaceae *Strychnos spinosa* Lam. Loganiaceae |
| 89 | *Massambala branco* | *Sorghum* sp. Poaceae |
| 90 | *Massambala rubro* | *Sorghum* sp. Poaceae |
| 91 | *Massango liso* | *Pennisetum* sp. Poaceae |
| 92 | *Massango barbado* | *Pennisetum* sp. Poaceae |
| 93 | *Milho (Mupungo)* | *Zea mays* L. Poaceae |
| 94 | *Mubafo* resin | *Pachylobus edulis* G.Don Burseraceae |
| 95 | *Goma copal de Benguela* (*Ocote* or *Cocote*) | |
| 96 | *Goma copal do Zenza de Golungo* | |
| 97 | *Goma Tragacantha* *Alquitiri* *Chixe* (tree) *Ici ià Chixe* (gum) | *Sterculia setigera* Delile Malvaceae |
| 98 | *Goma de Muance* | *Albizia welwitschii* Oliv. Fabaceae |
| 99 | *Goma de Mumango* | *Croton mubango* Müll.Arg. Euphorbiaceae |

(continued)

**Table 7.1** (continued)

| No. Synopsis | Common name | Accepted name |
|---|---|---|
| 100 | *Sangue de Drago*<br>*Mirahonde* (locals from Huíla)<br>*Ngillasonde* (locals from Pungo Andongo) | *Pterocarpus angolensis* DC.<br>Fabaceae |
| 101 | *Cabella* | *Xylopia aethiopica* (Dunal) A.Rich.<br>Annonaceae |
| 102 | *Marabú* feathers | Animal origin |
| 103 | *N-Bungo* (tobacco boxes)<br>*Quiambungo* | *Oxytenanthera abyssinica* Munro<br>Poaceae |
| 104 | Elephant tail (diadem) | Animal origin |
| 105 | Sieve made of *Súbi* | *Marantochloa conferta* (Benth.)<br>A.C.Ley<br>Marantaceae<br>*Marantochloa purpurea* (Ridl.)<br>Milne-Redh.<br>Marantaceae |
| 106 | *Bordão* Palm tree filaments<br>*Jimbusu* | *Raphia textilis* Welw.<br>Arecaceae |
| 107 | *Mateva de Pôrto Pinda* (with fruits) | *Hyphaene petersiana* Klotzsch ex Mart.<br>Arecaceae |
| 108 | Indigenous saddlebag, made from Imbondeiro's bark | *Adansonia digitata* L.<br>Bombacaceae |
| 108 B | Rope made from Imbondeiro's bark | *Adansonia digitata* L.<br>Bombacaceae |
| 109 | Imbondeiro's bark | *Adansonia digitata* L.<br>Bombacaceae |
| 110 | Workpieces of *Mabéla* (or *Mabella*) | *Raphia textilis* Welw.<br>Arecaceae |
| 111 | *Mabéla branca* | *Raphia textilis* Welw.<br>Arecaceae |
| 112 | *Bordão* fruits | *Raphia textilis* Welw.<br>Arecaceae |
| 113 | *Cachingas* (Sobas cap) | *Raphia textilis* Welw.<br>Arecaceae |
| 114 | Pineapple filaments | *Ananas comosus* (L.) Merr.<br>Bromeliaceae |
| 115 | Banana filaments | *Musa acuminata* Colla<br>Musaceae |
| 116 | *Mundondo* filaments and fruit | *Mondia whitei* Skeels<br>Asclepiadaceae |
| 117 | *Cachinga* (Sobas cap) made of banana filaments | *Musa acuminata* Colla<br>Musaceae |

(continued)

**Table 7.1** (continued)

| No. Synopsis | Common name | Accepted name |
| --- | --- | --- |
| 118 | *Subi* rods | *Marantochloa conferta* (Benth.) A.C.Ley Marantaceae *Marantochloa purpurea* (Ridl.) Milne-Redh. Marantaceae |
| 119 | *Quibosa* | *Triumfetta* sp. Tiliaceae |
| 120 | *Cairo da palmeira Dendem* | *Elaeis guineensis* Jacq. Arecaceae |
| 121 | *Calolo* haulm to start a hat | *Phoenix reclinata* Jacq. Arecaceae |
| 122 | *Sabugo* (marrow of the papyrus) | *Cyperus papyrus* L. Cyperaceae |
| 123 | *Lã de palmeira* (*Ucúcu*) | *Elaeis guineensis* Jacq. Arecaceae |
| 124 | *Cola* (Coleira fruit) | *Cola acuminata* (P.Beauv.) Schott & Endl. Sterculiaceae |
| 125 | *Riamba* *Liamba* *Diamba* | *Cannabis sativa* L. Cannabaceae |
| 126 | Powder of *Caseque* (or *Caseco*) | *Milletia* sp. Leguminosae |
| 127 | *Mucôco* | *Triclisia sacleuxii* Diels Menispermaceae |
| 128 | *Mûcua* | *Adansonia digitata* L. Bombacaceae |
| 129 | *Orucú* fruits *Urucú* *Quisafu* | *Bixa orellana* L. Bixaceae |
| 130 | Bark of *Mungo* *Mohambo* | *Hallea stipulosa* (DC.) J.-F.Leroy Rubiaceae |
| 131 | *Bálsamo de S.Tomé* | Burseraceae |
| 132 | *Goma arábica* | ? *Acacia* sp. Mimosaceae |
| 133 | *Goma elástica de Hungo* *Mupapata* | ? *Ficus elastica* Roxb. ex Hornem. Moraceae |
| 134 | *Goma elástica de Golungo Alto* *Licongue* | *Landolphia owariensis* P.Beauv. Apocynaceae |
| 135 | *Goma de Cajueiro* | *Anacardium occidentale* L. Anacardiaceae |
| 136 | *Castanhas de Caju* (from Luanda) | *Anacardium occidentale* L. Anacardiaceae |

(continued)

**Table 7.1** (continued)

| No. Synopsis | Common name | Accepted name |
|---|---|---|
| 137 | *Mutúge* fruits<br>*Moscardeira Brava de Angola* | *Pycnanthus angolensis* (Welw.) Exell<br>Myristicaceae |
| 138 | *Cassia fistula de Angola* (*Cannafistula*) | *Cassia fistula* L.<br>Leguminosae |
| 139 | *Cassia fistula de Huíla* | *Cassia fistula* L.<br>Leguminosae |
| 140 | Seeds of *Nocha* (or *Noxa*) | *Parinari curatellifolia* Planch. ex Benth.<br>Chrysobalanaceae |
| 141 | *Dendem* fruits | *Elaeis guineensis* Jacq.<br>Arecaceae |
| 142 | *Amêndoas de Disanha* | *Treculia africana* Decne. ex Trécul<br>Moraceae |
| 143 | *Cambundo* | *Coix lacryma-jobi* L.<br>Poaceae |
| 144 | *Fel da terra de flor roxa* | *Faroa salutaris* Welw.<br>Gentianaceae |
| 145 | *Fel da terra de flor amarela* | *Sebaea brachyphylla* Griseb.<br>Gentianaceae |
| 146 | Stems of *Mobiro* | *Adenia lobata* Engl.<br>Passifloraceae<br>*Adenia lobata* subsp. *rumicifolia* (Engl.)<br>Lye<br>Passifloraceae |
| 147 | *Salsaparrilha de Angola* | *Smilax anceps* Willd.<br>Smilacaceae |
| 148 | *Cahémbia-émhbia* | *Wissadula rostrata* (Schumach.) Hook.f.<br>Malvaceae |
| 149 | *Múbafo* bark | *Pachylobus edulis* G.Don<br>Burseraceae |

There are also examples of species that were used and experimented by Welwitsch on his patients and himself. In the case of *losna-de-Humpata—Artemisia afra* Jacq. (no. 57), Welwitsch confirmed the good results when using it for fever and stomach and *encotahóte* –probably *Andropogon stypticus* Welw. (no. 61) was applied with good results on female patients with gynecological problems (Welwitsch 1862).

The case of *Cachinde-Candange—Myrothamnus flabellifolius* Welw. (no. 58) is one of many examples of plants adopted by the Portuguese settlers and denominated *alecrim-das-paredes*. The locals used it for headaches, rheumatic pains, and slight paralysis. The Portuguese took into account these medicinal properties and also used it in place of rosemary (*alecrim* in Portuguese) to perfume houses (Welwitsch 1862).

Unfortunately, it was not possible to trace the colonial objects that were on display at the 1862 London Exhibition. Considering that these objects remained silent in the newspapers and official reports, the only way to bring these objects back was through the descriptions at *Synopse*. These descriptions provided insights into not only the uses of the plants but also the cross-cultural encounters between the explorer and the locals. Table 7.1 lists all the Angolan objects (149) organized by Welwitsch for the London Exhibition.

## Final Remarks

In the 1860s, the organization and participation in international exhibitions were essential instruments of political, economic, social, cultural, and scientific diplomacy. The participation of the different nation-states in these exhibitions meant the affirmation of national autonomies as a counterpoint to an increasingly global and internationalized context, taking into account the universal exhibitions as a "very powerful source for the most serious reflections, for the most important calculations, for the most important analyses, for the most productive studies."[50]

At this point, the colonial question was particularly complex with the expansion of European imperialism in Africa, marked not only by the political dimension but also by a strong economic, scientific, and technological component. As mentioned by the newspaper *A Revolução de Setembro*:[...] if we appear poorly before the assembly of peoples meeting in London, we will continue to be slighted and judged as unskillful in maintaining our autonomy [metropolitan and colonial] for lack of our own resources [...].[51]

The centrality of politics in these events largely explains the composition of the Portuguese Permanent Central Commission, marked by personalities related to Regeneration and central and colonial administration. Despite current criticism from opposition newspapers such as *A Revolução de Setembro*,[52] and generalist newspapers, such as *Jornal do Commercio,* consider the commission as elitist and unskilled,[53] it played an important role in the good Portuguese results in London, especially in the area of colonial products. The action of the Royal Commissioner and the study commission, appointed by the Portuguese government, was indispensable in assuming an ideal that considered the presence in these events as an indispensable means in the context of Portuguese *Regeneration*, related to the urgent need for material and immaterial progress. According to João de Andrade Corvo, the exhibitions presented themselves as a central instrument of civilization, promoting human "material improvement," collaboration and dialogue. Corvo considered

---

[50] "Commissão Portugueza para a Exposição Universal de Londres no anno de 1862". *Jornal do Commercio*, n.° 2280, 08-05-1861, p. 1.

[51] "Exposição Universal de Londres. Commissão Central Portugueza", *A Revolução de Setembro*, n.° 5805, 13-09-1861, p. 3.

[52] "Exposição Universal de Londres. Commissão Central Portugueza", *A Revolução de Setembro*, n.° 5805, 13-09-1861, p. 3.

[53] "Portugal na exposição Universal de 1862". *Jornal do Commercio*, n.° 2530, 11-03-1862, p. 1; "Exposição Universal de Londres". *Jornal do Commercio*, n.° 2664, 24-08-1862, p. 1.

the gathering and intellectual exchange between nations indispensable as a way of "patenting the productive forces" "and teaching each other the methods of creating wealth and satisfying not only the physical needs but also the intellectual needs of society."[54]

Most surprising would be the result of the collections sent by Welwitsch. Although the Portuguese representation was severely criticized by the press, Welwitsch was awarded four gold medals for the colonial objects presented, as a result of hard work done in Angola and funded by the Portuguese government. It is a successful example of the articulation of imperial strategic objectives, with particular research agendas by naturalists and collectors that would mark the scientific expeditions in Africa in the 1860s and 1870s. This result is impressive compared to the results obtained by institutions such as the Overseas Council [*Conselho Ultramarino*], which would only obtain two medals in London: one in the 2nd Class—section A (chemicals and pharmaceuticals) and another in the 4th Class—section C (animal and vegetable substances used in manufacturing).

This research brought to light objects that remained invisible since the London Exhibition of 1862. Welwitsch's publications were crucial for understanding which colonial objects from Angola were on display. Although it was not possible to trace these objects, the descriptions at the *Synopse* gave us a glimpse of tropicality: fruits, seeds, roots, gums, resins, and manufactured objects such as hats, tobacco boxes, saddlebags, and sieves. The descriptions included not only medicinal properties but also different uses of the species. The species listed at *Synopse* were chosen for display because of their fine quality, durability, and medicinal properties. These species, besides being used for woodwork (furniture and construction), also had diverse uses in which different parts and forms of the plants were used (oil, powder, leaves, and fruits) (Ficalho 1947). In addition, it is possible to notice that most of the species referred to at the *Synopse* were used by Welwitsch not only on his patients but also as a remedy to heal his own illnesses. On the other hand, through the Report by Royal Commissioner at the London Exhibition, it was possible to understand the spaces in which the Portuguese section was inserted, in particular, the space where the Angolan objects were displayed (Fig. 7.3).

The publication of *Synopse*, following a common practice within the nineteenth-century scientific community, contributed to the circulation, transmission, and dissemination of the knowledge acquired during the *Iter Angolense* expedition, simultaneously disseminating exotic objects and products in the European economic, scientific, and cultural context. Familiarizing the public with new realities and spatialities was essential in the effort to scientifically recognize colonial empires while, simultaneously, justifying the political and strategic interest of European nation-states in Africa, to reaffirm the economic potential of these territories—e.g., wood, cotton, and coffee.

---

[54] "Estudos sobre a Exposição de Londres". *Jornal do* Commercio, n.° 2725, 06-11-1862, p. 2.

**Acknowledgments** We thank the staff members of the Archives of the University of Coimbra, particularly Ana Margarida Dias da Silva. This work was supported by the Portuguese Foundation FCT (Fundação para a Ciência e a Tecnologia) (10.54499/DL57/2016/CP1372/CT0018) under the projects KNOW. AFRICA | KNOWledge networks in nineteenth-century AFRICA: A Digital Humanities approach to colonial encounters and local knowledge in the narratives of Portuguese expeditions (1853-1888) | ref. FCT - 2022.01599. PTDC (10.54499/2022.01599.PTDC) and PHONLAB—Phonetics Laboratory: Coimbra - Harvard. Rethinking twentieth-century scientific centres and peripheries | ref. FCT - 2022.06811. PTDC (10.54499/2022.06811.PTDC). It was also financed through national funding from the FCT under projects UIDB/04209/2020, UIDP/04209/2020, and LA/P/0132/2020 (10.54499/LA/P/0132/2020).

# References

Albuquerque S, Brummitt RK, Figueiredo E (2009) Typification of names based on the Angolan collections of Friedrich Welwitsch. Taxon 58(2):641–646

Alexandre V (1998) Situações Coloniais: I – A Lenta Erosão Do Antigo Regime. In: Bethencourt F, Chaudhuri K (ed) História Da Expansão Portuguesa, vol IV. Círculo de Leitores, pp 143–181

Alves JF (2006) As Exposições Industriais No Porto Oitocentista. In: Mourão JA, Matos AC, Guedes ME (eds) O Mundo Ibero-Americano Nas Grandes Exposições, Vega, pp 165–176

Areias MD (2012) Viagens e Expedições Científicas Dos Portugueses Ao Continente Africano Durante o Século XIX. Contributos Para o Conhecimento Da Geologia Africana. In: Diogo MP, Amaral I (ed) A Outra Face Do Império. Ciência, Tecnologia e Medicina (Sécs. XIX-XX). Edições Colibri, pp 31–48

Arnold D (1996) Inventing tropicality. In: Arnold D (ed) The problem of nature: environment, culture and European expansion. Blackwell Publishers, Oxford/Cambridge, pp 141–168

Arquivo Científico Tropical/Digital Repository (2023) Instituto de Investigação Científica e Tropical. https://actd.iict.pt/view/actd:XYLD0322. Accessed 21 Sept 2023

Bennett T (1988) The exhibitionary complex. New Formations 4:73–102

Blanchard P, Boëtsch G, Snoep N (2011) Exhibitions – L'invention Du Sauvage. Musée du Quai Branly, Actes Sud, Paris

Cantinho M (2006) O Museu Colonial de Lisboa (1870–1892). In: Museu Etnográfico Da Sociedade de Geografia de Lisboa: Modernidade, Colonização e Alteridade. Fundação Calouste Gulbenkian, pp 81–96

Castro MJ (2017) Art and Progress. Portuguese colonial representations in the great world exhibitions. In: Kong M, Monteiro MR, Neto MJP (eds) Progress(Es)—theories and practices. CRC Press, pp 309–314

Cayolla J (1945) Introdução. In: Mendonça A (ed) Colectânea de Escritos Doutrinários, Florísticos e Fitogeográficos de Frederico Welwitsch Concernentes Principalmente à Flora de Angola. Divisão de Publicações e Biblioteca e Agência Geral das Colónias, Lisboa

Çelik Z, Kinney L (1990) Ethnography and exhibitionism at the expositions Universelles. Assemblage 13:34–59

Dolezal H (1974) Friedrich Welwitsch Vida e Obra. Junta de Investigações Científicas do Ultramar, Lisboa

Ficalho C (1947) Plantas Úteis Da África Portuguesa – 2ª. Edição Prefaciada Revista Pelo Prof. Ruy Telles Palhinha. Agência Geral das Colónias, Lisboa

Figueiredo E, Smith G (2017) Common names of Angolan plants. Protea Boekhuis

Gilbert J (1994) World's fairs as historical events. In: Rydell RW, Gwinn N (eds) Fair representations: World's fairs and the modern world. VU University Press, Amsterdam, pp 13–27

Global Biodiversity Information Facility (2023) GBIF. https://www.gbif.org/pt/. Accessed 25 Sept 2023

Govaerts R (2017) World checklist of selected plant families. In: Bánki O, Roskov Y, Vandepitte L et al (eds) Catalogue of life checklist. https://doi.org/10.48580/d4sd-38c

Greenhalgh P (1991) Ephemeral vistas: the expositions Universelles, great exhibitions and World's fairs, 1851–1939. Manchester University Press, Manchester

Hiern WP (1896) Catalogue of the African plants collected by Dr. Friedrich Welwitsch in 1853–1861: Dicotyledons, vol I. British Museum (Natural History) – Dept. of Botany, London

Hoffenberg PH (2001) An empire on display: English, Indian, and Australian exhibitions from the crystal palace to the great war. University of California Press, Berkeley/Los Angeles/London

Huntley BJ (2008) Foreword/Prefácio. In: Figueiredo E, Gideon FS (eds) Plants of Angola/Plantas de Angola. South African National Biodiversity Institute, pp iv–v

International Plant Names Index (2023) The Royal Botanic Gardens – Kew, The Harvard University Herbaria, The Australian Nationa Herbarium. https://www.ipni.org/. Accessed 30 Sept 2023

Jerónimo MB (2018) Portuguese colonialism in Africa. In: Oxford research Encyclopedia of African history, pp 1–30

Kananoja K (2015) Bioprospecting and European uses of African natural medicine in Early Modern Angola. Portuguese Studies Review 23(2):45–69

Mackenzie J (2008) The Imperial exhibitions of Great Britain. In: Blanchard P, Bancel N, Boëtsch G et al (eds) Human zoos: science and spectacle in the age of colonial empires. Liverpool University Press, Liverpool, pp 259–268

Matos AC (2010) Les Musées Tecnhiques Portugais et les expositions Universelles au XIX Siècle/ Portuguese technical museums and universal exhibitions in the nineteenth century. In: Matos AC, Gouzévitch I, Lourenço M (eds) Expositions Universelles, Musées techniques et Societé Industrielle/world exhibitions, technical museums and industrial society. Edições Colibri, Lisboa, pp 71–96

Matos AC, Gouzévitch I, Lourenço M (2010) Introduction. In: Matos AC, Gouzévitch I, Lourenço M (eds) Expositions Universelles, Musées techniques et Societé Industrielle/ world exhibitions, technical museums and industrial society. Edições Colibri, Lisboa, pp 19–27

Natural History Museum (2014) Collection specimens [Data set]. Natural History Museum. https://doi.org/10.5519/0002965

Plants of the World Online (2023) The Royal Botanic Gardens – Kew. https://powo.science.kew.org/. Accessed 30 Sept 2023

Pratt ML (1992) Imperial eyes: travel writing and transculturation. Routledge, London/New York

Qureshi S (2011) Peoples on parade: exhibitions, empire, and anthropology in nineteenth-century Britain. University of Chicago Press, Chicago

Souto MH (2010) The Portuguese section of the histoire Du travail at the 1867's universal exhibition: ornamental art and museology in Portugal. In: Matos AC, Gouzévitch I, Lourenço M (eds) Expositions Universelles, Musées techniques et Societé Industrielle/ world exhibitions, technical museums and industrial society. Edições Colibri, Lisboa, pp 97–112

Souto MH (2011) Portugal Nas Exposições Universais 1851–1900. Edições Colibri, Lisboa

Vicente FL (2003) Viagens e Exposições: D. Pedro V Na Europa Do Século XIX. Gótica, Lisboa

Welwitsch F (1859) Carta Sobre a Cultura Do Algodão Na Província de Angola. Bol Offic Angola Parte Não Official: 694

Welwitsch F (1861) Cultura Do Algodão Em Angola. Imprensa Nacional, Lisboa

Welwitsch F (1862) Synopse Explicativa Das Amostras de Madeiras e Drogas Medicinaes e Outros Objectos Mormente Etnographicos Colligidos Na Provincia de Angola Enviados á Exposição Internacional de Londres Em 1862 Incluindo Os Que Foram Offerecidos Ao Gabinete Pharmacologico. Imprensa Nacional, Lisboa

Welwitsch F (1866) Observations on the origin and geographical distribution of the gum copal in Angola, West Tropical Africa. Bot J Linn Soc 9:287–302

Yallop J (2011a) Exhibition road, London, 1862. In: Magpies, squirrels & thieves: how the Victorians collected the world. Atlantic Books, London, pp 7–14

Yallop J (2011b) The promise of the east. In: Magpies, squirrels & thieves: how the Victorians collected the world. Atlantic Books, London, pp 331–351

# Chapter 8
# Developing Botany—Photography During the Victorian Era

**Vera Gonçalves and Filipe Barroso**

**Abstract** The Victorian era witnessed a period defined by an increasing turn towards progress and scientific interest, particularly the study of natural history. In addition, the emergence of photography resulted in significant changes in Victorian art. The natural world attracted great interest from nineteenth-century photographers. One of the main inventors of photography, Fox Talbot, had a great scientific interest in nature and natural phenomena, including botany. The study of botany and photography increased in popularity during the Victorian era. Moreover, many scientists and academics were interested in improving and developing photographic techniques and taking advantage of these processes, as they could be especially useful for documenting and displaying scientific knowledge, as in the field of botany.

This chapter has two objectives regarding the link between botany and photography. The first objective is to find botany among photographs, namely, the botanical elements and the garden represented across the different processes throughout the history of photography. Also, highlight some relevant scientific articles, books, and academic papers in which photography had a very significant presence. Second, the importance of botany in the development of photography should be assessed by evaluating the discreet botanical elements present in the chemistry of each photographic process.

**Keywords** photography · botany · Victorian era · nineteenth century · gardens · botanical books · photomechanical photo printing

V. Gonçalves (✉) · F. Barroso
Independent Researcher, Five Historic Photography Studio, Sintra, Portugal
e-mail: info@fivestudiosintra.com

## Introduction

The study of photography in the Victorian era (1838–1901) reveals the birth and evolution of a new art and science, with all the elements that characterize an era and the evolution of a technique. Chronologically, the evolution of photography has focused on a large part in the Victorian era, which was very rich in knowledge and progress. This development and advance were largely due to the influence of photographers from the United Kingdom as well as photographers and scientists interested in this art from all over the world. The Great Exhibition of 1851, held in London, gave photography a strong boost, as the first examples of photography were shown there, which also influenced Victorian art. Queen Victoria is also considered to have been the first British monarch to be photographed (Lyden 2014).

With the advent of photography, many academics and researchers were interested in investigating and taking advantage of their scientific knowledge to improve photographic techniques by associating photography with science. There are many examples and studies that show the importance of photography for the study of science, for example, in medical studies (Peres 2013). In addition, if we think of other sciences, such as zoology or botany, which were in constant progress in those times, they were certainly influenced and fortunate by the development of photography.

The true history of botanical photography began well before the nineteenth century and even before the invention of photography. The desire to be able to reproduce nature is ancient (Browne 2008). However, during the nineteenth century, a time when gardens, flowers, and botany, in general, were part of the interests of a large group of society (Burns 2017), it is not surprising that botanical elements and gardens (as a whole) had a very significant presence in the first photos. Many inventors and early proponents of photography were actively engaged in producing photographic images of botanical subjects. Such immediate entwining was the result of common interests among practitioners of botany, photography, and illustration. When photography was introduced to the public in 1839, it immediately began to displace the record-making function of other art forms, such as drawing and painting. At the time, photographs seemed to be a direct transcription of reality, precisely recording what was put in front of the camera or in contact with the photographic materials. In creating these early transcriptions of reality, it is not surprising that the first photographers turned to their gardens for inspiration (Allen and McNear 2018) because several British scientists and pioneer photographers either kept a garden or studied botany.

As photography means 'sun writing', it is not surprising that some of its earliest subjects were botanical. A British scientist and one of photography's earliest practitioners, William Henry Fox Talbot (1800–1877), used flowers during his first experiments. As he wrote to a botanist friend about his discovery, '*I believe that this new art will be a great help to botanists…especially useful for naturalists since one can copy the most difficult things with a great deal of ease…I have practiced this art since the year 1834*.' Botany and gardens were the perfect subjects for long photographic exposures typical of early techniques: they were inexpensive, and they did not move (Browne 2008), which was something important at those times. Therefore, the garden had a significant presence in early photography during the Victorian era (Kocol 2010). In addition, when we speak of gardens, we mean gardens and all its elements. Thus, from an attentive analysis of the experiences and works presented by different photographers, all over the world during the Victorian era, it is not difficult to find botany as a subject of photography. The garden was reflected in different elements, such as leaves, flowers, trees, floral arrangements (still life), and the capture of the garden as a whole. Still life was a popular theme for photographers during the nineteenth century for several reasons. Technically, the long exposure times required to capture an image on a light-sensitive surface meant that moving subjects were impossible to register until at least the 1860s (Rhodes 2008). The most immediate relationship between photography and botany was the use of botany and its elements as objects to be photographed; thus, we can easily identify these elements in early photographs.

However, if we consider the influence of botany in photography, in a much more detailed way, we will find that the raw materials used in different photographic processes had a plant origin, mainly the wood used in most cameras of the time, without forgetting oils, resins, or pigments. In fact, much chemistry, used in experiments and inventions introduced throughout the history of photography, is based on organic elements or compounds extracted from plants.

On the other hand, it is also necessary to point out that photography offered botanists a better definition of their objects of study, and so became part of the research and recording of observations, as well as in the description of the species. Photography has also become very important in books on botany and played an important role in spreading science to the public.

## Botany as a Subject of Photography

### *The Representation of the Garden and Botanical Elements*

In the nineteenth century, there was an explosion of interest in gardening at all levels of English Society, including the new middle classes. Just as simple botanical elements were represented in photography, gardens were also a very common element in Victorian-era photography. In certain respects, the history of photography is not dissimilar to the history of gardens. The pioneers of photography were scientists who understood chemistry and optics. Many, such as Fox Talbot and Julia Cameron (1815–1879), were from the upper class, educated, and artistically motivated (Allen and McNear 2018). Talbot and Cameron were great pioneers of photography and were both enthusiastic about botanical and garden trends and botanical discoveries of their time. As a result, botanic elements and gardens were represented prominently in early photographic works (Kocol 2010) (Fig. 8.1).

**Fig. 8.1** 'Leaf' by
William Henry Fox Talbot
(ca. 1839). Salted paper
print. Metropolitan
Museum Collections

Early photographs reflected this myriad of applications, many of which relied on existing artistic and scientific conventions, and botany appeared across all of them. As Henry Fox Talbot suggested, the potential of photography to facilitate the scientific study of botanical specimens was recognized immediately upon its invention (Allen and McNear 2018). Just as botanicals from Fox Talbot garden provided material in some of his first photographs, they continue to be closely connected with photography and gardens remain a site of experimentation and play for photographers (Allen and McNear 2018). Talbot later wrote, '*The first kind of objects I attempted to reproduce with this process were flowers and leaves, either fresh or selected from my herbarium* (Figure 1)' (Kocol 2010).

Following the public announcement of the method developed by Louis Jacques Mandé Daguerre (1787–1851) for capturing a unique image on a silver-coated copper plate, in 1839, Andreas Ritter von Ettingshausen (1796–1878), an Austrian academic, became interested in the application of the daguerreotype in science (Cull 2016), eventually making daguerreotype photomicrographs of cuts from plant tissues (1840). It was through microphotography that daguerreotypes began to benefit science studies, and many photographers began to use Daguerre's invention in their studies and observations (Table 8.1).

However, even before the invention of the first photographic processes, the pioneers of photography had already begun their investigations with photograms and photogenic drawings, which are not yet considered true photographs but were very important in the research process of inventing photography. Some were even invented after the first true photographic processes emerged, as is the case with the cyanotype process, showing interest in creating photographs without a camera, which used to be called '*cameraless*' photography. The object to be photographed and the surface of the paper needed to be in direct contact with each other, producing an image of the botanical specimen in light and shadow. The translucent parts of the object result in a darker impression, and the more transparent parts produce a lighter colouration. This method of placing a plant in direct contact with a surface to make an impression was a common technique in botany known as '*nature printing*' (Steidl 2012). The desire for very precise and accurate details, which are necessary for the classification and identification of plants, posed problems for scientific

**Table 8.1** Botany and photography—Botany revealed throughout the history of photography

| Botanist/Photographer | Year | Botanical Reference | Photographic Process/ Photomechanical Photo Printing |
|---|---|---|---|
| Thomas Wedgwood (British, 1771–1805) with Sir Humphry Davy (1778–1829) | 1802 | Plants leaves Managed to create photographic images, but not yet devised a way of 'fixing' them or making them permanent | Salt printing image - contact print |
| Sarah Anne Bright (British, 1793–1866) | 1839 | *'Quillan leaf'* | Contact print |
| William Forrester (Scottish, 1804–1866) | 1839 | Photographic impressions of ferns | Photo-lithograph |
| Alfred Donné (French, 1801–1878) | 1840 | First photographic pictures of natural objects, e.g. pollen grains | Daguerreotype |
| Andreas Ritter von Ettingshausen (Austrian, 1796–1878) | 1840 | Microscopic cross-sections of botanical specimens | Daguerreotype |
| Joseph-Philibert Girault de Prangey (French, 1804–1892) | 1841/1842 | 'Palm study'/ *'palm tree near the Church of Saints Theodore'* | Daguerreotype |
| Henry Fox Talbot (British, 1800–1877) | 1838–1839 | Some experiments titled 'botany' / 'botanical specimens' 'leaf of a plants' (...) | Photogenic drawings on salted paper and calotype |
| | 1839–1840 | *The Bertoloni album* | Photogenic drawings on salted paper |
| | 1844–1846 | *The pencil of nature* | Print images from calotype negatives |
| | 1858 | Experiments with flowers and leaves for the development of a new printing technique | Photoglyphic engraving |
| Golding Bird (British, 1814–1854) | 1839 | *'Observations on the application of heliographic or photogenic drawing to botanical purposes'* | Paper published in the *Magazine of Natural History*, with reference to new photographic techniques |
| Richard Willats/Theodoro Redman (British, 1820-?)/(British, 1811-?) | 1839 | *'Richard Willats album'* | Photogenic drawings (some with flowers and leaves) |
| Michael Pakenham Edgeworth (British, 1812–1881) | 1839 | Photogenic drawing as an aid to his botanical studies in India | Photogenic drawings |
| Johann Carl Enslen (German, 1769–1849) | 1839 | *'Spring leaves'* | Photogenic drawings on salted paper |

(continued)

**Table 8.1** (continued)

| Botanist/Photographer | Year | Botanical Reference | Photographic Process/ Photomechanical Photo Printing |
|---|---|---|---|
| Levett Ibbetson (British, 1799–1869) | 1840/1852 | *Le premier livre imprimé par le soleil** (contact prints of ferns, grasses, and flowers) | Photogenic drawings/ calotype |
| Hippolyte Bayard (French, 1801–1887) | 1839–1841 | *'Dessins photographiques Sur papier'* | Photogenic drawings on salted paper |
| | 1847 | *'In the garden'* (self-portrait) | Salted paper print |
| | 1850 | *'Bouquet of flowers'* (still life) | |
| Mathew Carey lea (Philadelphia-born, 1823–1897) | 1841 | *Photogenic drawings of plants indigenous to the vicinity of Philadelphia** | Photographic process based on potassium dichromate |
| Robert Hunt (British, 1807–1887) | 1839 | *'Photographic drawing of leaves'* | Direct-positive contact photogenic drawing using bleaching process |
| | 1844 | *'Botanical specimens'* | Chromotype (experimental photogenic drawing) |
| Anna Atkins (British, 1799–1871) with Anne Austen Dixon (1799–1877) | *1843–1853* | *British algae: Cyanotype impressions** | Cyanotype photograms |
| | 1852–1864 | *Cyanotypes of British and foreign flowering plants and ferns ** | Cyanotype photograms |
| Bessie Rayner Parkes Belloc (British,1829–1925) | 1848 | *'Photogenics drawings of various objects'* | Photogenic drawing negatives and prints |
| Blanche Shelley (British, 1841–1898) | 1854 | *'Ferns and daffodil'* | Photogram |
| Alois Auer (Austrian, 1813–1869) | 1855–1856 | *'Physiotypia plantarum Austriacarum'* a collection of natural self-prints of the *'vascular plants of the Austrian Imperial state'* | Natural self-prints |
| Constantin von Ettingshausen (Austrian, 1826–1897) with Alois Pokorny (Austrian, 1826–1886) | 1855–1856 | *'Physiotypia plantarum Austriacarum'* | Natural self-prints |
| | 1864 | *'Photographisches album der Flora Osterreichs'* | Heliogravures |
| Caleb Burrell Rose (British, 1790–1872) | 1859 | *Two specimens of fern leaf* (an early attempt at photography from Norwich) | Photogenic drawing of two plant specimens |

(continued)

**Table 8.1** (continued)

| Botanist/Photographer | Year | Botanical Reference | Photographic Process/ Photomechanical Photo Printing |
|---|---|---|---|
| John Dillwyn Llewelyn (British, 1810–1882) | 1842 | Images of rare orchids | Daguerreotype |
| | 1854 | *'Thereza Dillwyn Llewelyn with her microscope'* | Salted paper print from glass negative and ferns photogram |
| Mary Dillwyn (British, 1816–1906) | 1845 | *'The Mary Dillwyn album'* (some flower studies) | Salted paper print |
| Thereza Llewelyn (British, 1834–1926) | 1853 | Album with 12 contact prints of marine algae | Photogenic drawings |
| Cecilia Louisa Belville Glaisher (British, 1828–1892) | 1854–1856 | *'The British ferns represented in a series of photographs'* | Salted paper photoghenic drawings |
| Adolphe Braun (French,1812–1877) | 1854 | *'Fleurs Photographiees'* (studies of botanical specimens) | Collodion-albumen |
| Louis-Alphonse de Brébisson (French,1798–1872) | 1855 | Microphotography applied to botany, mainly with algae | Wet plate collodion |
| John Wheeley Gough Gutch (English, 1808–1862) | 1857 | Decorating the title page of each album with a photographic collage with leaves and flowers | Salted paper photograms |
| Henri Le Secq (French, 1818–1882) | 1856 | *'Bouquet of flowers'* (still life) | *Unidentified process* |
| Francis Bedford (British, 1815–1894) | 1856 | *'A study of plants'* (still life) | *Unidentified process* |
| Franz Antoine (Austrian, 1815–1886) | 1857 | *Die Cupressineen-Gattungen: Arceuthos, Juniperus u. Sabina*\* | *Unidentified process* |
| Roger Fenton (British, 1819–1869) | 1859 | *'Lily house, botanic garden'* (still life) | Albumen silver print |
| | 1860 | *'Group of Fruit and Flowers'* | Albumen silver prints, stereograph |
| Eduard Moritz Lotze (German, 1809–1890) | 1859 | *Saggio fotografico di alcuni animali e piante fossili dell'Agro Veronese*\* (paleophotography) | *Unidentified process* |
| Louis Godefroy Lucy-Fossarieu (French, 1822–1892) | 1860 | Album photos for the new zoological and botanical acclimatisation garden in bois de Boulogne | Albumen prints |
| Charles Hippolyte Aubry (French,1811–1877) | 1864 | *'Studies of leaves'* (still life) | Albumen print from glass negative |

(continued)

**Table 8.1** (continued)

| Botanist/Photographer | Year | Botanical Reference | Photographic Process/ Photomechanical Photo Printing |
|---|---|---|---|
| Julia Margaret Cameron (British, 1815–1879) | 1863 | *'My printing of ferns'* | Albumen print photogram |
| | 1862 | Portrait of Kate dore with photograms of ferns | Albumen print from glass negative and ferns photogram |
| | 1867 | *'The Gardener's daughter'* | Unidentified process |
| Eugène Charles de Gayffier (French, 1830–1910) | 1867 | *Herbier Forestier de la France** | Heliogravures / phototype reproduction |
| Charles F. Himes (American, 1838–1918) | 1868 | *Leaf prints: Or glimpses at photography** (publication to teach the photogram technique) | (with a albumen silver print of a photogenic drawing of leaves) |
| Arthur-Louis Ducos du Hauron (French, 1837–1920) | 1869 | *Les Couleurs en Photographie - solution du Problème** | Trichome process |
| Anna K. Weaver (American, 1847-?) | 1874 | *'No cross, no crown'* | Albumen silver print of photogrames |
| Eugène Chauvigné (French, 1837–1894) | 1875 | *'Roses'* (still life) | Albumen silver print from glass negative |
| Pietro Guidi (Italian,? -?) | 1870–1874 | *Flora fotografata delle piante piu pregevoli e peregrine di SanRemo e sue adiacenze per Francesco Panizzi** | Albumen print on cardboard |
| Amelia Bergner (American, 1853–1923) | 1877 | *'Photogram of leaves'/'study of leaves'* | Gum dichromate photograms |
| Herbert Boucher Dobbie (New Zealander, 1852–1940) | 1880 | *New Zealand ferns** | Cyanotype photograms |
| Louis Olivier (French, 1854–1910) | 1881 | *Recherches Sur l'appareil tégumentaire des racines** | Microphotography on glass negatives (collodion and gelatin) printing on the book in 'photoglyphie' (Woodburytipia) |
| Heinrich von Waldner (?-?) | 1883 | *Deutschlands Farne** | Phototypes |
| Eugéne Trutat (French, 1840–1910) | 1884 | *La photographie appliquée à l'histoire naturelle** (botanical examples, namely, algae) | Phototypes and figures engraved in the text |
| Geraldine Moodie (Canadian, 1854–1945) | 1898 | Studio photograph of blooms | (mounted on photographer's card. Inscribed in negative) |

(continued)

**Table 8.1** (continued)

| Botanist/Photographer | Year | Botanical Reference | Photographic Process/ Photomechanical Photo Printing |
|---|---|---|---|
| John Joly (Irish, 1857–1933) | 1895 | Lantern slides with botanical specimens, *the National Library of Ireland* collection | Glass photographic plate with fine vertical red, green, and blue lines less than 0.1 mm wide printed on them |
| Katharine Gordon Breed (American, 1871–1915) | 1896–1907 | Lantern slides with botanical specimens, *Mira dock forestry lantern slides* collection | Silver-albumen positive transparencies (hyalotype) and handcolored |
| Mira Lloyd Dock (American, 1853–1945) | 1896–1907 | Lantern slides with botanical specimens, *Mira dock forestry lantern slides* collection | Silver-albumen positive transparencies (hyalotype) |
| Ogawa Kazumasa (Japanese, 1860–1929) | 1895 | *Some Japanese flowers** | Book with collotypes of flowers |
| Robert Demachy (French, 1859–1936) | 1899 | *'Young woman among flowers'* | Proof printed in bichromated gum |
| Henri Gadeau de Kerville (French, 1858–1940) | 1890 | *Les vieux arbres de la Normandie—Étude Botanico-Historique** | Photocollographie (phototype) |
| Karl Blossfeldt (German, 1865–1932) | 1890–1898 | Teaching materials for arts and crafts classes | Gelatin silver negatives / slides/gelatin silver prints |
| Moritz Meurer (German, 1839–1916) | 1896–1899 | *Meure's Pflanzenbilder (Studie of plants)** | (with Karl Blossfeldt photos) |
| | 1896 | *Die Ursprungsformen des griechischen Akanthusornamentes und ihre natürlichen Vorbilder** | (with Karl Blossfeldt photos) |
| Louis Braeme (French, 1858–1935) with Armand Suis (French, 1858-?) | 1900 | *Atlas de photomicrographie des plantes médicinales** | Similigravure (autotypie) |
| Henry Troth (American, 1859–1945) | 1900 | *'Lady Fern'/'white-fringed Orchis Hab'* | Gelatin silver print |
| | 1891–1907 | Lantern slides with botanical specimens, *Mira dock forestry lantern slides* collection | Silver-albumen positive transparencies (hyalotype) |
| | 1900 | *'Nature's Garden'* (in collaboration with Arthur Radclyffe Dugmore 1870–1955)* | Unidentified process |
| | 1911 | *'Phytogeographic survey of North America'* (botanical study of John William Harshberger's) | Unidentified process |

**Fig. 8.2** 'Photographs of British Algae: Cyanotype Impressions (1843–1853)'. First page and *Rhodomenia polycarpa cyanotype printing*. Anna Atkins. Metropolitan Museum Collections

illustrators. Both nature printing and photogenic drawing were claimed to be more exact because they produced images made directly from actual specimens. Anna Atkins (1799–1871), considered by many to be the first who attempted the practical marriage of photography and botanical science, employed the cyanotype process for the first book in the world illustrated entirely with photographs (photograms): *British Algae: Cyanotype Impressions* (1843–1853). The first book to be printed photographically (Fig. 8.2) stands as a testament to this Victorian woman's imagination and perseverance (Schaaf 1979). Atkins's project lacked a rigorous method, but it demonstrated that she enjoyed the freedom to work '*at the outer limits of the patriarchal conduct of normal science*' in a way that might be seen to problematize '*the system of positivist classification and the apparatus of the illustration*' that would dominate much of the scientific literature of the nineteenth century. The beauty and uniqueness of her publication continued to impress photographic practitioners and students of botany even though, by the 1850s, other books using drawings of specimens, dried and mounted specimens, or images produced by Alois Auer's (1813–1869) nature printing technique rendered Atkins's work obsolete (Browne 2008). With the decade-long project completed, she continued to make cyanotype photograms in collaboration with her close friend Anne Dixon (1799–1864). Between 1852 and 1864, the two women produced three albums: *Cyanotypes of British and Foreign Ferns*, *Cyanotypes of British and Foreign Flowering Plants and Ferns*, and an unnamed third album, the latter with the possible assistance of Herschel's daughter, Isabella Herschel (1831–1893) (Browne 2008). The cyanotype photographic printing technique used by Anna Atkins was invented in 1842 by Sir John Herschel (1792–1871) shortly after photography was

**Fig. 8.3** 'Photogenic Drawing of a Plant' William Henry Fox Talbot (1839–1840). Salted paper print. Metropolitan Museum Collections

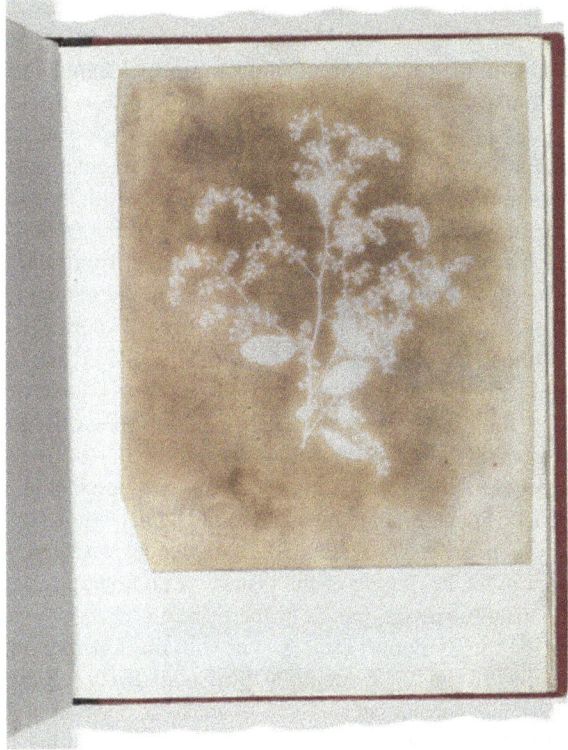

announced to the world in 1839 (Herschel 1842). However, as mentioned earlier, Fox Talbot was among the first to state the potential of photography for botanical studies, and that was not limited to his photogenic drawings of botanical elements or his contributions to the evolution of photography. Around 1839, Talbot sent early photogenic drawings (Fig. 8.3) to Antonio Bertoloni (1775–1869), based in Bologna, and to William Jackson Hooker (1785–1865), at the Royal Botanic Gardens, Kew, near London, with the suggestion that his process could be useful for sharing information, reproducing the form of plants, and solving the problem of transporting botanical specimens, since plants could be left behind, and lightweight, thin paper photographs could be carried home (Cull 2016). However, the legibility of the photographic image was frustrating for botanists. Hooker, according to Smith (1993), was less than impressed with the lack of detail in Talbot's images as a method of collecting information from nature, suggesting instead that his process could be more effective if it was used to reproduce botanical drawings. Talbot *photogenic drawings* of plants were publicly displayed in exhibitions, published in scientific journals, sent to leading botanists, and included in his famous book on the use of photography, *The Pencil of Nature* (1844–1846) (Cull 2016). Talbot was also the English inventor of photography. Encouraged by the frustration of his inability to draw accurately, he saw the potential for using new photographic processes to capture botanical details and the garden as a whole. He developed a technique called photoglyphic engraving, intended to overcome the fact that some of his early experimental photos faded. Photoglyphy is a process in which gelatin is used to fix an image to the surface of a metal plate, which can then be etched, inked, and printed onto paper (Nadeau 2018). He tried this process using a range of English flora.

Talbot and Atkins accurately predicted the role that photography would play across the sciences, and their photograms revealed early on how new modes of representation could be used to interpret the world; in that respect, their contribution to the art of photography proved to be substantial (Allen and McNear 2018).

Also important to highlight are the self-portraits (printed on salted paper) of Hippolyte Bayard (1801–1887) sitting in the garden and with gardening tools—*In the Garden* (1847). In five of the seven self-portraits, he placed himself in garden settings. This was, in part, a practical decision since natural light was required to make photographs at the time. However, his choice of setting also reflects his passion for plants. He came from a family of gardeners—his maternal grandfather worked in the extensive grounds of the abbey—in Breteuil, the village where Bayard grew up. His father, a magistrate, was a passionate amateur gardener who grew peaches in an orchard attached to his family home (Getty Museum 2020).

There are many examples of experiences with photograms and photogenic drawings in the beginning of the history of photography, using different techniques and printing processes, but all of them are motivated towards the same inspiration, that is, the use of plants (Table 8.1) or with the concrete objective of using them in their studies (Ryan 2017; Steidl 2019a; Steidl 2019b). Similarly, Talbot's technique was used in approximately 1854 to make a portrait of the Welsh scientist Thereza Dillwyn Llewelyn (1834–1926) looking through her microscope (Fig. 8.4). In her portrait, seemingly made by her father (John Dillwyn Llewelyn), she is encircled by a wreath of contact printed ferns, perhaps intended as an ocular metaphor that aptly reiterates her own act of enhanced seeing (Batchen 2016). The image combines the two principal means of photographic picture-making invented by Talbot: an image made by direct contact with an object and a camera image printed from a negative. The picture presents two characteristic pursuits of educated Victorian women, art and science, the dual poles that also characterized amateur photography itself (Metropolitan Museum 2020). In 1853, she herself had put together an album of 12 of her photogenic drawing contact prints of marine algae gathered at Caswell (Batchen 2016). There are many references to images created by John Dillwyn Llewelyn (1810–1882) and their relationship with botany. An attentive analysis of the collection of the National Museum of Wales allows us to understand the work of this photographer and evaluate the relationship between botany and photography. Talbot himself considered Llewelyn the first botanical photographer. In 1842, Llewelyn used the daguerreotype process to send images of rare orchids to Kew for identification (Morris 2008).

An English album maker named Blanche Shelley (1841–1898) made a similarly delicate contact print using Talbot's photogenic drawing process of a daffodil and ferns, as late as 1854. From 1853 to 1856, another English woman, Cecilia Glaisher (1828–1892), compiled an album with photogenic drawings of ferns of unparalleled beauty. Meticulously done, they impressed the natural history publisher Edward

**Fig. 8.4** 'Thereza' by
John Dillwyn Llewelyn
(1853–1856). Salted paper
print. Metropolitan
Museum Collection

Newman (1801–1876), so he intended to introduce these photogenic drawings into the highly competitive botanical print market (Batchen 2016). He proposed the publication of an album of salted paper impressions with the title *The British Ferns – Photographed from Nature by Mrs Glaisher, represented in a series of photographs* (Batchen 2016). Nevertheless, the book project was abandoned (Marten 2008). Reflecting on the links between botany and photography, Cecilia Glaisher wrote, '*the process of Photography is admirably adapted to making faithful copies of Botanical Specimens, more especially to illustrate the graceful and beautiful class of Ferns: it pursues the advantage over all others hitherto employed of displaying, with incomparable exactness, the most minute characteristics: producing absolute fac-similes of the objects, perfect both in artistic effect and structural details*' (Tucker, 2005).

Louis-Alphonse de Brébisson (1798–1872) was one of the first botanists to discover microscopic algae (and shoot them), and he published several works on this subject. He was the author of *Flore de la Normandie* (1836) and contributed to the monumental *Flore Générale de France* (1828–1829). Brébisson made important contributions to the development of photography, pioneering the art of photography in Normandy. He was one of the first in France to use collodion negatives, and in 1849, built a camera adapted to a microscope for use in his botanical research (Frizot 1998).

**Fig. 8.5** 'Spring' (1865).
Julia Margaret Cameron.
Albumen print. The J. Paul
Getty Museum

Additionally, the photographer Julia Margaret Cameron (1815–1879) made photograms of botanical subjects (Table 8.1). The portrait of Kate Dore (1815–1879), in collaboration with the fellow photographic pioneer Oscar Gustave Rejlander (1813–1875), features a border decorated with photograms of ferns. Using Rejlander's glass plate negative, Cameron placed ferns on the glass during the exposure of light-sensitive albumen paper (Kocol 2010). The great part of the work of this photography pioneer is marked by her predilection for flowers (Fig. 8.5), and their Victorian symbolism is evident in her photographs, which included ivy (representing fidelity), sunflowers (adoration), camellias (graciousness), and magnolias (dignity). Lilies-of-the-valley, laurels, daisies, morning glories, magnolias, grapevines, and wheat ears were also featured (Kocol 2010). Many Cameron's photos are portraits in which the subjects are in the garden. We must not forget that the garden was a romantic place, an ideal setting. A good example is Cameron's photograph *The Gardener's Daughter* (1867).

In the early 1850s, Adolphe Braun (1812–1877) began photographing flowers to design new floral patterns. Making use of the recently developed collodion process, which allowed printing reproductions of glass plates, he published over 300 of his photographs in the album *Fleurs Photographiées* (1855). These photographs caught the attention of the Paris art community, and Braun produced a second set for display at the Paris Universal Exposition that same year (Monovisions 2017).

The representations of still lifes reflected the interests and material resources of the wealthy Victorian middle classes, namely, the coded language of flowers

(Roberts 2004). Several were the photographers who portrayed still lifes with flowers, fruits, and leaves: Hippolyte Bayard (1801–1887), '*Bouquet of Flowers*' (1850); Henri Le Secq (1818–1882), '*Bouquet of Flowers*' (1856); Francis Bedford (1815–1894), '*A Study of Plants*' (1856); Roger Fenton (1819–1869), '*Lily House, Botanic Garden* (1859); Eugène Chauvigné (1837-1894), '*Roses*' (1875).

Early botanical and floral studies represented a combination of art and science (Allen and McNear 2018), as in the example of the American photographer Henry Troth (1859–1948), who speaks to his curiosity and fondness for accuracy: each composition '*dissects*' a botanical specimen so that every one of its components— flower, leaf, stem, and root—comes to life. He often mimicked an approach typical of botanical drawing—he cut the specimen into pieces to show all parts on the same page (Peterson 2012). Botany held sway for Troth, prompting him to make hundreds of straightforward photographs of flowers. These were used for several books, e.g., *Nature's Garden* (1900), in collaboration with Arthur Radclyffe Dugmore (Table 8.1). He also provided original photographs to the *Botanical Division of the Academy of Natural Sciences of Philadelphia* (Peterson 2012). We can also find Troth's works in the field of photography with botanical elements on slides for magic lanterns (Fig. 8.6) and photos on glass, such as those in the *Mira Dock Forestry Lantern Slides Collection* (mostly on Pennsylvania flora). We can also find works by photographers Katharine Gordon Breed (1871–1915), John Horace McFarland (1859–1948) and Mira Lloyd Dock (1853–1945), a renowned Pennsylvania environmentalist, botanist, and educator. This collection includes 468 glass lantern slides dating from 1897 to approximately 1902, apparently used by Mira Lloyd Dock in her lectures at the Mont Alto Forest School (*Mira Dock Forestry Lantern Slides Collection*) (Fig. 8.7). This collection emphasizes the importance that photography had in the dissemination and study of flora in certain areas. Some of the glass photographs were hand-painted because there was a desire for colourful

**Fig. 8.6** *Aralia trifolia.* Dwarf sinsing plants showing roots, leaves, and flowers. Henry Troth (1891/1907). Lantern slide from Mira Lloyd Dock Glass Lantern Slides Collection. Pennsylvania State University

**Fig. 8.7** *Sarracenia flava*
plants in vase (1898). Mira
Lloyd Dock. Lantern slide
from Mira Lloyd Dock
Glass Lantern Slides
Collection. Pennsylvania
State University

images. This was a long-standing desire to go beyond monochrome photography to colour. This was achieved by Louis Ducos du Hauron (1837–1920) in 1869, using a sequence of green, orange, and violet filters printed on thin sheets of bichromated gelatin containing carbon pigments of red, blue, and yellow, thus allowing the production of *naturally coloured* photographic images. In the same year, as he made his *cameraless* example, Ducos du Hauron described his results in *Les Couleurs en Photographie: Solution du Problem*. Although a commercially viable solution was still some way off, the possibility of obtaining colour photographs could no longer be denied (Batchen 2016).

Almost 40 years after the cyanotype works of Anna Atkins, another cyanotype work emerged, with ferns from New Zealand. Herbert Dobbie (1852–1940), a railway station master and amateur botanist who emigrated to New Zealand, from England, in 1875, made cyanotype contact prints of specimens of all 148 known species of fern in his new country (1880) and published them in the album—*New Zealand Ferns* (I and II) (McCraw1988). The result is a group of images that hover somewhere between science and art, between popular aesthetic enjoyment and commercial profit (Fig. 8.8) (Batchen 2016).

However, quickly evolving technology, which both simplified photographic processes and made them more affordable, helped to commercialize production and professionalize the trade. Then, just as quickly, with the introduction of the first Kodak box camera in 1888, photography became the pastime of the amateur, anyone with time and money to spare (Allen and McNear 2018). English landscape gardens became popular guidelines for many landscape photographers, professionals, and amateurs. With the evolution of photography and cameras, amateur photographers began to capture their interest and the people around them, including those who were enjoying themselves in the garden, with family and friends. These images go far beyond the work of the great photographers already mentioned.

**Fig. 8.8** *Nephrodium decompositum* var. A, Kaipara. From the album: 'New Zealand ferns'. 148 varieties (1880). Herbert Dobbie. Museum of New Zealand Collection

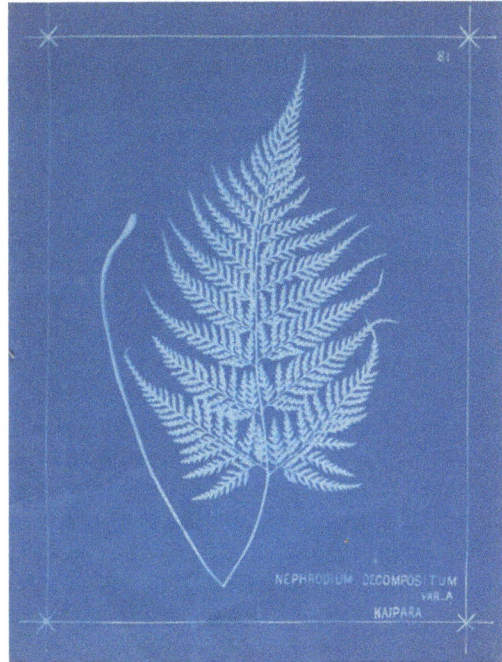

In our search for Victorian Age botanical photographs, we found that most photographers had a particular interest in botany. Victorians are known for their many interests and studies in different sciences, so it is easy to accept that most botanists were also interested in photography, intending to create images for their own publications. The examples of botanical elements represented in nineteenth century photography are many, and certainly, the numerous collections found in Botanical Museums and Botanical Gardens will tend to increase our knowledge. A non-exhaustive survey of photographers, techniques used, and botanical records is shown in Table 8.1.

## Photography and Botanical Publications

The introduction of images in books or scientific publications occurred much earlier than the presentation of photography (Peres 2013). Talbot was the first to consider that photography could benefit scientists, particularly in cataloguing their observations.

With the introduction of photography, an attempt to reproduce images recorded by photography emerged (Peres 2013), leading to the emergence and advancement of photomechanical printing processes.

For scientists, and particularly for naturalists, an image could be a working document, but it was also of greater importance if it was used to illustrate their work. Some examples have already been mentioned above. Many examples of botanical elements were photographed by scientists (Table 8.1). Thus, it is natural that this new possibility of representation was a reason for their new interest and study. With the invention of photography, several technical manuals were published with the aim of teaching and spreading this new technique and its fields of application, and the sciences and natural history were no exception (Fieschi 2008). In 1867, Albert Moitessier (1833–1889) published a manual to explain how photography could be applied to micrographic research, *La Photographie Appliquée aux Recherches Micrographiques*, providing all the technical data necessary for scientists to record images under a microscope (Moitessier 1867). A few years later, an illustrated manual entitled *Photography Applied to Natural History* (Trutat 1884) was published by Eugéne Trutat (1840–1910) to explain the application of photography in natural sciences, including botany. However, long before the publication of these technical manuals, there were several books with photographs. For example, the aforementioned (first photography book to be published) with the photograms of Algae by Anna Atkins (1843) was used as an herbarium, similar to publications on botany with illustrations (Schaaf 1979). It was a book with little botanical information but functioning as a complement to the book *Manual of the British Marine Algae* (1841) by the botanist William Henry Harvey (1811–1866) and with a more artistic purpose, intended to be offered to his circle of friends and scientists. This book was important for demonstrating to scientists, on a practical level, how education and science could benefit from photography. It was the precursor of many attempts to illustrate books with photographs that were to follow. Above all, it foretold the democratizing effect that photography would have on illustration and publication (Schaaf 1979).

With a somewhat similar typology, even if the technique was more advanced, because, in this case, it is a book made up of photos of the leaves of an herbarium (52 phototypes) accompanied by the description of these same species Heinrich Waldner published *Deutschlands Farne* (1883) (Fieschi 2008). In 1867, the book *Herbier Forestier de la France* was published by Eugène Charles de Gayffier (1830–1910). This was noteworthy because it was one of the first such books illustrated with photography (200 phototypes) rather than with lithographs or engravings. It was one of the first to photograph true forest foliage and to document it rather than the large parks near Paris (Fieschi 2008).

Initially, photographers, with an interest in botany, used plants as objects for their images, but for scientists, these photographs had a more specific purpose. Photography emerged as an instrument for research and science communication, where scientists used photography for their work; therefore, photography was introduced in laboratories as a tool for research and recording observations (Fieschi 2008). The evolution of photography facilitated its use by botanists and naturalists, especially since 1880, with the advent and commercialization of plates with *gélatino-bromure d'argent* (gelatine silver-bromide), which were easier to use in amateurs' photography (Fieschi 2008).

The use of photomicrography allowed recording observations made under the microscope. These observations could be tiring and do not allow perfect observation. However, capturing the image under the microscope allowed a calmer observation later. Louis Olivier (1854–1910), who published *Recherches sur l'Appareil Tégumentaire des Racines* (1881), stated that his discoveries were only possible thanks to the photographs made under the microscope (Fieschi, 2008). The photographer and botanist Alphonse de Brébisson, previously mentioned, considered his microphotographs precious for his botanical work. Another good example of photography used to record the results of experiments is the work of George Ville (1824–1897) *Recherches Expérimentales sur la Végétation*, published in 1857.

Similigravure (autotypie) was a photomechanical process of extreme importance for the illustration of botanical books with photographs (Fieschi 2008). An important work, also in the field of microphotography illustrated with similigravure, was the *Atlas de Photomicrographie des Plantes Médicinales,* published by Louis Braemer (1858–1935) and Armand Suis (1859–?), in 1900 (Braemer and Suis 1900). It was in this time that the use of photographs in the illustration of books on botany underwent great development. In 1890, Henri Gadeau de Kerville (1858–1940) published *Les Vieux Arbres de la Normandie - Étude Botanico-Historique*, a photographic inventory of 113 notable Normandy trees [40 photogravures and 62 photocollographies (phototypes)] (Kerville 1890), another approach to photography associated with botany (trees) and publications.

Photography was not used solely for scientific study or the presentation of botanical results, but also illustrated nature guides and essays, such as *Nature's Garden; an aid to knowledge of our wildflowers and their insect visitors* (1900) with coloured plates and many other illustrations photographed directly from nature by Henry Troth and Arthur Radclyffe Dugmore (1870–1955) with text by Neltje Blanchan (1865–1918).

Collection (survey) of references to botanical elements present in the history of photography (1802 to 1900), highlighting some works developed in the field of botany and photography, as well as the photographic processes used in the registration of these botanical elements.

(*collection constantly updated*)

**Note 1** The ***Botanical Reference*** column can be a publication (marked with *), some of which were discussed in the previous section ***Photography and Botanical Publications*** or an example of a photograph of the aforementioned botanist/photographer with reference to botanical elements. Many of these photographers have other images with botanical specimens.

**Note 2** This table was made, in part, with the research of the online collections of institutions and museums, from 2019 until 2024, such as the *Bibliothèque Nationale de France; Fitz Museum; George Eastman Museum; J. Paul Getty Museum; National Museum Wales; The Metropolitan Museum of Art; SFMOMA Museum; Musée d'Orsay; Victoria and Albert Museum; The New York Public Library; Mira Dock Forestry Lantern Slides Collection.*

## Botany as a Material for Photography

One of the first objects in which we can identify plant raw material is the wood used in large format nineteenth century cameras. The wood used came from several plant species, according to the place of manufacture (Osterman *pers. com.*). Cameras were made with standard plate sizes—five French sizes and English four sizes –, which were reduced to seven sizes by the end of the 1840s. Most cameras were made by optical, clock or mirror makers, and they used different woods depending on where and when the cameras were made. The British makers favoured mahogany or rosewood and lacquered brass; the continental Europe makers preferred walnut and unlacquered brass (Pritchard 2008). The earliest cameras were often made with an inexpensive 'core' and then veneered with an expensive work, as in the Giroux camera made for Daguerre. American daguerreotype cameras were also fabricated using rosewood and mahogany veneers. Early continental cameras could also be made with white maple. Some cameras were made with oakwood, but this was never a popular wood for camera making (Osterman *pers. com.*). By the mid-1850s, British cameras were made of mahogany and cherry, and French cameras were made of local walnut, although this was not a very good wood for the cameras (Osterman *pers. com.*).

Obtaining a photograph is a complex chemical process. This is why the history of photography is full of inventions, with different materials and techniques. Changes were made in the support, the photosensitive material, and the way the components were mixed. This evolution was closely related to that of botany because some chemicals used in photography have a direct plant origin or, at least, some components from plants. Thus, the history of photography is 'sprinkled' with elements of plant origin, which were of great importance in experiments that allowed the evolution of photographic processes, some even called organic processes (Table 8.2).

The oldest organic process that provided the oldest image that has survived to this day was the Heliographic process, which was invented by Joseph Nicéphore Niépce (1765–1833), in approximately 1822, and is also known as the Asphalt or Bitumen process. *Bitumen of Judaea* dissolved in the *lavender oil* was thinly coated onto a metal plate or stone; the layer hardened selectively in the light and was developed by the same solvent (Ware 2008). However, perhaps the photographic printing process of the Victorian era that was closely linked to botany was the *anthotype* (or phytotype), which has been put aside and almost forgotten. This process, first conceived by Sir John Herschel in 1839, was a photographic printing process that used the photosensitivity of the pigments contained in the different parts of the plant. When UV light falls on the vegetable pigments emulsified in paper, it causes whitening, leaving an impression on unexposed areas.

In the experiments described in his article, Herschel (1842) used several plant species, including *Matthiola annua*, *Papaver orientale*, *Curcuma longa*, *Bulbine bisulcata*, *Cheiranthus cheiri*, *Ferranea undulata*, *Viola odorata*, *Sparaxis tricolor*, *Papaver rheum*, and *Senecio splendens*. Unfortunately, the ephemerality of the

**Table 8.2** Botany in Photography—Botany as a Raw Material in the History of Photography Collection (survey) of references to elements of plant sources in the chemical composition of historical photographic processes, with reference to the inventors of the processes

| Photographer | Year | Photographic Process/ Photomechanical Photo Printing | Botanical Reference |
|---|---|---|---|
| Joseph Nicéphore Niépce (French, 1765–1833) | 1822 | Heliographic or sun drawing | **Bitumen of Judaea,** a fossil compound, obtained from ancient plants |
| Niépce and Louis Daguerre (French, 1787–1851) Jean-Louis Marignier | 1832 | Physautotype | Images were produced using **lavender oil** residue dissolved in alcohol as the photographic agent |
| | 1995 | | Usually replaced with **colophony resin,** obtained mainly from the turpentine of Masson's pine (*Pinus massoniana* lamb.) and slash pine (*Pinus elliottii* Engelm.). |
| Henry Fox Talbot (British, 1800–1877) | 1839 | Calotype | **Gallic acid** used as a developer; organic acid found in different plants |
| Sir John Herschel (British, 1792–1871) | 1839 | Anthotype also Phytotype | Is a photographic printing process that uses the photosensitivity of the pigments contained in the different parts of a plant |
| Abel Niépce de St-Victor's (French, 1805–1870) | 1847 | Albumen negative on glass | **Starch** was used as a binder for albumin to glass |
| Frederick Scott Archer (British,1813–1857) | 1851 | Wet plate collodion | Negatives varnished with **sandarac gum.** To this varnish was added **lavender oil** |
| Jean Marie Taupenot's (1822–1856) | 1855 | Dry collodion | Use of **Indian rubber** latex, *Hevea brasiliensis* (Willd. Ex A.Juss.) Müll.Arg. |
| Alphonse Louis Poitevin (French, 1819–1882) | 1855 | Carbon printing | Use of **arabic gum** and **charcoal powder** in prints |
| John Pouncy (British, c.1808–1894) | 1858 | Gum Bichromate | Use of **arabic gum** and pigments in printing |
| John Mercer (British, 1791–1866) | 1858 | Chromatic photograph | Use of some **vegetable dyestuffs** |
| Hermann Vogel (German,1834–1898) | 1873 | Sensitize photographic plates to other colours | **Coralina** pigment, obtained from the red algae *Corallinales* |
| Jakob Schmid (Swiss, 1856–1924) | 1880 | Photochrom | Image developed with **turpentine oil** |

images produced by plant pigments does not allow us to appreciate some of the works developed at that time. When we study the scientific production of Herschel era, we see a great interest on understanding the interactions between light and dyes and between light and organic compounds (Eder 1945).

Plant secretions, such as resins and oils, had a great importance in the history of photography. In 1832, Niépce and Louis Daguerre (1787–1851) developed a new process called the *physautotype* (a neologism meaning nature's self-image). Instead of bitumen, this process used a whitish resin extracted from lavender oil. (Petersen 2008; Watson and Rappaport 2013). This process was rediscovered only recently (ca. 1995) by Jean-Louis Marignier. The exposure of *colophony resin* (abietic acid) to light can cause it to become insoluble, even without the presence of metallic salts (Ware 2008). Colophony resin is obtained mainly from the turpentine of Masson's pine (*Pinus massoniana* Lamb.) and slash pine (*Pinus elliottii* Engelm.).

However, there is a gum that stands out in the history of photography—the Arabic gum—because it had a great influence on several photographic processes. Arabic gum is a mixture of polysaccharides and glycoproteins extracted from the species *Acacia senegal* (L.) Willd. and *Acacia seyal* Delile (Peres 2013). According to Lerebours (1843), Joseph Berres, in 1840, published the first process for engraving daguerreotype plates. On top of the daguerreotype, Berres placed a layer of varnish, exposed the plate to nitric acid vapours and, finally, covered it with Arabic gum. When negative dry collodion emerged (in 1855), was considered one of the greatest innovations in photography after wet plate collodion, since it did not require sensibilization and development at the time the photograph was taken. To do this, it was necessary to maintain a humid collodion for a long period. Therefore, a layer of liquid gelatin or Arabic gum was added to the collodion (Pavão 1997). However, the greatest difficulty was that the film tends to peel from the glass when it dries, and to avoid this issue, the glass was covered with albumen, gelatin, or with a layer of Indian rubber (Meldola 1889) obtained from the latex of the *Hevea brasiliensis* tree (Hofmann 1989). Arabic gum was also used to fix on paper an emulsion containing ammonium dichromate or potassium dichromate, creating a light-sensitive surface. For this reason, mixtures of Arabic gum and dichromate were widely used in coal tests (Peres 2013). The photographic process with bichromated gum was more attractive because of the possibility for the photographer to introduce pigments into the viscous material and add colour to the print, especially in the Victorian era when painting with watercolours was a popular hobby. Gum printing allowed the photographer to choose textures, colours, and papers closer to what was possible for painters or etchers (Pinheiro 2008). In techniques based on the use of Arabic gum, an aqueous solution of pigments is thickened with the addition of powdered gum until it reaches the desired consistency. When drying, the gum acts as a binder, keeping the pigments fixed in the support and preventing the oxidation of the pigments and consequent changes in their colours, so it was also useful in photomechanical processes (Peres 2013). In the photomechanical process of lithography, Arabic gum is used to protect the engraved plates during the image transfer process, preventing the ink from filling the spaces that are intended to be white. After a positive photograph is finished, it is often glued to a card to give it greater resistance. Dextrin or Arabic gum (mixed in a starch paste) was usually used to stick photographic paper to a card (Bentes 1866). Starch was also used as a thickener or to obtain matte photographs. It was also used as a binder in glass by Niépce de Saint-Victor, in 1847, in an attempt to create the first glass negatives (Chicandard 1909).

In the first and most widely used negatives, in wet collodion on glass, as well as in ambrotypes and ferrotypes, after obtaining the negative or positive image and for it to remain, it was necessary to protect the glass with varnish. The varnish used was made from sandarac gum, a natural resin extracted from *Tetraclinis articulata* (Vahl) Mast. Lavender oil (*Lavandula* spp.) was also added to this varnish (Bordin 2015).

In 1873, Hermann Vogel (1834–1898) tried to obtain photographic plates sensitive to other colours (yellow, orange, and green) and found that by adding certain coloured substances to silver bromide, as was the case with coralline (pigment obtained from the red algae *Corallinales*), the sensitivity to green radiation increased, creating orthochromatic plates. Vogel concluded that with the addition of the appropriate pigments, he could reproduce the correct tones of coloured objects (Peres 2013).

## Conclusion

It is easy to find botany, flowers, leaves, and images of gardens in a multitude of photographs from the Victorian era. A very pleasant exercise, in constant update. There are many examples of period experiences that reveal how important the gardens and their elements were for photographers, as part of a society very interested in this area. A way to produce art without forgetting passion and the subjects of study, or even by reflecting the interests of the time, which were often associated with collections, namely, of dried leaves in herbarium—the Victorian obsession for the classification of the natural word. Thus, we can find botany and its elements in a wide variety of photographic processes, and as photography has evolved, plants continue to be an object of study and elements to be photographed and experienced.

By necessity, the early photos of plants were photograms, printed by contact, and thus, images were obtained by transmitted light, not our normal way of seeing plants. Later, photographers were able to take photographs of plants in the camera by reflected light, images that had a much greater degree of familiarity. Perhaps this was less of a concern for Atkins because her *'flowers of the sea'* were depicted against the cyanotype's naturally blue background, much as they might have been seen by a diver facing the skylight.

Another problem in the photographic illustration of plants was that botanists were used to engravings that typically had one or more detailed drawings in addition to the main one. Atkins emulated this in some of her plates by arranging more than one specimen on a sheet, but this still did not allow for magnification. Another major drawback of photography for botanical illustration was the flip side of its very strength. Photography excelled at depicting a real-world object very precisely. However, botanists consulting an illustration wanted to observe what was typical for a type of plant, not what was specific to an individual specimen. In the end, this lack of ability to generalize the image was perhaps the single largest drawback of botanical photography.

When authors write about the history of Victorian photography, only the main photographers are commonly highlighted—Talbot and Anna Atkins—as examples of photographers who related botany and photography, especially Talbot's fascination with botany, which is very well represented in his inspiring work and is now found in several museums. It could have been possible to focus our research only on the botany present in Talbot's photographic work, as it is so vast and interesting. However, slightly more in-depth analysis can easily take us to a new world of photography associated with botany.

Botany as a subject in photography is still an area that requires much research. Considering that throughout the history of photography, many of these processes were invented and experienced, not all were analysed in this chapter because some ended up having no practical application. Photography in botany publications is a very interesting topic and deserves more attention. In the analysis of botanical publications, it is not always easy to identify, looking only at the images, which photomechanical processes were used in the printing process of the image, and even more difficult if the negative form was used to create the image.

We should not forget that plants were decorative and symbolic elements in photo albums of the time — the art of Victorian album photocollage. These elements embellished the pages of countless photographic albums, which were considered a respectable hobby, especially for women. The delicate drawings, sometimes watercolour, with a predominance of botanical elements, for example, the albums of Lady Frances Jocelyn (1820–1880) and the wonderful album of Madame B (1831–1906) (Siegel 2009). In the elaboration of photocollages, we can also highlight the work of John Wheeley Gough Gutch (1808–1862). He mainly worked with wet collodion glass negatives and printed his photographs on salted paper. He mounted his work into slim, soft-cover albums where he titled and dated the images, often decorating the title page to each album with a photographic collage. The collages were made by placing leaves, flowers, and feathers in printing frames (in contact with salted paper) to produce negative images (photograms), which were then cut up, made into patterns and pasted onto the page (Summer 2008). Was this done to have a mere decorative and aesthetic purpose?

Botany was an inspiration for photography, and photography was very important for the dissemination of botanical knowledge to scientists and as a symbolic representation in art. In addition to the spread of botanical images through publications, we can also find examples of plant presence on slides of the magic lanterns used in the education and teaching of botany. These collections deserve to be the object of more research due to their role in the history of photography and botany.

Finally, we cannot forget the importance of the plant raw materials used throughout the development of different photographic processes because much photographic chemistry was dependent on plants, and in this way, the evolution of photography was largely due to botany.

# References

Allen JM, McNear SA (2018) The photographer in the garden. Aperture and the George Eastman Museum, New York

Batchen G (2016) Emanations. The art of the cameraless photograph. DelMonico Books-Prestel, New York/London

Bentes JA (1866) Tratado theórico e pratico de photographia. Livraria de A M Pereira, Lisboa

Bibliotéque Nationale de France (2024). https://gallica.bnf.fr. Accessed 17 February 2024

Bordin G (2015) Fascino e rigore del collodio. Il collodio umido positivo e negativo. Antiche Teccniche Fotografiche, Guaradilab, Rimini

Braemer DL, Suis DA (1900) Atlas de photomicrographie des plantes médicinales. Vigot Frèrs Éditeurs, Paris

Browne LK (2008) Botanical photography. In: Hannavy J (ed) Encyclopedia of nineteenth-century photography. Routledge, New York/London, pp 193–195

Burns M (2017) Printing and publishing the illustrated botanical book in nineteenth century Great Britain. Cogent Arts & Humanities 4:1364058. https://doi.org/10.1080/2331198 3.2017.1364058

Cull B (2016) Early canadian botanical photography at the exposition universelle, Paris 1867. Sci Can 39(1):27–50. https://doi.org/10.7202/1041377ar

Chicandard G (1909) La Photographie. Octave Doin et Fils Éditeurs, Paris

Eder JM (1945) History of photography. Columbia University Press, New York

Edwards S (2008) Great Britain. In: Hannavy J (ed) Encyclopedia of nineteenth-century photography. Routledge, New York/London, pp 606–612

Fieschi C (2008) Photographier les plantes au XIXe siècle. La photographie dans les livres de botanique. CTHS Sciences n° 4. Éditions du Comité des travaux historiques et scientifiques, Paris

The Fitzwilliam Museum (2024) Cambridge https://www.fitzmuseum.cam.ac.uk/collections. Accessed 13 March 2024

Frizot M (1998) A new history of photography. Könemann, Köln

George Eastman Museum (2024) Photography collection online, Rochester NY https://collection-seastmanorg/objects. Accessed 23 February 2024

Herschel JFW (1842) On the action of the rays of the solar spectrum on vegetable colours, and on some new photographic processes. Philos Trans R Soc Lond 132:181–214

Hofmann W (1989) Rubber technology handbook. Hanser, New York

J. Paul Getty Museum (2024). https://www.getty.edu/art/collection Accessed 14 February 2024

Kerville HG (1890) Les vieux arbres de la Normandie. Étude botanico-historique. J-B Baillière et Fils, Paris

Kocol M (2010) The garden in early art photography. Gardens Illustrated 167:74–77

Lerebours NP (1843) A treatise on photography. Longman, Brown, Green and Longmans, London

Lyden AM (2014) A royal passion. Queen Victoria and photography. The J. Paul Getty Museum, Los Angeles

Marten C (2008) James Glaisher (1809–1903). In: Hannavy J (ed) Encyclopedia of nineteenth-century photography. Routledge, New York/London, pp 592–594

McCraw JD (1988) H B Dobbie - fern enthusiast. N Z J Bot 26(2):171–178. https://doi.org/10.108 0/0028825X.1988.10410109

Mira Dock Forestry Lantern Slides Collection (2024). Pennsylvania https://libraries.psu.edu/about/collections/mira-dock-forestry-lantern-slides. Accessed 1 March 2024

Moitessier A (1866) La photographie appliquée aux recherches micrographiques. J-B Baillière et Fils, Paris

Meldola R (1889) The chemistry of photography. Macmillan and Co, New York

The Metropolitan Museum of Art (2024). https://www.metmuseum.org/art/collection/search. Accessed 12 February 2024

Monovisions, Black & White Photography Magazine (2017) Biography: 19th Century photographer Adolphe Braun. https://monovisions.com/adolphe-braun-biography-19th-century-photographer. Accessed 13 March 2024

Morris R (2008) John Dillwyn Llewelyn (1810-1888). In: Hannavy J (ed) Encyclopedia of nineteenth-century photography. Routledge, New York/London, pp 866–868

Musée d'Orsay (2023). https://www.musee-orsay.fr/fr/collections/catalogue-des-oeuvres/recherche-simple.html Accessed 10 December 2023

Nadeau L (2018) Photoglyphic engraving. In: Hannavy J (ed) Encyclopedia of nineteenth-century photography. Routledge, New York/London, pp 1080–1081

National Museum Wales (2024) Collections & research https://museum.wales/collections Accessed 1 March 2024

The New York Public Library (Digital Collections) https://digitalcollections.nypl.org Accessed 12 March 2024

Pavão L (1997) Conservação de Colecções de Fotografia. Dinalivro, Lisboa

Peres IM (2013) Fotografia científica em Portugal, das origens aos séc. XX: Investigação e Ensino em Química e Instrumentação. Doutoramento em Química. Universidade de Lisboa, Faculdade de Ciências, Departamento de Química e Bioquímica

Peres IM (2004) Processos fotográficos históricos. In: Costa FM, Jardim ME (eds) 100 anos de Fotografia Científica em Portugal (1839–1939). Edições 70, Lisboa, pp 15–34

Peterson S (2008) Joseph Nicéphore Niépce (1765–1833). In: Hannavy J (ed) Encyclopedia of nineteenth-century photography. Routledge, New York/London, pp 1003–1006

Peterson CA (2012) Pictorial Photography at the Minneapolis Institute of Arts: history of exhibitions, publications, and acquisitions with biographies of all 243 pictorialists in the collection. Christian A Peterson: Privately printed

Pinheiro NA (2008) (Leon) Robert Demachy (1859–1936). In: Hannavy J (ed) Encyclopedia of nineteenth-century photography. Routledge, New York/London, pp 408–409

Pritchard M (2008) Camera Design: 1 (1830–1840). In: Hannavy J (ed) Encyclopedia of nineteenth-century photography. Routledge, New York/London, pp 244–245

Rhodes K (2008) Still Lifes. In: Hannavy J (ed) Encyclopedia of nineteenth-century photography. Routledge, New York/London, pp 1343–1346

Roberts P (2004) 'The exertions of Mr Fenton': Roger Fenton and the founding of the photographic society. In: All the mighty world. The photographs of Roger Fenton, 1852–1860. Yale University Press, New Haven/London, pp 211–221

Ryan JR (2017) Placing early photography: the work of Robert Hunt in mid-nineteenth century Britain. Hist Photogr 41(4):343–361. https://doi.org/10.1080/03087298.2017.1357268

Schaaf L (1979) The first photographically printed and illustrated book. Papers Bibliograph Soc Am 73(2):209–224

SFMOMA Museum (2024) San Francisco https://www.sfmoma.org/museumfromhome/ Accessed 14 March 2024

Siegel E (2009) Playing with pictures: the art of Victorian Photocollage. Art Institute of Chicago, Chicago

Smith G (1993) Talbot and botany. Hist Photogr 17(1):33–48. https://doi.org/10.1080/0308729 8.1993.10442590

Steidl K (2012) Leaf prints early cameraless photography and botany. PhotoResearcher 17:26–35

Steidl K (2019a) Natur als Bild—Bilder der Natur. In: Steidl K (ed) Am Rande der Fotografie. Eine Medialitätsgeschichte des Fotogramms im 19. Jahrhundert. De Gruyter, Berlin, pp 189–257

Steidl K (2019b) Experiment, Sammelobjekt, Frauenkunst. In: Steidl K (ed) Am Rande der Fotografie. Eine Medialitätsgeschichte des Fotogramms im 19. Jahrhundert. De Gruyter, Berlin, pp 259–305

Summer I (2008) John Wheeley Gough Gutch (1808–1862). In: Hannavy J (ed) Encyclopedia of nineteenth-century photography. Routledge, New York/London, pp 627–628

Trutat E (1884) La photographie appliquée à l'histoire naturelle. Gauthier-Villars, Imprimeur-Libraire, Paris

Tucker J (2005) Nature exposed: photography as eyewitness in Victorian science. The Johns Hopkins University Press, Baltimore

Victoria and Albert Museum (2020). https://www.vam.ac.uk. Accessed 18 March 2020

Ville G (1857) Recherches expérimentales sur la Végétation. Mallet-Bachelier, Paris

Ware M (2008) Positives: minor processes. In: Hannavy J (ed) Encyclopedia of nineteenth-century photography. Routledge, New York/London, pp 1154–1161

Watson R, Rappaport H (2013) Capturing the light: a story of genius, Rivalry and the birth of photography. Pan Macmillan, London

# Chapter 9
# The Victorian Return to Nature and the Simple Life

Vicky Albritton and Fredrik Albritton Jonsson

**Abstract** In the nineteenth century, industrialization, pollution, parliamentary enclosure acts, and the degradation of natural landscapes due to large-scale agriculture increasingly disconnected people from nature. This led to a spiritual disaffection notably expressed in the Romantic vision of William and Dorothy Wordsworth and John Clare. In the 1860s and 1870s, Ruskin urged readers to reject the industrialized marketplace in favor of the simple life and the holistic cultivation of art, community, and nature through gardening, reviving handicrafts, preserving landscapes, and implementing pedagogies fortified by the close observation of the environment. Susanna Beever, Albert Fleming, H.D. Rawnsley, and W.G. Collingwood were some of the most energetic practitioners of Ruskin's ideas. Ruskin himself went so far as to counter what he saw as the narrow, soulless, Darwinian science of botany with a spiritual form of classification drawn in part from mythology and art. Despite his dark forebodings about anthropogenic climate change late in life in The Storm Cloud of the Nineteenth Century, Ruskin's ideas reemerged in the twentieth century in Arthur Ransome's nature-oriented children's books, Gustav Stickley's American Arts and Crafts furniture, and Back-to-the-Land movements like those led by Charles Robert Ashbee and Julius Augustus Wayland. Ruskin's writing on the sanctity of green spaces, their vulnerability and potential in the face of industrialization and consumer excess, both reflected and shaped the Romantic impulse to return to nature.

**Keyword** Ruskin · Simple life · Back to the land · Gardening · Arts and crafts

---

The Sections "The Simple Life in Practice", "Ruskin's Storm-Cloud of the Nineteenth Century", and "Legacy of the Simple Life" include excerpts from *Green Victorians: The Simple Life in John Ruskin's Lake District*. Chicago: The University of Chicago Press, 2016. I am grateful to the University of Chicago Press for permitting me to print these previously published passages.

---

V. Albritton (✉)
Independent Researcher, Ogden Dunes, IN, USA

F. Albritton Jonsson
History Department, University of Chicago, Chicago, IL, USA

In nineteenth century Britain, urbanization and industrialization provoked powerful social criticism in many corners of society. This essay considers one strand of this critique, in particular, the romantic and poetic fascination with the simple life, which in most cases involved living close to nature. In a time when industrialization increasingly devastated landscapes, devalued labor, diminished leisure time, and bombarded consumers with advertisements for cheap, poorly made goods, the ideal of the simple life offered an alternative. Perhaps best articulated by the social and art critic John Ruskin (1819–1900), the simple life called for a disavowal of consumer excess, an appreciation of arts and crafts, experimental pedagogy, landscape history, sustainable farming, and the aesthetic appreciation of nature. Botanical pursuits were of particular importance as a spur to poetic and literary creativity and the cultivation of ethical comportment toward the natural world. For disciples of the simple life, a newfound concern for the environment reoriented pleasures and desires away from cities back toward the increasingly fragile countryside and traditional work and lifestyles.

We begin with William Wordsworth and his sister Dorothy who moved to Grasmere in the English Lake District in 1799. There Dorothy fell in love with the local flora, taught herself botany and transformed her cottage garden into a veritable museum of Lakeland plants. She and her brother deliberately sought to live simply and close to nature while William wrote arguably the first environmental, anti-industrial poems proclaiming the beauty and sublimity of the landscape. We then consider the "peasant poet" John Clare, a commoner with rather closer ties to his community in Helpston. In the 1820s and 1830s he recorded the loss of free access to the countryside as the rich began fencing off private property, clearing forests, tilling the soil, and destroying ecosystems. Next, we touch on the ways working men accessed nature within cities through various social enterprises. From here, we turn to the work of John Ruskin, the famous art and social critic, beginning in 1860 with his first full scale attack on industrialization and immoderate consumption and its effect on the landscape and laborers. Since many of Ruskin's theories were most thoroughly carried out by his followers, we then survey several of their efforts. This includes, Susanna Beever, a spinster and woman gardener; Albert Fleming, a solicitor and promoter of revivalist artisanal linen weaving that prized good labor conditions and products deeply connected to nature; H.D. Rawnsley, Cannon and founder of the National Trust; and W.G. Collingwood who with his wife and children engaged in experimental pedagogy and imaginative nature-oriented endeavors far from the city.

The happiness and relative success of Ruskin's disciples stood in marked contrast, however, to the apocalyptic proclamations of their Master in his later years. To this end, we touch on Ruskin's 1884 lectures on the Storm Cloud of the Nineteenth Century. There he announced a new era of terrifying terrestrial despoliation and anthropogenic climate change, which he believed was already damaging plants and trees all around the globe. Yet Ruskin's legacy in the late nineteenth and early twentieth century was far less dire, as we see in his influence on Arthur Ransome's famous children's books, the Arts & Crafts and Back-to-the-Land movements, as well as community agricultural projects and a museum devoted to Ruskin's ecological principle of the simple life.

## William and Dorothy Wordsworth: The Romantic Return to the Countryside

> The world is too much with us; late and soon,
> Getting and spending, we lay waste our powers:
> Little we see in Nature that is ours... (Wordsworth 2014, 403).

So begins William Wordsworth's famous sonnet (ca. 1802–1804) on excessive consumption in the industrial age. Many city dwellers escaped to the countryside, seeking a brief reprieve from slums and relentless commerce, foul air, and polluted waterways. Tourism flourished and those who could afford it spent holidays in quaint natural landscapes. This included the picturesque English Lake District with its slate mining and agricultural towns nestled beneath massive, heathered fells. A few desired to make their move permanent. To this end, William Wordsworth (1770–1850) and his sister Dorothy Wordsworth (1771–1855) moved to the village of Grasmere in 1799, trading urban conveniences for poetic inspiration and the close connection to nature that came to define the Romantic period.

Dorothy and William grew up in Cockermouth about 25 miles northwest of Grasmere. Their father was James Lowther, first Earl of Lonsdale and a solicitor. In a way, it was like coming home when they moved into Dove Cottage, though their education and culture set them apart from the local farming community. Nonetheless, they each tried to embrace their simplified lifestyle with great thoughtfulness. William wrote poems in a relatively casual iambic pentameter that recalled the natural rhythms of vernacular speech. He expounded on the sublime joys of the rural landscape, finding in nature a model of simple satisfaction. As he wrote in *Lines Written in Early Spring* (1798), the birds and trees seemed happy and "every flower/ Enjoys the air it breathes" (Wordsworth 2014, 30). He lamented that modern man did not follow nature's example. By contrast, peasants' more traditional lives blended seamlessly into the natural world and William and Dorothy looked often to locals for inspiration. In the poem "Michael," Wordsworth praised the shepherd, noting his fondness for "these fields, these hills/Which were his living Being, even more / Than his own Blood" (Wordsworth 2014, 147). Such a man had no need of urban luxuries. He needed only the pleasure of wholesome activity and plainest fare. Admittedly, Wordsworth never tilled the fields himself and his praise of peasant life tended to downplay the hardships of rural subsistence living. In any case, he felt there was much to be gained from living so close to nature. As he wrote in "Tintern Abbey," he learned "To look on nature" and rejoice in "elevated thoughts; a sense sublime / Of something far more deeply interfused" (Wordsworth 2014, 68). Jonathan Bate (2001, 148) explains that for Wordsworth the poem is "a meditation on the networks which link mental and environmental space."

Where William described his connection to nature in cerebral, meditative poems, his sister Dorothy trained her eye on more practical matters. In her *Grasmere Journals*, she recorded the humble details of daily life, pleasures small but profound and activities that embody the art of simple living. She made her own shoes, mended stockings, gathered seeds and copied her brother's poems. One day she and William sat before a window awaiting rain, "deep in Silence & Love, a blessed hour. We drew to the fire before bed-time & ate some Broth for our suppers" (Wordsworth 1991, 104). Such was the simplicity of her contentment. In May of 1800 she longed for a book of botany to help plan her flower garden at their home in Grasmere. Her interest in botany grew and "Throughout the spring of 1802," one scholar says, she drew "her brother's attention to beauties that by his own admission he had not heeded in all his thirty years" (Mahood 2008, 24). She gathered peas and gooseberries from her own garden and ventured into the hillside for wild mosses, thyme, and columbine roots. As her botanical vocabulary grew, she sometimes mingled scientific nomenclature with common plant names, as when she "brought home heckberry blossom, crab blossom—the anemone nemorosa—Marsh Marygold—Speedwell" (Wordsworth 1991, 98). She was obviously pleased with her local plant collecting though it was possible to order exotic plants by mail. Her consumer restraint was a consequence of both limited finances and a love of the landscape. Humility and wonder suffused her description of the Coniston Fells "in their own shape & colour—not Man's hills but all for themselves the sky & the clouds & a few wild creatures" (1991, 90). Yet she understood the impact humans might have on the environment. She recorded that Miss Hudson of Workington sowed flowers "in the Parks several miles from home." "This," remarked Dorothy, showed that "Botanists may be often deceived when they find rare flowers growing far from houses" (1991, 111).

The Wordsworths' frugality and unpretentious habits reflect the influence of the local population as well as their innate desire to protect the landscape from destructive technological improvements. In the 1840s Wordsworth sought to safeguard this rural way of life by campaigning to prevent the railways from laying tracks deep into the Lake District. He feared it would scar the land and allow hordes of tourists to unsettle the fragile balance between humans and nature. He failed, but his celebrity and the force of his sentiments expressed so eloquently in his poems inspired generations of preservationists (Ritvo 2009, 21–22).

## The "Peasant Poet" John Clare: Enclosure of the Countryside

Although William and Dorothy lived humbly in comparison to their own class, William enjoyed a secure living from genteel work and became Poet Laureate in the 1840s. They eventually left the rustic simplicity of Dove Cottage for a more impressive house at Rydal Mount. The rewards were great for the Romantic poet who moved to the country by choice. Commoners, of course, lived and worked there by necessity—often in harsh conditions. Yet their views on plants and trees also bore the mark of a new Romantic perspective on nature, as is seen in the work of Northamptonshire commoner and poet John Clare (1793–1864), who witnessed up close how industrialization threatened rural life. Clare had been born to a farm laborer and received a limited education within his parish. At 13, he began to write poetry after reading James Thomson's "The Seasons" and in 1819, this "peasant poet" was discovered by a London bookseller. His earnings were small so he continued work as a manual laborer. Subsequent volumes of poetry won praise but did not sell well, he struggled to support his growing family, and his relationship with his publishers soon fell off. In 1835, the mental illness with which he had long struggled worsened, and he spent his last 20 years in the Northampton General Lunatic Asylum (Bate 2003).

Twentieth century critics became fascinated by Clare's combination of rural labor and literary imagination. His poems grew out of his experience of the changing Victorian landscape, which had always been a resource for the poor. The "commons" provided food, grass for cattle, and wood-fuel to anyone in need. But in 1807 an Act of Parliament led to the enclosure of Helpstone, Clare's village. He and other villagers suddenly found themselves barred from using common land (Paulin 2004, xviii-xix). Far from preserving field and forest from the marauding poor, enclosure inflicted environmental devastation of a kind never seen before. Woodlands were leveled to make way for agriculture. In "Remembrances" (ca. 1832–1837) Clare recalled a landscape from his childhood now utterly despoiled. The commons were "gone" and the area was "All leveled like a desert by the never weary plough"; the "axe of the spoiler" had felled the woods "With its hollow trees like pulpits" and "Inclosure like a Buonaparte let not a thing remain" (Clare 2004, 260). In another poem, "Helpstone" (ca. 1812–1831), Clare laid blame squarely on the rich. "Accursed wealth o'er bounding human laws / Of every evil thou remainst the cause" (Clare 2004, 4).

Like most commoners, Clare obtained some knowledge of nature and plants through traditional lore and homeopathic medicine. After reading Hill's *Family Herbal* (1754) and James Lee's *An Introduction to Botany* (1760) he even developed an interest in botanical classification. But ultimately, it failed to enthrall him. He preferred to view each individual flower in its particular environmental and sociological context (Bate 2003, 102–103). One day in 1825, he took a walk and wrote down what he saw, blending local dialect and poetic description to convey a markedly holistic understanding of how plants, trees, humans, and wildlife could coexist peacefully. The woods were covered with "white anemonie which the childern call Lady smocks." He mentioned "hare bells" daisies and Celadine and a young boy out "peeping for pootys" (i.e. snails). The bees sang "a busy welcome to spring" and he wondered how they found their way home "from the woods and solitudes were [sic] they journey for wax and honey." The bees were "fond" of their flowers and, likewise, the wood pigeon "fluskering among the Ivied dotterels on the skirts of the common" was "very fond" of the "Ivy Berrys." He conversed with "woodman" out planting about a bird's song, weaving human, animal and botanical life together in intricate harmony. Moreover, it had long been this way. "I could almost fancy that this blue anemonie sprang from the blood or dust of the romans for it haunts the roman bank in this neighbourhood and is found no were else." Sadly, he noted, it does not "grow in great plenty" anymore, for "the plough that destroyer of wild flowers has rooted it out of its long inherited dwelling" (Clare 2004, 474–476). After a thousand years, the relationship between humans, plants, and animals had begun to unravel, but it's obvious that Clare himself was fully immersed in these simple natural pleasures.

By contrast, Wordsworth sometimes fretted that his cerebral, highly developed linguistic skill distanced him from the natural world. Nevertheless, in "Tintern Abbey," he argued that this distance is impossible if one communes with nature properly. Nature could inspire "lofty thoughts" to protect man against the "sneers of selfish men" and the "dreary intercourse of daily life." Wordsworth believed that in the country the mind of the poet "can be part of nature" (Bate 2001, 148). Yet as we see in "Nutting" (ca. 1798) this entry may at times have been difficult for city trans- plants. Wordsworth recounts how he gathered nuts, an activity he had enjoyed as a schoolboy. But now he went "Trick'd out in proud disguise of Beggar's weeds" to preserve his fine attire. After gathering so much that he felt "rich beyond the wealth of kings," he saw he had damaged a tree. The "shady nook / Of hazels, and the green and mossy bower / Deform'd and sullied, patiently gave up / Their quiet being." Remorseful, he implored the reader to be more careful, "for there is a spirit in the woods" (Wordsworth 2014, 130–131). Clare also wrote a poem called "Nutting," but it told quite a different story. He evoked the simple, pleasant pastime of search- ing for nuts with companions in a shaded wood, already immersed in nature and connected to each other. Fatigued, they made their way to a "velvet bank of short sward pasture" to enjoy the fruits of their labor in a scenic landscape. The fragrant field "Showed the dimmed blaze of poppys still in flower" (Clare 2004, 131). All was well. But the poem's abrupt end—"Now the thymes in bloom but where is pleasure gone"—subtly underscored on the cruel abolishment of this pastime after Enclosure. Where Wordsworth sometimes felt that he stumbled into nature and then tried to make amends, Clare always sensed he was being pushed out of it.

Clare empathized with the banishment of nonhuman creatures from the land- scape as well. In his poem "To the Snipe," he noticed the fearful birds who retreated to a "dreary spot" in the marshes to find peace. He was moved, not by the sublime fells, but by this wasteland, "Giving to all that creep or walk or flye / A calm and cordial lot." Recognizing the importance of marginal lands to all manner of life was part of his poetic mission. Seemingly insignificant things might hold the greatest power. As Bate (2001, 161) notes, Clare's poetry often "catches at little things" in nature and shows how "each thing here is known and loved because it is small."

## John Ruskin: The Idea of the Simple Life

The poems of Wordsworth and Clare both vividly convey the importance of an indi- vidual's intimate and even solitary connection to a natural world that is wild or at least remote. But working men living in cities without any means to travel, much less leave for good, had limited access to green, ecologically diverse spaces. Victorian novels depicted wretched slums where nature was nowhere to be seen and morality had all but collapsed. Social critics, religious leaders, politicians, and poets alike debated how to counteract threats posed by modern industrialization to both humans and nature—even in urban centers. The Kew Gardens in London opened to the public as early as 1840 and other gardens followed suit all over Britain. With the passing of the 1848 Public Health Act, civic leaders recognized that fresh air taken in public parks would enhance the lives of citizens otherwise subjected to dismal, polluted urban conditions (Lawes 1849, 117). As John Clare's countryside became increasingly closed to commoners, new city parks opened up to them. Garden manuals were written for every class. In 1838 John Loudon published *The Suburban Gardener and Villa Companion* to great acclaim, which contained advice even for mechanics, whose terraced houses had usually only "a meagre fourteen foot frontage...and a small back garden" (Uglow 2005, 179). Gardens had become crucial to even the poorest laborer. As Winters (1999, 189) explains, "the greening of cities" was not a fair compensation for mounting pollution and crowding, but the Victorians "humanized the urban landscape" as they could.

The urgency of the situation was not lost on Victorian intellectuals. "THERE IS NO WEALTH BUT LIFE," declared John Ruskin in *Unto This Last* (1860). The Oxford don who made his name promoting the work of J.M.W. Turner's art in *Modern Painters* (ca. 1843–1860) turned now more resolutely to social criticism and concern for the environment. He argued that "not greater wealth, but simpler pleasure" would bring joy (Ruskin 17.112). "A truly valuable or availing thing is that which leads to life with its whole strength" (17.84). The idea of consuming wisely by making good use of well-designed objects went hand in hand with promoting the dignity of labor, preserving landscapes and imbuing life with greater meaning. In this, he echoed Wordsworth's sentiments. However, as Craig (2006, 13) notes, Ruskin "shifts the responsibility for culture and the locus of critical judgment from the poet, genius, and hero to individual consumers." Further, Ruskin covered far more intellectual territory. He was the first great Victorian figure to link the burning of coal and subsequent pollution of the atmosphere to anthropogenic climate change. (Albritton & Albritton Jonsson 2016, 14). Coal powered the manufacturing industry, bringing large quantities of cheap, poorly made goods to the masses. Further, excessive consumption began to ravage the environment, fouling the air and water and covering plants and trees with soot. The desire for cheap goods had to be curbed. Wider society needed to learn what poor labor conditions produced them and how landscapes and wildlife suffered as a result. To this end, Ruskin used his considerable influence to promote the cultivation of desires more in keeping with sustainable production and wise consumption. The art of living, he believed, depended on learning to ask how much a person should consume and what was most needful in life (Albritton and Albritton Jonsson 2016, 34).

In 1872, Ruskin moved to Brantwood in the Lake District, not far from Wordsworth's home. The house overlooked Coniston Water with the 5000-foot high Old Man on the other side. Here he explored on his own terms what it might mean to curb overweening desires and seek satisfaction instead. He turned to nature for inspiration. His estate became a kind of laboratory with experiments involving outdoor work. He revived certain modes of land use from the pre-industrial era, being concerned to work with nature rather than against it. This meant looking carefully at what nature provided ready-to-hand and considering how best to shape it gently into useful form. He considered how nature and human society could coexist and thrive under each other's care. This was all the more relevant given that Brantwood was situated in a farming and slate mining community, which had worked the land for centuries without decimating it. Something could be learned from that relationship. He designed a small garden to assess what exactly a working man might gain, spiritually and physically, from maintaining it. There he grew vegetables and herbs and more common flowering vines instead of useless, showy exotics unsuited to their environment. He used his own hands in the garden, often chopping his own wood to burn instead of coal. It is true his house was poorly insulated and needed all the fuel he could find. But there was a moral value in testing out the benefits of physical labor in a natural setting (Albritton and Albritton Jonsson 2016, 17–18).

Ruskin sought to bring this ethos of material simplicity to readers of all classes, including the "workmen and laborers of Great Britain." In his newsletter, *Fors Clavigera* (published between 1871–1884), he detailed his activities at Brantwood and argued for the importance of reining in consumer desires and cultivating skilled, artisanal work while respecting the environment. He supported many of his followers who consequently revived older forms of handwork such as spinning and weaving, carpentry, metalworking, and lace-making. Gardening offered everyone a chance to work with their hands and step into nature now and then. As *The Gardener's Chronicle* wrote in 1841, gardening had become an "indispensable part of the domestic establishment of every person who can afford the expense" (Constantine 1981, 388). But what of those who could not afford it or had no experience with it? With such questions in mind, Ruskin established the Guild of St. George in 1878, a working-men's organization for laborers, shopkeepers and those without any schooling that they might learn about the benefits of art and community agriculture. As Frost (2014, 6) notes, Ruskin's plans for the Guild were in many ways ill-conceived, mismanaged, and even elitist. But the message—bringing the working man back to nature—was eagerly received by many.

Ruskin's utopian critique of society, then, was visionary but not without flaws. He often ignored the better aspects of modern cities and new technology and science. For instance, he failed to foresee the benefits of railway travel which he, like Wordsworth, felt marred the landscape and sped travelers too quickly across it, depriving them of the spiritual uplift of scenic beauty. He also criticized the theory of evolution. The idea of competition among species seemed to reflect and celebrate unrestrained capitalism. Like John Clare, he found little use for Darwinian botany. Its depiction of sexual reproduction seemed distasteful. As Smith (2010, 120) notes, in Darwin's system, "Crosses are 'unions' and even 'marriages,' while the resulting seedlings are 'offspring' and 'children.'" Hybrid offspring, Darwin noticed, generated more vibrant, beautiful colors and structures, which in turn attracted insects to them and further increased pollination and fertility. This suggested that the beauty of flowers was merely the effect of evolutionary efficiency and not a symbol of God's majesty on earth. To Ruskin, it was scandalous, vulgar, and even blasphemous.

In *Proserpina*, Ruskin's study of common flowers, he took an anti-scientific approach to botany. He was not alone in this. As Shteir (1996, 30–31) notes, "the Linnaean system of plant classification and nomenclature came under attack, for example, from romantic poets, essayists, and scientists who developed an approach to nature often guided by holistic and emotional concerns." For instance, Ruskin questioned the value of scientific systems of naming plants, scoffing at the term "chlorophyll," which in Greek literally means "greenleaf." This struck him as false knowledge. Apparently unimpressed by the fact that the scientific term referred to a precise chemical reaction, he argued instead for the value of looking at a leaf without the impediments of scientific knowledge. "Think awhile of its dark clear green, and the good of it to you"; he bade his reader recall "the general fact that leaves are green when they do not grow in or near smoky towns," noting that "very soon there will not be a green leaf in England, but only greenish-black ones" (Ruskin 25.232). What good was a science that neglected the plight of the plants it sought to describe? Ruskin instead sought out deeply felt connections—between plants, humans, literature, political economy etc.—where others saw none. Unfortunately, as O'Gorman (2010, 276–277) notes, *Proserpina* "could never be to the new empirical, experimental and 'materialist' analysts of the Victorian laboratories less than eccentric and anachronistic." Consequently, it has been little read. In its own time a reviewer in *The Gardeners' Chronicle* called it "in its way interesting," but added, "the seeker of botanical information must glean warily" (A.G. 1875, 358–359). It is ironic that close by was an advertisement for hot-houses made with Ormson's "Steam-Power Machines" and offering "Boilers and Heating Apparatus of the best description" (354)—coal-powered, polluting machinery designed to grow useless, showy exotics in the hot-houses Ruskin decried.

Clearly, "botanical information" was never the point. For his part, Ruskin vowed to ignore "the recent phrenzy for the investigation of digestive and reproductive operations in plants," and the scientific impulse to analyze "every possible spur, spike, jag, sting, rent, blotch, flaw, freckle, filth, or venom" of plant matter (Ruskin 25:390–91). Far more important was the aesthetic and spiritual value of flowers, a point he drove home memorably with a brief description of a single poppy:

> I have in my hand a small red poppy which I gathered on Whit Sunday on the palace of the Caesars. It is an intensely simple, intensely floral, flower. All silk and flame: a scarlet cup, perfect-edged all round, seen among the wild grass far away, like a burning coal fallen from Heaven's altars. (Ruskin 25.253–254).

Ruskin eloquently contended that the poppy's beauty was not the soulless consequence of an evolutionary drive to attract more insects and increase fertility. It was instead a sign that the "poppy-Maker" wished us to admire it. Beauty was a moral good in itself rather than a biological instrument. He challenged his followers to create beauty in their daily lives and appreciate it in nature. In this way, they might live the simple life—the good life—and protect the environment too.

## The Simple Life in Practice

Ruskin's followers in the Lake District developed his concept of the simple life in their own unique ways. As the following case studies show, they distanced themselves from cities and commercial temptations, immersed themselves in nature, arts and crafts, writing and painting, and let beauty and the spiritual good of humanity direct their thoughts and actions.

**Susanna Beever: Women and Gardening** Susanna Beever (b.1806) moved to Coniston in the Lake District in 1827 with her father, one of her two brothers and three sisters (none of whom ever married). Thirty-four years before Ruskin moved into Brantwood across the lake, she was to some degree already embracing a simple life shaped by gardening and a love of wild and native plants. This was not always easy for a genteel woman in this period. While Dorothy Wordsworth cultivated her home and garden out of the wilderness with freedom, this was unusual for one of her status. In 1811 Jane Loudon wrote a practical manual for lady gardeners, *The Ladies' Flower Garden of Ornamental Annuals*. There she defined types of garden-ing that might be relatively suitable for women. Ornamental annuals, for instance, were "peculiarly suitable for a feminine pursuit":

> The pruning and training of trees, and the culture of culinary vegetables, require too much strength and manual labor; but a lady, with the assistance of a common laborer to level and prepare the ground, may turn a barren waste into a flower-garden with her own hands. (Loudon 1811, i).

A woman was not to be seen straining, sweating, or laboring. Beever mentioned that she dug in the earth with her feet so as not to be seen stooping in public (Albritton and Albritton Jonsson 2016, 79). For women, the garden was something of a "transitional or liminal zone" (Page and Smith, 2011, 1). Their place within it was not always clearly defined. On the one hand, they risked soiling their hands or appearing too mannish. On the other, the garden was associated with the beauty and serenity of Eden and was thought to imbue peace and gentility in women (Page and Smith 2011, 3).

As for the science of botany, Victorian society grew more and more uneasy regarding women's involvement. Women had been writing about herbal cures since at least the seventeenth century and had written books on botany from the late eigh-teenth century. Flowers, being small and delicate and always near the home, were deemed "the most 'feminine' of God's creations"; as long as women avoided scan-dalous descriptions of sexual reproduction in plants, botany was, "the scientific study considered most suited for women" (Page and Smith 2011, 55). But in time women began to be excluded "from science and science writing as women were pushed to the margins of an increasingly masculinized science culture" (Shteir 1996, 103). Working within these parameters for female involvement, Beever and her last remaining sister, Mary, nevertheless acquired significant knowledge of Lake District plants and were cited in botanical studies. Mary even had a plant named after her: *Lastraea felix-mas* var. *Beevorii*, a type of fern (Albritton and Albritton Jonsson 2016, 73). Beever found comfort in this work after her father died and, one by one, her siblings too. Before Ruskin's arrival, she bore her mounting loneliness by painting, writing poetry, gathering quotations from Shakespeare and establishing a garden that would later win minor fame as an inspiration to Ruskin.

When Ruskin moved into Brantwood in 1872, he had already conceived of the Guild of St. George and had many plans to experiment with the simple life. He was also overcoming the death of his parents and the first attack of a mysterious mental illness that would plague him for the rest of his life. Yet he was drawn to the older spinster (now 66) who lived across Coniston Water above a gently sloping garden. Damask plum trees and gooseberry bushes grew at the base of it. The garden path began near the water and led up past vegetable and herb beds and patches of roses, nasturtiums, and convolvulus. Farther up the slope the clematis spread over an arbor marking the entrance to a steeper, terraced hillside garden that culminated in a stand of apple and pear trees (Albritton and Albritton Jonsson 2016, 70–71). Beever's friend the Reverend William Tuckwell gave a detailed description of the plants she grew, noting (to name just a few) the "Rock-rose...Bitter-vetch, Barren-worth, Trillium, endless varieties of Saxifrage... a hazel-alley...[and] an outcrop of native stone planted here and there with rock-plants" (Tuckwell 1891, 109–113).

Like Ruskin, Beever prized common, hardy, and native plants rather than anything exotic that might require a coal-fired hothouse. They both took an interest in native plants and natural methods of cultivation. Ruskin admired William Robinson's *The Wild Garden* (1870), which argued for the use of hardy plants (even exotics) well adapted to their location. Beever's garden possessed such plants suitable to the Lake country. In fact, Tuckwell wrote that she hoped her garden might be preserved for public use after her death and that it might include a section devoted to "the more characteristic or rare Lake plants" (Tuckwell 1891, 112). Some were native to Northern England and were locally frequent in Upper Teesdale and the Lake District, like *Gentiana verna* and *Potentilla fruticosa*. Her correspondence mostly mentions plants naturalized in Britain, such as scarlet rhododendrons, Herb Robert geraniums, Travellers Joy (clematis), *Lithospermums* (gromwells), saxifrage, rock roses, and sweet brier. The few obvious transplants in the list—*Schizostylus* (Kaffir Lily), *Senecio pulcher* (ragwort)—were hardy and preferred the rocky soils found in the Lake District. In a letter she once mentioned heliotrope (one of the least suitable plants for the northwest), but perhaps only because it was a dear old friend's favorite. Beever could have ordered exotic plants by mail, but she seems to have lacked any interest in them. Her letters mention only two orders she made from a nursery in Hyères. Thus the Thwaite gardens were humbler and more local than they might have been (Albritton and Albritton Jonsson 2016, 75, 85).

Beever's unassuming ways and concern for nature were a balm to Ruskin's increasingly dark warnings of anthropogenic climate change. His correspondence with her lasted ten years and amounted to at least 900 letters, which were eventually published by a mutual friend, Albert Fleming, as *Hortus Inclusus* in 1887. They offer a rich, detailed record of simple life ideals centered on gardening, which flourished under his influence. She studied his work more carefully and in 1874 began collecting quotations from Ruskin's *Modern Painters*. Her selections in what was titled *Frondes Agrestes* reveal a preference for themes relating to ethical consumption, sufficient living, and the beauty of nature (Albritton and Albritton Jonsson 2016, 82–84). She never longed for expensive goods or cheaper novelties. Instead, she shared her observations with cheerful fascination. For instance, she wrote of sparrows squabbling in the ivy "like many human beings." But they were safe in this "*fashionable* place of resort," and free to tell each other "the events of the day." (Albritton and Albritton Jonsson 77). When Ruskin grumbled about pollution or the heartlessness of evolution, he found solace in Beever's imaginative tendency to anthropomorphize nature. Ruskin praised her for delighting in all common plants and wildlife, exclaiming, "What infinite power and treasure you have in being able thus to enjoy the least things, yet having at the same time all the fastidiousness of taste and fire of imagination which lay hold of *what is greatest in the least*, and best in all things!" (Albritton and Albritton Jonsson 2016, 81). Her humble perspective on consumption and the natural world so impressed him that he once told her, "you know, you really represent the entire Ruskin school of the Lake country" (Albritton and Albritton Jonsson 2016, 72). No such school formally existed, but its clever students kept migrating to the region to try their own hand at the simple life.

**The Revival of Artisanal Crafts** John Clare had seen peasants' use of the land curtailed. But as consumers began to purchase ever cheaper machine-made goods, skilled and educated craftsmen and women found less work as well. The connection to nature afforded through careful handwork in wood carving and joining, spinning and weaving, and potting became more limited. Labor became more mechanized as each step in the manufacturing process was cut off from the others. Workers began to suffer the physical pain and spiritual despair of repetitive actions. Ruskin described in *The Stones of Venice* (1851–1853) how Venetians made glass beads without joy. They first created a glass rod and then "the men who chop up the rods sit at their work all day, their hands vibrating with a perpetual and exquisitely timed palsy" (Ruskin 9.197). Worse, they were made for frivolous consumers who never thought about the conditions of their making. To counter this trend, through the Guild of St. George, Ruskin supported business ventures that promoted artisanal craftsmanship and avoided coal- and steam-powered machinery, injurious chemical processing, and other working conditions harmful to people and nature (Albritton & Albritton Jonsson, 50).

William Morris (1834–1896), a socialist and activist, had been known to "read Ruskin out loud to his friends" (MacCarthy 1995, 69) as a student at Oxford. As MacCarthy (1996, 168) notes, Morris was captivated by Ruskin's *Unto this Last* (1860) which laid out his attack on industrialization, capitalism, and excessive consumption. Morris said that "It was through him that I learned to give form to my discontent" (MacCarthy 1995, 169). Like Ruskin, Morris became interested in the revival of medieval handicrafts and in 1861, he founded a decorative arts firm with the Pre-Raphaelite artists Edward Burne-Jones and Dante Gabriel Rossetti as well as the Neo-Gothic architect Philip Webb (MacCarthy 1995, 166). They produced highly influential pieces including tapestries, wallpaper, fabrics, carved and painted furniture, and stained glass. He wrote poetry and translated literature such as the medieval Icelandic sagas to great acclaim. But he became perhaps best known for his textile arts, which often displayed intricate botanical motifs such as the "Kelmscott Tree" or "Strawberry Thief" patterns. "By the 1890s," Bingham (2009, 115) says, "Morris and his friends had attracted a new generation of followers, and the Arts and Crafts movement had been born."

Indeed, even from the early 1880s, the Arts and Crafts had already begun to flourish in the Lake District where, under Ruskin's close watch, it took on a particularly nature-oriented aspect centered on the wise and careful use of local natural resources.

**Albert Fleming, Barrister at the Wheel** When one of Ruskin's disciples, the barrister Albert Fleming, moved into a home not far from Brantwood he decided to revive the spinning and weaving industry. Ruskin encouraged him. Fleming aimed to give work to elderly women in the area who might benefit from extra income. He wanted to teach consumers to shun inferior, mass-produced goods by offering compelling alternatives. In 1883, with the help of his energetic housekeeper, Marian Twelves, they located their first spinning wheel and got it working again. Fleming wagered that if people could only be made to understand what they were buying and why, they would be willing to pay a little more for the product. Handwoven linens, after all, were far more durable and would last a long time. He advertised his new business, the Langdale Linen Industry, as a symbol of healthy, happy labor steeped in ancient, environmentally gentle traditions. Sun-bleaching and natural dyes were used despite the popularity of new artificial "aniline" dyes. "Lichens, grasses, heather and logwood" provided alternative sources of color (Brunton 2001, 76, 79). His venture was part of the larger Lakeland Arts and Crafts movement, which was known for its nature-inspired regionalism (Haslam 2004, 31–33). When consumers bought their hand-spun and hand-woven linens, they were to think of the countryside where it was made. No factory-made items could ever boast of qualities so redolent of the natural world.

Fleming's efforts found unusual and satisfying expression in 1889 when his ally H.H. Warner produced a book titled *Songs of the Spindle & Legends of the Loom*. It brought together poems dating back to ancient times about spinning and weaving, including one by Wordsworth, along with images of looms and spinning wheels in use. The book was bound in Langdale linen and sewn together by hand. Admittedly, they were forced to import flax for the cloth as they had been unable to grow it locally. Nonetheless they believed the cover would still remind the reader of the Langdale region's hills and grasses because the linen had been bleached "by no deleterious chemicals, but by the pure mountain air and sunshine" (Warner 1889, 7). Just 250 limited-edition copies were printed on a hand-press and the name of every single person engaged in the book's production was listed in the first pages, from spinner to folder and sewer of pages, thus underscoring the union of satisfying labor, craftsmanship, and an idyllic landscape (Albritton and Albritton Jonsson 2016, 6–7).

The Langdale Linen industry ultimately proved a moderately stable business until the 1920s largely because Fleming and Twelves successfully linked business savvy directly to the nature-oriented ideals of the simple life. It was but one of many Arts and Crafts enterprises in the late nineteenth century in Britain. Another notable disciple of Ruskin's, H.D. Rawnsley, started a separate organization in the nearby town of Keswick.

**H. D. Rawnsley: Tending the Flock, Founding the National Trust** In his youth, Canon Hardwicke Rawnsley (1851–1920) went up to Oxford in 1870 where he studied under Ruskin. There he acquired a vision of leading rural laborers to a more enriching art- and nature-oriented life. Like Fleming, he moved to a town not far from Ruskin and also supported local artisanal industries, eventually founding the Keswick School of Industrial Arts in 1883. This included workshops on linen weaving and design, woodwork, as well as metalwork in brass, copper, and silver. Like most Arts & Crafts era items, their products bore simple, stylized, often botanical patterns stitched into cloth, carved into wood or beaten into the metal. The KSIA became so popular it held a large handicraft exhibition in 1885, the first of many in the region. This allowed the wider world to see and purchase a share of the simple life or at least become familiar with it (Albritton and Albritton Jonsson 2016, 104–107).

As a Vicar, Rawnsley led his parishioners to celebrate nature by overseeing so-called Rushbearing ceremonies. Reeds were gathered from the land and strewn across the church's floor. Based on the Roman festival of Floralia, the ceremony was intended to underscore his rural parishioners' connection to the landscape. He urged them to follow Christ's example and eschew the temptation of modern luxuries in favor of the simple pleasure in honest labor—especially "simple labor of the fields." In an undated "Harvest Sermon" on "the Dignity of Manual Labour," Rawnsley emphasized that a man never stands "in nearer relation to the creator...than when he tills," echoing the sentiments of Ruskin's Guild of St. George (Albritton and Albritton Jonsson 2016, 112–113). Because he encouraged poor people to embrace a level of simplicity to which he did not himself aspire, it is hard not to sense his paternalistic politics and understand his version of the simple life as merely vicarious (Albritton and Albritton Jonsson 2016, 98). Nevertheless, this religious ritual was wholesome, evidently enjoyable for all, and was part of a larger effort to safeguard the landscape.

In 1883, Rawnsley took up the fight against the encroachment of railways into the Lake District. With several allies, he founded the Lake District Defence Society after successfully blocking a plan to build a railway near Derwentwater. Not long after, he joined others in a fight against the proposal to turn Thirlmere's lake into a reservoir to supply Manchester with water, a plan that would drown an entire village and change the uniquely rugged landscape forever (Ritvo 2009). This fight was lost. Nevertheless, he continued to speak out against unsightly telephone lines and to promote the benefits of public footpaths even as property owners sought to enclose ever more fields and prevent ramblers from hiking freely through the hillside (Albritton and Albritton Jonsson 2016, 108).

Rawnsley's most famous work began in 1895 when he became a founding member of the National Trust for Places of Historic Interest and Natural Beauty. As Ritvo (2009, 171) says, "The National Trust...was an outgrowth of the movement to preserve access to open spaces, but instead of campaigning to persuade others to maintain scenic landscapes and historic buildings, it proposed to acquire them itself." The best chance to provide the public with accessible green space was to buy the land and limit commercial building. Within 20 years, 1387 acres were acquired and preserved. Today it reports owning more than 610,000 acres. This too was a continuation of the work begun long ago by the Guild of St. George, which had also sought to preserve places of great beauty and historical interest for the common man and woman to enjoy long into the future.

**W.G. Collingwood and Family: Imagination and Experimental Pedagogy**
William Gershom Collingwood (1854–1932) had been a brilliant student at Oxford

under Ruskin's tutelage as well. He took part in one of Ruskin's social experiments, the "Hinksey Dig," in which privileged young students were invited to help build a road. Taking on the poor community's labor with their own hands, they learned about the benefits of outdoor physical activity and working for a common social good. The lesson hit home. Collingwood decided to forego the possibility of a secure, conventional life as a scholar. He married and took a dilapidated cottage in the Lake District in 1883. He became Ruskin's private secretary, painted, and wrote to cobble together a living. He later moved to a house just down the road from Brantwood in Coniston (Albritton and Albritton Jonsson 2016, 29–30).

Like Rawnsley, Fleming and Beever, he too took up the simple life by engaging in a wide variety of activities. He wrote books on Ruskinian topics, including *The Philosophy of Ornament* (1883), in which he argued for labor-friendly artisanal goods over mass-produced items. Working closely with a local carpenter, he helped design hand-crafted furniture for his home though he could barely afford it. He also engaged in archaeological and antiquarian research in order to illuminate the region's past, find a deeper connection to the countryside and bolster local responsibility toward landscape preservation (Albritton and Albritton Jonsson 2016, 121). Like the Wordsworths, Collingwood and his wife, also an accomplished painter, left behind many urban conveniences to pursue their dream. They raised their children—Dora (b. 1886), Barbara (b. 1887), Robin (b. 1889), and Ursula (b. 1891)—to practice self-reliance and seek satisfaction in simple things, cultivating a complex and sensitive bond with their rugged natural surroundings.

A childhood spent in the city, far from nature, struck Collingwood as impoverished and uninspiring. In 1887 he wrote a poem complaining that such children had "nothing to see, / But bricks, and a waterspout." His own children, by contrast, took in the trees and hillside every hour of the day. The family employed a few governesses but more often pursued their own, unique educational endeavors. The children's days were filled with the music of Mozart or Mendelssohn—Dorrie played for them—and lessons in Greek, Latin, archaeology, history, literature, astronomy, evolutionary theory and germ theory. But of paramount importance was their exploration of the countryside. They hiked up fells, rambled through woods and sketched local flora and fauna. They compiled their observations and pictures, as well as stories, poems, letters, and humor in a monthly family "magazine" that was then sent off to their subscribers (family and friends), each recipient reading it and then sending it to the next on the list. These magazines—*Nothing Much* and *What Ho!*—are today held by Abbot Hall and the Cumbria Archive Centre in Kendal. They constitute a thoroughly original record of a pedagogical orientation toward the natural world (Albritton & Albritton Jonsson 2016, 149, 153–154).

In *Nothing Much* the children often mocked the overzealous claims of unscrupulous advertisements, demonstrating a preference for the priceless wonders of nature (Albritton and Albritton Jonsson 2016, 160–161). They scoured the woods in search of subjects for stories, paintings, and drawings of the landscape in all sorts of weather. Paintings of flowers and studies of birds, feathers, and butterflies abound; the landscape was a steady source of inspiration. Dora's watercolor of a dead chaffinch stands out for its accurate and unsentimental record of nature. Seven-year-old

Ursula's crude but lively painting of a siskin is accompanied by the motto "From Nature"— clearly in honor of Ruskin, who strongly advocated drawing directly from nature. Changes in weather and season were followed with great interest. "Flora's Time Table" was a recurring feature that gave the date of the seasonal first appearance of flowers. The children acquired a formidable knowledge of local plants, having grown familiar with "Speed-well," "Ramp and Vetch," "Potentilla," "Milkwort," and "Brooklime," to name just a few (Albritton and Albritton Jonsson 2016, 156–158). The Cumbria Archive Centre in Kendal holds many of the Collingwood family's private letters offering further glimpses of family life. Dora writes that she made her own little garden in the woods, describes a hike she took and mentions that her brother Robin made two slides of a geranium for his microscope (CACK, WDWGC/1/5). These carefully recorded observations indicate the family's deep appreciation of the countryside.

For Collingwood, natural history was never seen as divorced from human history. Therefore, they also endeavored to uncover the extensive human history inscribed on the land. The children were encouraged to think broadly about the physical, tactile reality of the landscape, including active quarries, abandoned buildings, archaeological sites dating back to the Romans, ancient pitsteads, disused bloomeries, and any other traces of local industries that operated centuries before. During their explorations, they acquired a sense of how nature can recover from human impact over time. They noticed, for instance, how coppice woods had been harvested deep in the past, but had not been destroyed—an early example of the sustainable use of resources (Albritton and Albritton Jonsson 2016, 159).

Collingwood's aim was to understand human history as a part of natural history. This required the active cultivation of the imagination—both his own and that of his children. To this end, he wrote a fascinating work of historical fiction set in the area around Lake Thirlmere. By 1894 the battle to save it had already been lost. Manchester had won the right to build a dam and pipe water back to the city. The hills had been covered with deciduous trees, but the Waterworks Committee decided they needed to be cut down. They were subsequently replaced with dark, rugged conifers uncharacteristic of the Lake District (Ritvo 2009, 126–128). Collingwood decided to set his historical novel here to tell the tale of what it must have been like before everything changed, back when Vikings settled in what was then a heavily forested wilderness. *Thorstein of the Mere* (1895) looked back to a time when Celtic, Anglo-Saxon and Viking societies intersected, but in a way, it also looked forward. For the Viking society depicted in the novel evoked the Ruskinian simple life. As Collingwood depicted them, these ancient people made beautiful, useful household goods, took pride in the manual skills they learned, engaged in spinning and weaving and woodworking, and prized their life in the country as opposed to those who lived in York, the largest, wealthiest nearby town full of inconceivable luxuries. Collingwood's novel acknowledged that Thorstein's way of life had been lost, but also implied that it was desirable to recreate (Albritton and Albritton Jonsson 2016, 134–140).

While optimism generally prevailed among the Collingwoods, they understood the landscape was always under threat. In *Nothing Much*, they discussed the possible source of soot that had fallen on and around the lake at one point, presuming it was from a nearby manufacturing town whose smoke stacks they could see (Albritton & Albritton Jonsson 2016, 156). Similarly, Collingwood complained in a private letter of slate rubble the miners left in unsightly piles, having cut down all the trees and spoiled the view (Albritton & Albritton Jonsson 2016, 143). Later in life, as the region filled up with motorcars and the roads grew crowded with littering tourists, as electric power lines and more railway tracks laced the landscape, Collingwood expressed some doubt that humans would choose to use nature wisely. In the preface to the second edition (1932) of his *Guide to the Lake Counties* he noted ruefully, "I suppose you must consider the greatest happiness of the greatest number, as the dogs told the fox" (Albritton & Albritton Jonsson 2016, 148). But then, living so near to Brantwood at this time, their embrace of the simple life could hardly escape some measure of pessimism as Ruskin's fears for the health of the natural world grew increasingly dire.

## Ruskin's Storm-Cloud of the Nineteenth Century

Even as the Collingwoods first moved to the Lake District in 1883, Ruskin's thoughts on the fate of the natural world had become increasingly gloomy. His prescription against mass consumption and industrialization had failed to take hold in the larger population. He felt the consequences would be grim. In 1884, Ruskin delivered two public lectures on "the Storm Cloud of the Nineteenth Century" in London. There he declared that a new epoch in natural history was underway. It was manifested in the form of a foul, polluted wind from which landscapes might never recover:

> This wind is the plague-wind of the eighth decade of years in the nineteenth century; a period which will assuredly be recognized in future meteorological history as one of phenomena hitherto unrecorded in the courses of nature, and characterized pre-eminently by the almost ceaseless action of this calamitous wind. (Ruskin 34.31).

The Storm Cloud was characterized by "a malignant *quality* of wind" with intemperate, changeful weather. Its effects grew more devastating every year. "It will blow either with drenching rain, or dry rage, from the south,—with ruinous blasts from the west,—with bitterest chills from the north,--and with venomous blight from the east" (Ruskin 34.34). It was already wreaking havoc with plants and trees. Ruskin declared he had seen it in Britain and Western Europe and as far away as Sicily. The "plague-wind" or "plague-cloud," as he called it, was "*always* dirty, *and never blue under any conditions*" (Ruskin 34.51).

One particularly eerie effect of the plague-wind was the way in which Ruskin felt it affected plants and trees. He had taken notes in his journal for years and had sketched the effects of the storm cloud in his sketchbook since at least 1845. He had grown hypersensitive to the suffering of vegetation. To his mind, the wind gave the trees "an expression of anger as well as of fear and distress" (Ruskin 34.34). In this he nearly echoed Wordsworth's and Clare's animistic sentiments. The tree leaves trembled in sickly syncopation, "as if they were all aspens...with a peculiar fitfulness." In his London lecture he read out an entry from his diary made at Brantwood on July 4, 1875:

> [A]n hour ago, the leaves at my window first shook slightly. They are now trembling *continuously*, as those of all the trees, under a gradually rising wind, of which the tremulous action scarcely permits the direction to be defined,—but which falls and returns in fits of varying force, like those which precede a thunderstorm—never wholly ceasing. (Ruskin 34.31).

Ruskin's religious upbringing and intermittent bouts of mental illness clearly played a role in shaping his bleak presentiments. Yet this concern for the environment began well before he suffered any prolonged mental or physical breakdown. Even in *Modern Painters*, in the fifth volume from 1860, he noted the catastrophic effects that industrial society would have on the natural world. There he compared Renaissance Venice to Victorian London, finding the industrial city already blighted by comparison. He took his evidence from the paintings of J.M.W. Turner, where the atmosphere routinely bespoke "dinginess, smoke, soot, dust, and dusty texture... dunghills, straw-yards, and all the soilings and stains of every common labour." Ruskin had long shared the concerns of J. S. Mill who, in *Principles of Political Economy* (1848), asked how society might come to terms with permanent physical limits to economic growth. What would happen if the population of Britain grew too large for the land to support it? What if the planet came to be so densely populated that every inch of it was under cultivation? All wilderness would vanish. Every "wild shrub" would be "eradicated as a weed" and wild animals would lose any and all habitat. Mill's answer was to embrace the notion of a stationary state, a stage of history when economic development was constrained by the physical limits of finite natural resources. Crucially, he advised embracing this state before population growth got out of hand. But where Mill had imagined deterioration in terms of overcrowding, extinction, and the loss of solitude, Ruskin instead saw environmental change as a product of spiritual decline, atmospheric pollution, modern warfare, and anthropogenic climate change. Curbing population was only one piece of the puzzle. For Ruskin, it had to be coupled with the moral rehabilitation attainable through the practice of simple living and through the cultivation of views and habits promoting the aesthetic appreciation of nature. (Albritton and Albritton Jonsson 2016, 31, 37–38).

It is a testament to the power of these earlier arguments, that amid his increasing mental illness, failing health, and general despair, his more devoted followers continued and even furthered what he began.

## Legacy of the Simple Life

**Arthur Ransome's Children's Books** As unsettling as Ruskin's prognostications are, practitioners of the Ruskinian simple life have tended to live and work with optimism and his legacy has been rosier than one might imagine. For example, the nature-oriented educational activities of the Collingwoods inspired a family friend to write a novel about them. Arthur Ransome met the Collingwoods in 1903 and developed an intimate bond with the whole family, finding a kindred spirit in Collingwood and falling in love with one daughter after another. Twenty-five years later, Ransome became the chronicler of their life in the Lake District in his beloved books for children, starting with *Swallows and Amazons*. This was published in 1930 to great acclaim in Britain. Much of the narrative centers on surprisingly technical descriptions of manual skills—sailing, camping, boatbuilding, and charcoal burning—the likes of which had seldom been seen in children's literature before. Everywhere, nature is a force to be understood and reckoned with in a wise and thoughtful way; work with the hands is prized over any use or thought of modern technology. The novel often calls to mind the Collingwood children's rambles through the fells so carefully recorded in *Nothing Much*. Ransome's talent was to translate the appeal of the simple life into a concrete language that children could understand (Albritton and Albritton Jonsson 2016, 172–173). The novel also popularized the idea of the Lake District as a site of leisure and play, drawing ever larger segments of the public, adults and children alike, to hike and sail in an area of renowned natural beauty (Welberry 2004, 82, 84). *Swallows and Amazons* and its film and TV adaptations imprinted the simple life on generations of English children on holiday. Admittedly, love of the countryside and sailing were not part of any radical critique of urban life and industrial production, only a joyous but temporary suspension of reality (Albritton and Albritton Jonsson 2016, 173).

**The "Return to Nature" and Arts and Crafts in America** The desire to return to nature sprang up in America in the nineteenth century as well, though it developed its own character. Transcendentalists like Henry David Thoreau and Ralph Waldo Emerson in the first half of the nineteenth century advocated spiritual improvement by living more simply and closer to nature. In this, they resembled British Romantics like Wordsworth, choosing to forego the conveniences and luxuries of modern urban life and industrial innovation in favor of escaping to an ennobling rustic natural setting. As Shi (1985, 125) notes, for Transcendentalists, "romantic simplicity was less a societal ethic than a spiritual ethic." A similarly spiritual and individualist strain was seen in the wilderness movement inspired by John Muir, a far more rugged practitioner of living simply, even primitively, in the wild than Thoreau. He did not expect people to live as he did. But he hoped they would experience now and then a "revivifying contact with nature." Thanks to Muir, America's first National Park at Yellowstone was founded in 1872 and the Sierra Club was founded in 1892 (Shi 1985, 196–197). Later in the century, elites like Oliver Wendell Holmes praised the benefits of wealth to those seeking contact with nature, noting shrewdly that money could buy access to green healthy spaces, which in turn would likely result in happier, kinder children than those who grew up in cramped, polluted urban back alleys (Shi 1985, 159). The plight of those who could not leave was a lesser concern.

The simple life and its reverence for nature also found unique expression in the American Arts and Crafts movement, which got under way in the late 1890s in Boston. Where Ruskin and William Morris (1834–1896), his younger counterpart, prized a revival of medieval handicrafts that gave meaning and satisfaction to common laborers, connecting them to nature, the American movement focused more on the middle classes and tended not to discourage industrial progress. As Shi puts it:

> The return to nature advocated by most [American] Arts and Crafts spokesmen was not that of a Thoreau or a Tolstoy but that of an Emerson, Downing, or Norton—a comfortable home and vegetable garden in the country, within commuting distance of the city, and blessed with the amenities of civilized living. (Shi 1985, 194).

Many American craftsmen, like Gustav Stickley who was known for the simplified "mission style" furniture still popular today, catered to these bourgeois middle-class city dwellers. He felt that he could "provide both therapeutic personal and regenerative social benefits" by selling them well-made items or teaching them to make them (Shi 1985, 191). He even created "Craftsman Farms" in New Jersey where, not unlike the Guild of St. George, he sought to teach them about the benefits of gardening, agricultural work, and handicrafts (Shi 1985, 194). He also founded *The Craftsman: An Illustrated Monthly Magazine for the Simplification of Life*, first published in 1901. Stickley had been inspired by both Morris and Ruskin and wished to promote the simple life. The entire second issue of the magazine was devoted to Ruskin's life and work, attending to his social and economic ideas. Each of the essays by Professor Irene Sargent, an art critic and moral reformer at Syracuse University, touch on Ruskin's influence in some way. The foreword to the issue mentions that the choice to include a review of a novel titled "Back to the Soil" came about because of "Ruskin's desire to improve the tenements and environment of the city poor."

Stickley's Craftsman farms and magazine were indicative of a more socially radical strain in the American Arts and Crafts movement. It prized quality, hand-made goods of simplified design over mass-produced machine-made goods and the use of durable materials in connection with nature. But, being more strongly influenced by Ruskin and Morris, it also drew attention to the plight of common city workers' dissatisfaction and disconnection from nature. In fact, all around the country, rural Arts and Crafts colonies cropped up—often explicitly inspired by Ruskin. These were "intended to attract disaffected artisans and craftsmen from the cities," and "usually combined subsistence agriculture with cottage industries to provide for their basic needs" (Shi 1985, 190). This "Back-to-the-Land" movement, which occurred in England as well, was a far cry from the more patrician ideal of the middle and upper classes commuting between an urban workspace and a peaceful country home.

**The Back-to-the-Land-Movement, Nature and Community** In England there were many experiments designed to remove workers from harsh urban environments and set them up in "artistic communities, some out on the edges of conventional society" in the countryside (MacCarthy 1995, 602–604). One of the more famous was led by Charles Robert Ashbee (1863–1942). He set up the Guild and School of Handicraft in London in 1888, then moved it in 1902 to the Cotswolds in Gloucestershire. There he and his wife Janet and his fellow craftsmen attempted to pursue the simple life according to the principles of Ruskin and Morris. Within months, "more than a hundred men, women and children" followed them (Bingham 2009, 122). Believing that cities gave rise to greed and disconnected humans from nature and beauty, this experiment in communal living reflected a nostalgic view of a pre-industrial life attuned to the natural world. Silversmiths, jewelers, cabinet-makers and woodcarvers lived and worked near to each other. They seemed to prosper in this idyllic rural setting. They developed a unique style that "combined country traditions with the influence of Art Nouveau (Bingham 2009, 123). But by 1906, things took an unfortunate turn, with younger craftsmen leaving the group and older craftsmen shortening their hours due to ill health. They could not compete with London shops. The simplicity of the Arts and Crafts style was its hallmark and its downfall, as it could easily be replicated (often mechanically) and sold for less. The colony disbanded in 1908 (Bingham 2009, 126). Similar colonies were established by H.V. Mills in the 1890s at Starnthwaite, near Kendal in the Lake District, which did not last long either due to disputes among colonists.

Americans tried their hand at moving back to the land as well. In a town named Ruskin, Tennessee, a utopian socialist colony was founded in 1894 by Julius Augustus Wayland. (Ruskin was not a socialist, but Morris and many of his followers were.) Wayland wished to establish an alternative to the greed and wastefulness of industrial city centers. As with Ashbee's colony, workers lived and worked together in the countryside. As Atwood (2016, 38) notes, Wayland followed Ruskin's tenets by insisting on the importance of education. All children received non-competitive lessons in "drawing, painting, modeling, writing, natural history, chemistry, oratory, physical exercise, geometry, and music" (Atwood 2016, 39). They spent their days "observing and engaging with nature and developing essential manual skills" (Atwood 2016, 39). In this, they distantly mirrored the activities of the Collingwood family. Unfortunately, the colony struggled with internal disputes as well. After moving a few more times, it too disbanded in 1901 (Atwood 2016, 39). Another similar colony based on the Guild of St. George arose in Ruskin, Florida with greater success, though World War I eventually depleted the town of its young people (Atwood 2016, 40).

The Garden Cities movement fared rather better. Beginning in 1899 with the first development at Letchworth, Ebenezer Howard conceived of these self-contained communities as a compromise between country and city living. Built for companies who wished to increase productivity by improving the health and well-being of their workers, town planners sought a balance between space for industry, residences, and agriculture. Community green spaces took center stage in the design of these "garden cities." Although many slowly took on the character of bedroom suburbs, falling short of its original aims, the garden city ideal continues to influence city planners today.

The simple life may have failed to reshape society on the whole, but it has nonetheless left a lasting mark. It continues to inspire those concerned about the environment. In cities and suburbs today, people sign up for various environmentally friendly groups such as CSAs - community supported agriculture cooperatives - which bring locally grown, organic produce directly to a person's house while enjoining ordinary non-farmers to get outside and help till the soil with their own hands (Shi 1985, 278–79). The role of plants and landscapes in peoples' lives continues to be expanded and broadened to include, for instance, ecotherapeutic activities whereby urban residents troubled by stress, trauma, or a variety of mental afflictions spend time in the country. This trend shows every promise of evolving to suit a variety of needs in a quickly changing, stressful world.

In 1951, Ruskin's beloved house in the Lake District was bought and a trust was established to turn it into a museum for future generations. Brantwood, with its thoughtfully curated gardens and unique natural setting, now abounds with eclectic expressions of the simple life rooted in nature. The entire wooded estate with its fields and numerous experimental gardens has been restored, as far as possible, in accordance with the principles of the simple life, which as Ingram (2014, 85) notes, constitute "idealism and open-mindedness--a willingness to embrace new, even challenging, ideas." Practices familiar to today's students of environmentalism include organic farming designed to produce a harmonious whole "in which the soil, plants, wild creatures, humanity and landscape are all in tune with one another" (Ingram 2014, 85). In this way, the ideals of the simple life and the Victorian desire to return to nature continue to be explored and adapted to our changing needs.

## Conclusion

The nineteenth century saw profound social change instigated by rapid industrial growth, degradation of the environment and spiritual disaffection at all levels of society. This in turn inspired numerous attempts to return to nature through literature and the arts in combination with socially progressive endeavors, some more successful than others. In the case of the Romantics William Wordsworth and his sister Dorothy Wordsworth, the impulse to flee the city constituted a privileged, conscious choice to reconnect with nature. In the case of John Clare, who grew up and worked in a rural community, life in the country was a matter of necessity, but one increasingly curtailed by industrialization and encroachment in the form of large-scale agriculture and parliamentary acts of enclosure which cut commoners off from the land and the sustenance it previously provided. Urban planners attempted to mitigate the pernicious effects of urbanization on the working class by promoting novel ways to access nature within city limits. This included the establishment of botanic gardens and large, public green spaces and promoting garden allotments and flower shows for horticulturalists of all levels.

Few Victorian social critics had more to say about the benefits of drawing people of all classes back to nature than John Ruskin. Through the establishment of the Guild of St. George for the working classes, voluminous writings on allotments, artisanal crafts, nature, literature, and art as well as his critique of poor labor conditions, excessive consumption, and the devastation of natural resources, Ruskin's name became synonymous with the ideal of the simple life. Ruskin's followers in the Lake District showed how the wise use of nature coupled with consumer restraint could free people from the hardships of the capitalist industrial economy while also safeguarding the countryside. They gardened, revived medieval artisanal crafts, wrote books and poems on nature, engaged in archaeology, painted, founded the National Trust, and engaged in experimental pedagogy to sensitize children to the ecology and culture of the local environment.

Despite Ruskin's dire warnings about the Storm-Cloud of the Nineteenth Century where he described anthropogenic climate change as a result of industrial overreach and thoughtless consumption, his ideas about the simple life and its ability to connect people to nature have found optimistic and beneficial expression in popular children's literature, the Arts & Crafts movement, nature-oriented residential planning, Back-to-the-Land movements and community agriculture. Even today, Ruskin's estate, Brantwood, in the Lake District ensures that these ideas will continue to expand our ability to access nature and drive home the need to preserve fragile landscapes.

# Bibliography

AG (1875) A beginner in botany. Gardener's Chronicle 90(4):358–359

Albritton V, Albritton Jonsson F (2016) Green Victorians: the simple life in John Ruskin's Lake District. The University of Chicago Press, Chicago

Atwood S (2016) Ruskin's educational ideals. Routledge, New York

Bate J (2001) The song of the earth. Picador, London

Bate J (2003) John Clare: a biography. Farrar, Straus and Giroux, New York

Bingham J (2009) The Cotswolds: a cultural history. Oxford University Press, Oxford

Brunton J (2001) The arts & crafts movement in the Lake District: a social history. Lancaster, University of Lancaster University

Clare J (2004) Major works. Oxford University Press, Oxford

Constantine S (1981) Amateur gardening and popular recreation in the 19th and 20th centuries. J Soc Hist 14(3):387–406

Craig D (2006) John Ruskin and the ethics of consumption. University of Virginia Press, Charlottesville

Foreword (1901) The craftsman: an illustrated monthly magazine for the simplification of life 1(2–6)

Frost M (2014) The lost companions and John Ruskin's Guild of St George: a revisionary history. Anthem, London

Haslam S (2004) John Ruskin and the Lakeland arts revival, 1880–1920. Merton Priory, Cardiff

Ingram D (2014) The gardens at Brantwood: evolution of John Ruskin's Lakeland paradise. Pallas Athene and the Ruskin Foundation, London

Lawes E (1849) The act for promoting the public health. Shaw and Sons, London

Loudon J (1811) The ladies' flower garden of ornamental annuals. William Smith, London

Mahood MM (2008) The poet as botanist. Cambridge University Press, Cambridge

MacCarthy F (1995) William Morris: a life for our time. Knopf, New York

O'Gorman F (2010) Ruskin, science, and the miracles of life. Rev Eng Stud 61(249):276–288

Page J, Smith E (2011) Women, literature, and the domesticated landscape: England's disciples of flora, 1780–1870. Cambridge University Press, Cambridge

Paulin T (2004) Introduction. In: Clare J Major works. Oxford University Press, Oxford

Ritvo H (2009) The dawn of green: Manchester, Thirlmere, and modern environmentalism. The University of Chicago Press, Chicago

Robinson W (1870) The wild garden. John Murray, London

Ruskin J (1903–1912) The works of John Ruskin. George Allen, London

Shteir A (1996) Cultivating women, cultivating science: flora's daughters and botany in England 1760 to 1860. Johns Hopkins University Press, Baltimore

Smith J (2010) Domestic hybrids: Ruskin, Victorian fiction, and Darwin's botany. In: Knoepflmacher U, Browning L (eds) Victorian hybridities: cultural anxiety and formal innovation. Johns Hopkins University Press, Baltimore, pp 117–126

Townend M (2009) The Vikings and Victorian Lakeland: the Norse medievalism of W.G. Collingwood and his contemporaries. Titus Wilson & Son, Kendal

Tuckwell W (1891) Tongues in trees and a sermon in stone. George Allen, London

Uglow J (2005) A little history of British gardening. Pimlico, London

Warner H (1889) Songs of the spindle & legends of the loom. N.J. Powell & Co, London

Welberry K (2004) Arthur Ransome and the conservation of the English lakes. In: Dobrin S, Kidd K (eds) Wild things: children's culture and ecocriticism. Wayne State University, Detroit, pp 82–96

Winters J (1999) Secure from rash assault: sustaining the Victorian environment. University of California Press, Berkeley

Wordsworth D (1991) The Grasmere journals. Oxford University Press, Oxford

Wordsworth W (2014) Wordsworth's poetry and prose. Norton, New York

# Chapter 10
# Violets in the Victorian Age

**Luís Mendonça de Carvalho and Francisca Maria Fernandes**

**Abstract** During the nineteenth century, violets were highly valued plants in Europe, especially in the courts of Napoleon III and Queen Victoria. Many new varieties were introduced, and some even won awards at events organised by the Royal Horticultural Society. In nineteenth-century English literature, violets were frequently mentioned, and they were grown outdoors or in greenhouses and sold on the streets of major cities. The French Riviera was a perfect location for cultivating violets because of the mild climate and soil. When Queen Victoria chose to spend part of her year there, the region became a popular holiday destination for the European aristocracy. The Grasse perfume industry used essential oils extracted from plants grown in the area, including violets, to make perfumes and other personal hygiene products. The Victorian language of flowers also gave violets a symbolic meaning related to discretion and humility. At the dawn of the Victorian Age, as society underwent significant changes, violets began to lose their cultural relevance.

**Keywords** Violets · Parma-violets · Queen Victoria · floriography · Victorian literature · French Riviera · Grasse

The history of violets (genus *Viola* L.) in Europe began with the classical Greeks, who praised violets for their sweet fragrance, unique colour, and medicinal uses. Violets were sung by Greek poets, such as Homer (eighth century BC) in the Odyssey (5.78) when he described the grotto where Odysseus (Ulysses) was kept prisoner by Calipso as full of '*beds of violets*' (Homer 1961). The poet Pindar (c.518–438 BC) also mentioned violets and associated these flowers with the

L. M. de Carvalho (✉)
Beja Polytechnic University, Beja, Portugal
e-mail: lmmc@ipbeja.pt

F. M. Fernandes
IHC/IN2PAST, Lisbon Nova University, Lisbon, Portugal

glorious city-state of Athens: '*shining and violet-crowned and celebrated in song, bulwark of Hellas, famous Athens, divine citadel*' (Pindar 1997).

In the treaty '*On Marvellous Things Heard*', attributed to Aristotle (c.384–322 BC), violets are reported to be so abundant in some regions of Sicily that their perfume filled the air and made dogs confused:

> '*In Sicily in the district called Enna there is said to be a cave, around which is an abundance of flowers at every season of the year, and particularly that a vast space is filled with violets, which fill the neighbourhood with a sweet scent, so that hunters cannot chase hares, because the dogs are overcome by the scent. Through this cave, there is an invisible underground passage, by means of which Pluto is said to have made the rape of Core*' (Aristotle 1955).

Violets were also a sign of spring, as written by Theophrastus (c.371–c.286 BC) in the *Enquiry into Plants* (6.8.1): '*Of the flowers the first to appear is the white violet*' (Theophrastus 2016). The poet Sappho (c.630–c. 570 BC), a Greek aristocratic that founded an academy for girls on the island of Lesbos, wrote many poems, now lost, but some fragments survived, such as '*You put on many wreaths of violets and roses and crocuses together by my side, and round your tender neck*' (Campbell 1990).

Violets were also present in the odes composed by the Roman poet Horace (65–8 BC), such as in the Ode XV: '*For elms, we plant the unwedded plane. Myrtles abound, and violet-beds, and every flower, that yields a scent, O'er olive-ground its perfume sheds*' (Horace, 1894). The Roman naturalist and writer Pliny, the Elder (c. 23/24–79 AD), dedicated a chapter to violets in his *opus magnum*, the Natural History (Book 21.14):

> '*Next in esteem comes the violet, of which there are several kinds, the purple, the yellow, the white, all of them planted as are vegetables, from cuttings. Of these kinds however the purple, which comes up wild in sunny, poor soils, springs up with a broader, fleshy leaf, coming straight from the root. It is the only one to be distinguished from the others by a Greek name, being called ion, from which ianthine cloth gets its name. Of the cultivated violets, the most highly esteemed is the yellow variety. The kinds called Tusculan and marine have a slightly broader but less perfumed petals. The Calatian variety however is entirely without perfume and has a very small petal; it is a gift of autumn, but all other kinds bloom in spring.* (Pliny 1961).

When the city of Rome fell and the classical world ended, violets indeed continued to be picked and used, although within very different economic and sociocultural contexts. In the following millennia—the Middle Ages—violets would gain new meanings, such as being linked with the Virgin Mary, as in a text attributed to Saint Bernard of Clairvaux (1090–1153), in which the mother of Jesus is called "*the violet of humility*" (Ancona 1977) (Fig. 10.1). This connection with humility was probably a consequence of the shade environment where most violets thrive and flourish and due to their morphological characteristics — discrete buds pointing downward, hidden flowers under the leaves — and other signs that Medieval Christian theologians saw as unmistakable signs of modesty. Close to our time, the Carmelite nun, later Saint Thérèse of Lisieux (1873–1897) [Little Flower of Jesus—*La Petite Thérèse*] wrote in her book *Histoire d'une Âme* [*The Story of a Soul*], the symbolic role that God attributed to violets:

**Fig. 10.1** Virgin Mary as the Violet of Humility. Turgis Lithography. Paris (late XIX/early XX century), postcard, private collection

'*Ainsi en est-il dans le monde des âmes, ce jardin vivant du Seigneur. Il a trouvé bon de créer les grands saints qui peuvent se comparer aux lis et aux roses; mais il en a créé aussi de plus petits, lesquels doivent se contenter d'être des pâquerettes ou de simples violettes destinées à réjouir ses regards divins lorsqu'il les abaisse à ses pieds*' [*And so it is in the world of souls, Jesus' garden. He willed to create great souls comparable to lilies and roses, but He has created smaller ones, and these must be content to be daisies or violets destined to give joy to God's glances when He looks down at his feet*] (Lisieux 1898).

The American priest and famous writer of Catholic devotional works, Francis Xavier Lasance (1860–1946), wrote a text in which the character of humbleness is also associated with violets:

'*As the wood-violets give forth their perfume from beneath the brushwood that conceals them from view, telling us of their unseen nearness, so kindness reveals to us the nearness of Jesus, the sweetness of whose spirit is thus breathed forth*' (Lasance 1908).

A high momentum in the history of violets began in the eighteenth century when, in Paris (France), violets began to be sold on the streets, probably from wild harvesting or small-scale cultivation. A letter from Madame Roland [née Jeanne Marie Phlipon] (1754–1793), a famous *salonnière* and writer who lived during the troubled times of the French Revolution and lost her life in the guillotine, reveals the joy that violets brought to her:

*'I always remember the singular effect produced on me by a bunch of violets at Christmas. When I received them, I was in that mood which a season favorable to serious thought induces. My imagination slumbered. I reflected coldly, the emotions were at rest. Suddenly, the color of the violets and their delicate perfume quickened my senses. It was an awakening to life. ... A rosy tinge suffused the horizon of the day'* (Roland 1901).

By the end of the eighteenth century, the highly scented Parma violets (Fig. 10.2) were also cultivated in frames. The history of Parma violets is somewhat unclear. Still, Malécot et al. (2007), using ITS (internal transcribed spacer) sequences and allozyme variation in 14 putative loci, concluded that the cultivars of Parma violets belong to the genetic pool of *Viola alba* Besser subsp. *dehnhardtii* (Ten.) W.Becker, with parental ancestors from the Eastern and the Western Mediterranean region, probably from Turkey and Italy. This is corroborated by the proto-scientific literature, such as in *Hortus Medicus et Philosophicus* (1588) by Joachim Camerarius (1534–1598), in which the author wrote (page 177) that travellers and writers saw

**Fig. 10.2** Parma Violets. Favourite Flowers of Garden and Greenhouse (1896). Eduard Step. Biodiversity Heritage Library

PARMA VIOLET

(VIOLA ODORATA—*var. parmensis*)

Nat. size

PL. 32

violets with many petals (pleiomerous), such as Damascene roses, but with a better fragrance, in the surroundings of Byzantium (Constantinople, now Istanbul):

> '*Costæus in Museun annotauit Byzantii Violæ specimen coli, cætera Martiæ similem, sed flore infinitis foliolis constructo & æquante Rosam Damascena, magnitudine & odore etiam nostrates superante*' [Costeo in (the book) Museun (Pseudo-Mesue, 1573) commented on a violet specimen from Byzantium, in all like *Martiæ* (*Viola odorata* L.), but having flowers with many petals, such as the Damascene Rose, surpassing even ours in size and fragrance] (Camerarius 1588).

The ancestors of Parma violets probably came from Turkey to Italy on an uncertain date, where they hybridised with Italian plants and originated the Parma violets. The history of these flowers is linked to the Bourbon royal family, which ruled the Kingdom of Naples and had dynastic ties with many royal families in Italy and Europe. In Naples, tradition states that Parma violets came from Portugal, hence its local name—*Violetta Portoghese* (Portuguese Violet); in Italy, these flowers were known as *Violetta di Napoli* (Naples Violet).

Although Parma violets were cultivated in many European countries during the late eighteenth and nineteenth centuries, as in Germany and Italy, it is in the French and British nursery catalogs that we find more data on the cultivation and trade of Parma violets, as well as the introduction of new cultivars into the market (Coon and Giffen, 1977, Coombs 2003). In France, the name *Violette de Parme* (Parma violets) is mentioned in *Le Bon Jardinier* (1805, page 471): '*Il y a depuis quelques temps une Violette semi-double, dont la couleur est extrêmement pâle. On l' appelle Violette de Parme. Elle a l' avantage de fleurir dès le mois d' octobre, sin on la tient en pots et dans l' orangerie, ou dans l' appartement*', but it can be older. This edition of the *Le Bon Jardinier* was dedicated and presented to Her Majesty, the Empress of France [*Impératrice des Français*], Joséphine de Beauharnais [born Marie Josèphe Rose Tascher de La Pagerie] (1763–1814), the first wife of Emperor Napoleon Bonaparte (1769–1821), which, since the beginning of the nineteenth century, maintained a collection of violets in the gardens of their country home—*Le Châteaux de Malmaison*, near Paris (Lack 2004).

Violets had a long association with this couple; Josephine was adorned with violets in their marriage, and she asked Napoleon to always give her violets on their wedding anniversaries (Coats 1977). After Napoleon's abdication and banishment to the island of Elba, in the Mediterranean Sea, violets became a symbol of the Bonaparte party because he promised to return to France when the violets were again in season. He did return and got the epithet *Le Père La Violette*. After his final defeat, and for the following decades, in France, violets were a symbol of sedition due to their liaison with the Bonapartists (Fig. 10.3).

The second wife of Napoleon, Marie Louise (1791–1847), daughter of the last Holy Roman Emperor, and mother of Napoleon's only son—the King of Rome—received the Duchy of Parma after Napoleon departed to the island of Saint Helena. Once in Parma, she patronised the violet fashion, cultivation, and trade (Sandrini 2008). Today, two of the most widely cultivated varieties of Parma Violets still evoke her name: *Viola* 'Duchess of Parma' (Fig. 10.4) and *Viola* 'Marie Louise' (Fig. 10.5).

**Fig. 10.3** A bunch of violets conceals the profiles of Napoleon, Marie Louise, and the King of Rome in 'The Pedigree of Corporal Violet', satirical print by George Cruikshank (1792–1878), published by H. Humphrey June 9th, 1815 – No. 27 St. James's St., London; Yale University Library

**Fig. 10.4** Parma violet 'Duchesse de Parma'. Courtesy of Clive Groves

**Fig. 10.5** Parma violet
'Marie Louise'. Courtesy
of Clive Groves

In the middle of the nineteenth century, violets were again in fashion under the Second Empire (1852–1870), ruled by Napoleon's nephew, Napoleon III (1808–1873), and his wife, Eugenia de Montijo (1826–1920). When the imperial couple was forced to leave France and found exile in England, they took their passion for violets with them, increasing the prestige of these flowers in the United Kingdom (Coombs 2003).

The emperor and the empress had one son only, Louis-Napoléon (1856–1879), who died in a Zulu ambush while serving the British Army in South Africa. The prince's body was recovered and brought to Europe, but his mother wanted to see the place where her son had lost his life, so she went on a trip to South Africa (Fig. 10.6). An apocryphal story recounts how the empress found the place where her son died, marked with stones but lost in the dense African brushwood. Decades later, when evoking the empress, who recently died, the journal *The King Islands News* November 23rd, 1921, describes it with the following words:

'(…) *After her son, the Prince Imperial, was killed in Zululand, the Empress* (…) *paid a visit to his grave. This spot had been marked by a cairn of stones, but by the date of the visit, the jungle had encroached so that even the Zulu guides, who had been among the Prince's assailants, could not find it. The Prince had a passion for violet scent; it was the only toilet accessory of the kind he used. Suddenly, the Empress became aware of a strong smell of violets. "This is the way," she cried, and went off on a line of her own* (…) *until, with a loud cry, she fell upon her knees, crying, "C'est ici!" [It is here]* (…).'

Parma violets were widely cultivated in Paris neighbourhoods, including Bourg-la-Reine, where the famous nursery of Armand Joseph Millet (1845–1920) was located. In 1898, describing the prominence of Parma violets in Paris, Armand Joseph Millet wrote:

"*everyone who has lived in Paris, even those who have merely passed through the city in March or April, have seen and admired those beautiful pots of violets which adorn our florist's shops. All the covered markets and street markets are filled with them, flattering the eyes and filling the air of the metropolis with their sweet perfume. Well, it is the Parma violet again that supplies these quantities of plants.*" (Millet 1898).

**Fig. 10.6** Empress
Eugénie kneeling on a
*prie-dieu* with a bouquet of
violets in her gloved hands
(c.1880). W. & D. Downey
Photographers, London.
Bibliothèque Nationale de
France Gallica 12,148/
btv1b530198589

It is estimated that, in Paris, approximately six million violet bunches were sold
yearly by the end of the century (Perfect 1996, Coombs 2003).

## Violets in the United Kingdom

In the 1840s and 1850s, all major centres for cultivating violets were located in
Middlesex. At Cranford, F.J. Graham raised the violet 'The Czar', the first to receive
an award from the Royal Horticultural Society, in 1865. This cultivar was very
important for breeding new hybrids, and most contemporary hybrids have this cultivar as an ancestor. In the *Extracts from Proceeding of the Royal Horticultural
Society and Miscellaneous Matter*, published in *The Journal of the Royal
Horticultural Society* (New Series Volume I, page 5), we can read the following:

> '*a very interesting collection consisted of a group of twenty-four plants of the new large-
> flowered Russian Violet raised by F. J. Graham, Esq., of Cranford, and named the Czar. The
> blooms of this new Violet, as well as the foliage, are very large, and the whole plant has a
> most robust-looking aspect. It had already gained a First-Class Certificate, and now fully
> maintained the high character thus given to it. With it was a seedling, called Princess
> Dagmar, with smaller flowers of a duller, paler purple, altogether inferior*' (RHS 1866).

As the development of the metropolitan area of London claimed new lands, nurseries were relocated to the counties of Kent, Sussex, and Devon, such as R.W. Beachey, the leading violet nursery in Devonshire in the 1880–1890s. Other famous violet nurseries included Henry Cannell's Swanley Nurseries (Swanley, Kent), Dillistone and Woodthorp (several locations), and Carles Turner, The Royal Nurseries (Slough, Buckinghamshire) (Coombs 2003).

Sir Joseph Banks (1743–1820), President of the Royal Society, an avid plant collector and 'unofficial' director of Kew Gardens under King Georg III (1738–1820), had 300 pots with Parma violets in his garden in Spring Grove House, Isleworth (Farrar 1989). In the first half of the nineteenth century, the Parma violets cultivated in the United Kingdom were from the cultivar known as *Viola* 'Pallida Plena' or *Viola* 'Suavis Pallida Plena Italica', probably the original Parma violet, which is now called *Viola* 'Neapolitan'. In 1873, a new violet listed by British nurseries and called 'de Parme' was probably the cultivar *Viola* 'Duchesse de Parme', which came from Florence (Italy) 3 years earlier. In 1875, the cultivar *Viola* 'Lady Hume Campbell' was introduced in the British markets and received the name of the lady who brought it from Milan (Italy) and who had it grown in Pinner (Middlesex). This latter cultivar has a long flowering period (up to May) and can produce many flowers; under ideal conditions, 100 plants can yield 600 to 1000 flowers weekly. In 1880, a new white Parma violet was introduced to the market, *Viola* 'Swanley White' (Fig. 10.7), by the Swanley Nurseries in Kent (United Kingdom). This violet was acquired from the Italian Count Filippo di Brazza, who had raised it in Udine (Italy) (Coombs, 2003). It is sold under these two names: 'Swanley White' or 'Conte di Brazza'.

**Fig. 10.7** Parma violet 'Conte di Brazza' or 'Swanley White'. Courtesy of Clive Groves

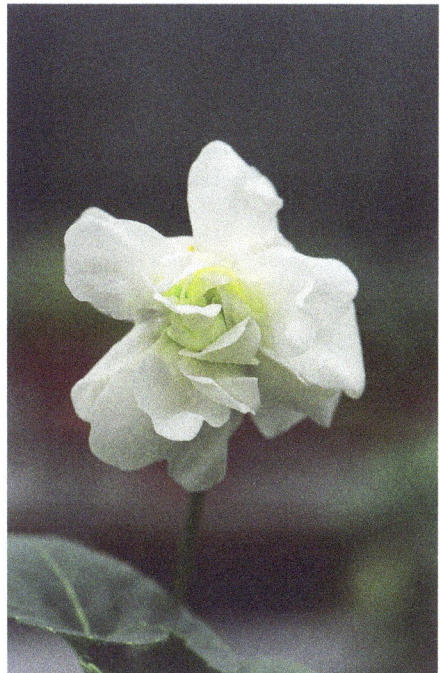

## Violet Cultivars in the UK (Nineteenth Century)

In the nineteenth century, many Parma violet cultivars were introduced to the British market, but not all were retained in the nurseries' catalogues. The following list describes some of the most common and sought-after scented single, double, and Parma violet cultivars; the dates indicate when they became available (Timbal-Lagrave 1862, Perfect 1996, Bertrand and Casbas 2001, Coombs, 2003):

> *Viola* 'Baronne Alice de Rothschild' (1894) giant flowers, mid-bluish-purple, long stems, early flowering [scented single violet]
>
> *Viola* 'Conte di Brazza' (1880) (synonymous with *Viola* 'Swanley White' or *Viola* 'White Parma') has bright white flowers with slight blue tints; this cultivar won the Royal Horticultural Society First Class Certificate in 1883 [Parma cultivar]
>
> Viola 'Devonensis' (ca. 1860) unknown origin, deep blue flowers, and mid to late-flowering; considered the hardiest winter blooming violet [scented single violet]
>
> *Viola* 'Double Rose' (1899) is a possible improvement of 'Double Rose de Bruant' [scented double violet]
>
> *Viola* 'Duchesse de Parme' (1870) pale lavender-blue flowers; very prolific and easy to grow [Parma cultivar]
>
> *Viola* 'Köningin Charlotte' (1890) blue flowers, very fragrant, free-flowering [scented single violet]
>
> *Viola* 'Lady Hume Campbell' (1875) (synonymous of *Viola* 'Gloire d'Angoulême') lavender mauve flowers [Parma cultivar]
>
> *Viola* 'Madame Millet' (1884) lilac rose flowers, slightly mauve, like little roses; obtained by Néant of Bièvres in Seine-et-Oise (1868) and introduced by the famous French nursery Millet et Fils of Bourg-la-Reine; very popular cultivar in the nineteenth century but no longer available, probably extinct [Parma cultivar]
>
> *Viola* 'Marie Louise' (1865, probably older) deep lavender mauve flowers [Parma cultivar]
>
> *Viola* 'Mrs. John J. Astor' (1895) rose lavender flowers; a cultivar originating in the USA that won the Award of Merit from the Royal Horticultural Society (1899); now a lost cultivar [Parma cultivar]
>
> *Viola* 'Princesse de Galles' [Princess of Wales] (1889) giant-flowered lilac-blue; probably the most grown scented cultivar in Europe and the UK [scented single violet]
>
> *Viola* 'Quatre Saisons' (1830) pale lilac-blue; semi-perpetual flowering (mid-summer to May), strongly scented [scented single violet]
>
> *Viola* 'Queen of Violets' (1865) rose-pink buds opening to white flowers [scented double violet]
>
> *Viola* 'St. Helena' (1897) pale lavender-blue; a highly praised cultivar [scented single violet]
>
> *Viola* 'The Czar' (1863) deep purple flowers, long stems, free flowering, strong scent; parent plant used for new cultivars; awarded by the Royal Horticultural Society in 1865 [scented single violet]

The association of violets with the ruling classes and the aristocracy was robust during the final decades of Queen Victoria's reign, and the reign of King Edward VII, when the gardens of Windsor Castle were supervised by John Dunn and circa 3000 violets were cultivated under the protection of frames. Only three cultivars were grown: two were Parma violets (*Viola* 'Marie Louise' and *Viola* 'Lady Hume Campbell'), and the other was *Viola odorata* L. 'Princess of Wales' (Cuthbertson, 1910). Queen Victoria (1819–1901) had a long association with violets; in the journals she kept all her life, the first reference to violets was made when she was still a princess, on Easter Day of 1834: '*Mamma gave me two very pretty little china baskets with violets…*' (Queen Victoria, unpublished manuscript). In her journal, Queen Victoria mentioned violets at least 105 times, many of which occurred during her sojourns at Osborne House (Figs. 10.8 and 10.9), the private residence of the Queen and Prince Albert on the island of Wight (Beasley 2017).

**Fig. 10.8**  Osborne House, the private marine residence of Queen Victoria and Prince Albert on the Island of Wight (c.1900). Library of Congress LC-DIG-stereo 1 s22298

**Fig. 10.9**  Osborne House Violet Collection. Viola 'Köningin Charlotte'. Courtesy of Toby Beasley

Floriography is a cryptological communication code that uses flowers or plants and their colours and arrangement to convey messages. This ancient language has changed according to society's cultural needs and values. A classic example of this codified use can be found in the Bible's book *Song of Songs*, where saffron, myrrh, roses, lilies, spikenard, calamus, cinnamon, aloes, frankincense, pomegranates, and some other plants and flowers have a strong symbolic meaning. During the long reign of Queen Victoria, many books on floriography were published, such as the popular *The Language of Flowers: An Alphabet of Floral Emblems* (Unknown author, 1857). Therefore, we can understand the meaning of violets: blue violet—*'faithfulness and love'*; purple violet—*'you occupy my thoughts'*; white violet—*'innocence and modesty'*; and wild violet—*'love in idleness'*. Victorians exchanged bouquets that *'talked'* and sent poetic messages to each other in a society that did not valorise the public expression of feelings (Fig. 10.10). Flowers' messages could have a positive meaning—love and friendship—or could send opposite signs—sadness, grief, jealousy, or ingratitude.

In France, Parma violets were first cultivated in warm southern provinces under the shade of olive orchards or as a main crop in open fields with partial shade (Figs. 10.11 and 10.12). These violets fulfilled the demand for cut violets from the

**Fig. 10.10** Violets as a symbol of discreet love (1905), postcard, private collection

**Fig. 10.11**   Picking violets in Hyères, La Côte de d'Azur (1900–1910), postcard, private collection

**Fig. 10.12**   Picking violets in olive groves, Côte de d'Azur (1900–1910), postcard, private collection

aristocracy and European royalty who had followed the pave of Queen Victoria who, after visiting Menton (Figs. 10.13 and 10.14), decided to spend, every year, several weeks in the French Riviera for its mild climate (Coombs 2003, Nelson

**Fig. 10.13** In the French Riviera, Queen Victoria visited Chiris's Perfume Manufacture, and M. Chrisis explained the distilling process to her. Illustrated London News (25 April 1891)

**Fig. 10.14** Marble monument (1912) in the memory of Queen Victoria by Louis Maubert (1875–1949), Nice (French Riviera); the four figures represent the cities where The Queen has sojourned (Nice, Cannes, Grasse et Menton), postcard, private collection

2007). She arrived at Menton on March 16, 1882, and stayed at the Chalet des Rosiers. The French newspaper *Petit Nicois* provided many details about this chalet, which were later reprinted in journals of the Empire, such as *The Port Adelaide News* (July 11, 1882):

> '(…) *The Chalet des Rosiers, which Queen Victoria has selected as her residence while at Mentone, is situated on the Baie de Ganvan, in the center of a vast semicircle, perpetually bathed in sunshine* (…) *a peculiarity of the house is that from the balcony of the royal chamber the eye looks down upon a thorough forest of orange and lemon trees, the perfume from which is mingled with that of roses and violets* (…).

By the end of the nineteenth century, a disease infected Côte d'Azur Parma violets, and flower growers began to cultivate *Viola odorata* L. cv 'Victoria', which is easier to maintain and more resistant to the common diseases that affect violets (Perfect 1996, Nanneli 2001).

The first record of Parma violets in Toulouse dates from 1854, and the production was mainly located in the villages of Lalande and Aucamville. The flowers were sold locally, in markets, on the streets (Fig. 10.15), or at the railway station

**Fig. 10.15** Seller of violets (in boxes), '*Types Toulousains – La Bouquetière*' (c.1904), postcard, private collection

**Fig. 10.16** Selling violets (bouquets and boxes) at Toulouse Railway Station (1900–1910), postcard, private collection

(Fig. 10.16), and by the end of the nineteenth century, they were also sent by train to other major French cities, including Paris (Figs. 10.17, 10.18, 10.19 and 10.20). Local confectionaries used Parma violets for the famous *fin de siècle* candied violets, which were exported to France and Europe and are still in production (Timbal-Lagrave 1862, Bertrand and Casbas 2001).

By reading *The Journal of Horticulture, Cottage Gardener and Home Farmer* (volume 30, III series, January–June 1895), we can observe not only the prices of the violet bouquets sold in London (Figs. 10.21 and 10.22), especially in the Convent Garden Market but also other information regarding the cultural uses of these flowers:

> '*Violets are also in much demand, sometimes because of their delightful perfume, at others because they were the favourite flower of the departed ones; but from an artistic point of view, wreaths made entirely of them do not compare favourably with others made from the various kinds of pure, rich, white flowers*' (page 66).

> '*Good Violets always sell well in country towns*' (page 189).

**Fig. 10.17** Souvenir postcard describing the main stages of violets' production and selling in Toulouse and their expedition to Paris (early twentieth century), private collection

**Fig. 10.18** Violets sellers in Paris, '*Les Petites Métiers de Paris – Marchandes de Fleurs*' (c.1900), postcard, private collection

**Fig. 10.19** Violet sellers in Paris, '*Scènes Parisiennes—Quartier de l'Étoile: Marchande de Violettes*' (c.1905), postcard, private collection

**Fig. 10.20** Selling violets on *Saint-Michel* bridge, Paris (c.1900), postcard, private collection

**Fig. 10.21** Covent Garden Flower Women. Street Life in London (1877) John Thomson. LSE Digital Library

'G. Nobbs, Royal Gardens, Osborne, Cowes, for which an award of merit was granted. A bunch of Violet Victoria was exhibited by Mr Jennings, gardener to Leopold de Rothschild, Esq., Ascot, Leighton Buzzard' (page 236).

'Marie Louise Violets—Mr J. Anderson sends us blooms of this Violet taken from plants grown in cold frames. They are large, rich in colour, and deliciously fragrant, and are highly creditable to their grower' (page 249).

**Fig. 10.22** Flower Seller
at Piccadilly Circus
(London). Types of
London Life, postcard,
private collection

*Types of London Life (4). Flower Seller at Piccadilly Circus.*

*It is questionable if any flowers give more pleasure at any time during the year than double Violets do in the winter. When we consider that Violets can be had with small cost and labour for over six of the dullest months of the year, it seems strange that they are not even more generally cultivated'* (page 289).

*'Battle of Flowers at Nice —Her Majesty viewed the recent battle of flowers from a space on the Rue du Congrès, placed at her disposal by the municipality. The Queen was received by Comte de Malaussena, the Mayor of Nice, and presented on behalf of the Felis Committee with a pink satin banner and a splendid basket of Violets. Her Majesty remained in her carriage during the proceedings, and the Royal equipage was soon filled with small bouquets thrown from the passing vehicles. The Queen appeared to enjoy the display, and from time to time, it returned the bunches of flowers, especially favouring the French officers of the garrison. Both on arriving and when leaving Nice Her Majesty was saluted with the British National Anthem. Miss Van Buren of New York won the first prize for the best decorated carriage, her Victoria resembling a cornucopia overflowing with flowers'* (page 291).

*'Twenty years ago, the Violet trade about London was a remarkable one, but we have rarely had good seasons for the flowers since, and whilst the foreign competition has been extraor-*

*dinary, the adverse seasons certainly have not been due to competition. But bunched Violets may be gathered fresh and run into London from remote places in a few hours; hence, it may be that our present abundant supply of sweet perfumed lich singles of the true Russian are home grown'* (page 293).

*'It is quite true that Violets like pure air and an open position in order that they may produce stout foliage, not large thin-textured leaves to fall a prey to their natural enemy, red spide'* (page 303).

*'One young lady boasts of never appearing in public without her bunch of Violets when it is possible to procure them, and in the garden here, we are expected to have Violets from October to May in a small but continuous supply'* (page 344).

*'The desire is to afford ladies an opportunity of gathering Violets in the winter, where litter for covering frames is objectionable'* (page 344).

*'Fatal Violets – A well-known botanist says that the root of the fragrant Violet is so poisonous that a very little of it causes nausea, interrupted heart action, difficulty of breathing, and other organic complications, which may result in serious illness. In this connection it is significant that an intimate relation has been discovered between the strong perfume of the fragrant Violet and the venomous qualities of its root, for the root of the scentless, or Dog Violet, is not venomous. More recently, cases of heart failure and defective circulation have been traced to the influence of Violet perfume. Since the fragrant Violet and its extracted perfumes are becoming all the vogue, it might be well, in view of these sinister reports concerning them, to consider the propriety of their indiscriminate and lavish enjoyment, particularly as they seem liable to injuriously affect not only those who use them but also susceptible persons who may be in their immediate neighbourhood'* (page 474).

The leaves and flowers of the violets were the raw materials used by the perfume industry, which, in Europe, was mainly located at Grasse (French Riviera), the world's capital of perfumes (Fig. 10.23). In the last quarter of the nineteenth century, the perfume industry began to use ionone, a molecule responsible for the violet fragrance, which is isolated from the rhizomes of *Iris* x *germanica* L. var. *florentina* (L.) Dykes (Nanneli 2001, Coombs 2003). This was a significant setback for violet farmers, and within a few years, in France, the cultivation area of violets was reduced by two-thirds (Tergit 1961).

By the late nineteenth century, prints depicting violets were trendy, as seen in late Victorian newspapers (Fig. 10.24). After Queen Victoria died (1901), the new consort sovereign, Queen Alexandra (1844–1925), continued to patronise violets, as we can read in the same journal (*The Journal of Horticulture, Cottage Gardener and Home Farmer* 1902, April, page 294):

*'In the Royal Conservatory Gardens in Windsor Great Park, a specially fine specimen of double Violets is being grown, which has gained the enthusiastic approval of the Queen. Her Majesty has said that she has seen none finer, and bunches are frequently forwarded to her. Each flower (says the Daily Mail) is very large and of a beautiful colour'.*

British literature is also full of violets, often serving as symbolic elements within the narrative, as in *Daniel Deronda* (1876), the last novel of George Eliot (1819–1880), and the only one set in Victorian society in which we can read a reflection on life that uses violets as a metaphor:

**Fig. 10.23** Violets leaves arriving at Roure-Bertrand Factory, in Grasse (c.1900), postcard, private collection

**Fig. 10.24** Girl with violet. Engraving from *The Illustrated London News* (12 January 1895), private collection

*'that dead anatomy of culture which turns the universe into a mere ceaseless answer to queries, and knows, not everything, but everything else about everything – as if one should be ignorant of nothing concerning the sense of violets except the sense itself for which one had no nostril'* (Eliot 1876).

In the acclaimed *The Picture of Dorian Gray* (1891) by Oscar Wilde (1854–1900), first published in *Lippincott's Monthly Magazine* (1890), violets are mentioned several times:

*'I once wore nothing but violets all through one season, as a form of artistic mourning for a romance that would not die'* (…) *'and in violets that woke the memory of dead romances'* (…) *'That evening, at eight-thirty, exquisitely dressed and wearing a large button-hole of Parma violets, Dorian Gray was ushered into Lady Narborough's drawing-room by bowing servants'* (Wilde 1891).

In *The Happy Prince* (1888), also by Oscar Wilde, the protagonist not only had *'violet-eyes'* but also asked the little swallow to give a precious stone to the writer who had nothing to eat nor to keep him warm:

*'Leaning over a desk covered with papers, and in a tumbler by his side there is a bunch of withered violets. His hair is brown and crisp, and lips are red as pomegranate, and he has large and dreamy eyes (…) the young man had his head buried in his hands, so he did not hear the flutter of the bird's wings, and when he looked up, he found the beautiful sapphire lying on the withered violets'* (Wilde 1888).

Many poems of Alfred Tennyson (1809–1892), the *Poet Laureate of the United Kingdom*

(from 1850 to 1892)—an honorary position appointed by Queen Victoria –evoke violets, such as:*To die before the snowdrop came, and now the violet's here.*
*0 sweet is the new violet, that comes beneath the skies*
*(…)*
*The smell of violets, hidden in the green,*
*Pour'd back into my empty soul and frame*
*The times when I remember to have been*
*Joyful and free from blame*
*(…)*
*She keeps the gift of years before,*
*A wither'd violet is her bliss:*
*She knows not what his greatness is;*
*For that, for all, she loves him more* (Tennyson 1912).

Charlotte Brontë (1816–1855), an English novelist and poet from the early Victorian period and the eldest of the three Brontë sisters, wrote novels that became classics, such as *Jane Eyre* (1847) and *Villette* (1853). In the latter, violets are present in critical moments of the plot to denote a contrast between women, as a means of communication between the main characters (Paul and Lucy), and as a symbol of a short life (Uhara 2024). In this novel, besides human connection and mortality, violets symbolise security through the house Paul gives Lucy for her career, allowing her to achieve autonomous security and making her a model for women in Brontë's desired society (Bidner 2023).

Violets and Parma violets continued to be cultivated and used (Figs. 10.25, 10.26, 10.27 and 10.28) throughout the late nineteenth century and early decades of the

**Fig. 10.25** Couple of lovers holding hands over a violet bouquet (c.1905), postcard, private collection

**Fig. 10.26** Violets' corsage (c.1910), postcard, private collection

**Fig. 10.27** Woman
holding several bouquets
of violets (c.1910),
postcard, private collection

**Fig. 10.28** Woman
holding a bouquet of
violets (c.1908), postcard,
private collection

twentieth century, and new cultivars were introduced, such as *Viola* 'Mrs. Arthur' (1902) and *Viola* 'Queen Mary' (1915), but after the First World War and the changes in fashion that occurred after the 1920s, the public interest in these plants steadily decreased, and violets never recovered the high status they once had. They are still being cultivated, and periodically, new amateurs, enthusiasts, and horticulturists have attempted to revive the public interest in these enchanting plants with unique perfumed flowers.

# References

Ancona ML (1977) The garden of the renaissance: botanical symbolism in Italian painting. L. S. Olschki, Firenze

Aristotle (1955) Minor Works (trans. W.S.HETT). Harvard University Press, Cambridge, MA

Beasley T (2017) In search of Queen Victoria's favourite flower. https://www.english-heritage.org.uk/visit/inspire-me/blog/blog-posts/in-search-of-queen-victorias-favourite-flower. Accessed 18 March 2024

Bertrand B, Casbas N (2001) Une Pensée pour la Violette. Le Compagnon Végétal, vol. 11. Editions de Terran, Sengouagnet.

Bidner C (2023) Wilted petals that loved: flowers and humanity in Charlotte Brontë's Villette. The Albatross 13:69–77

Camerarius J (1588) Hortus Medicus et Philosophicus. Johannem Feyerabend, Frankfurt am Main

Campbell DA (ed & trans) (1990) Greek lyric, volume I: Sappho and Alcaeus. Harvard University Press, Cambridge, MA

Coats A (1977) The empress Josephine. Gard Hist 5(3):40–46

Coombs R (2003) Violets: the history and cultivation of scented violets. Batsford, London

Coon N, Giffen G (1977) The complete book of violets. A. S. Barnes and Company, New York

Eliot G (1876) Daniel Deronda. William blackwood and sons, Edinburgh/London

Farrar E (1989) Pansies, Violas and Sweet Violets. Hurst Village Publishing, Hurst

Homer (1961) The odyssey (the Fitzgerald translation). Anchor Press/Doubleday, Garden City, NY

Horace (1894) Odes (W.E.Gladstone trans). Charles Scribner's Sons, New York

Lack WH (2004) Jardin de la Malmaison: Empress Josephine's Garden. Prestel, London

Lasance FX (1908) My prayer book—happiness in goodness: reflections, counsels, prayers, and devotions. Brezinger Brothers, New York

Lisieux T (1898) Histoire d'une Âme. Librairies de L'Oeuvre de Saint-Paul, Bar Le Duc

Malécot V, Marcussen T, Munzinger J, Yockteng R, Henry M (2007) On the origin of the sweet-smelling Parma violet cultivars (Violaceae): wide intraspecific hybridization, sterility, and sexual reproduction. Am J Bot 94(1):29–41

Millet A (1898) Les Violettes, Leurs Origines, Leurs Cultures. Librarie Agricole, Paris

Nanneli I (2001) Viole. Idea Books, Milano

Nelson M (2007) Queen Victoria and the discovery of the Riviera. Tauris Parke Paperbacks, London

Perfect EJ (1996) Armand Millet and his violets. Park Farm Press, High Wycombe

Pindar (1997) Fragments, Nemean Odes, Isthmian Odes, Volume II [William H. Race (ed & trans)]. Harvard University Press, Cambridge, MA

Pliny (1961) Natural history VI, Libri XX-XXIII (W.H.S. Jones trans). Harvard University Press, Cambridge, MA

RHS [Royal Horticultural Society] (1866) Extracts from proceeding of the Royal Horticultural Society and miscellaneous matter [the journal of the Royal Horticultural Society, new series volume I], Royal Horticultural Society, London

Roland M (1901) The private memoirs of Madame Roland [Edward Gilpin Johnson (edi.)]. A.C. McClurg & Co., Chicago, IL

Sandrini F (2008) Maria Luigia e le Violette di Parma. Quaderni del Museo n.10, Museo Glauco Lombardi Fondazione Monte di Parma, Parma, Italy

Tennyson A (1912) Poems of Tennyson (1830–1870). Oxford University Press, London

Tergit G (1961) Flowers through the ages. Oswald Wolff Publishers, London

Theophrastus (2016) História das Plantas [Maria de Fátima Sousa Silva and Jorge Paiva (trans)]. Imprensa da Universidade de Coimbra, Coimbra

Timbal-Lagrave E (1862) De la Culture de la Violette à Toulouse. Annales de la Société d'Horticulture de la Haute-Garonne 4:1–6

Uhara M (2024) Symbolic meanings of violets in Villette. Brontë Studies 49(1–2):116–128. https://doi.org/10.1080/14748932.2024.2317158

Unknown Author (1857) The language of flowers: an alphabet of floral emblems. T. Nelson and Sons, London

Wilde O (1888) The happy prince and other tales. David Nutt, London

Wilde O (1891) The picture of Dorian Gray. Ward, Lock, and Company, London

## *Periodicals*

Le Bon Jardinier (1805)

The Illustrated London News (2714, 25 April 1891)

The Journal of Horticulture, *Cottage Gardener and Home Farmer* (January–June 1895)

The Journal of Horticulture, *Cottage Gardener and Home Farmer* (April 1902)

The King Islands News (23 November 1921)

The Port Adelaide News (11 July 1882)

9 783031 687587